T0296837

Statistical Aspects of the Microbiological Examination of Foods

Statistical Aspects of the Microbiological Examination of Foods

Third Edition

Basil Jarvis

AMSTERDAM • BOSTON • HEIDELBERG • LONDON
NEW YORK • OXFORD • PARIS • SAN DIEGO
SAN FRANCISCO • SINGAPORE • SYDNEY • TOKYO

Academic Press is an imprint of Elsevier

Academic Press is an imprint of Elsevier
125 London Wall, London EC2Y 5AS, United Kingdom
525 B Street, Suite 1800, San Diego, CA 92101-4495, United States
50 Hampshire Street, 5th Floor, Cambridge, MA 02139, United States
The Boulevard, Langford Lane, Kidlington, Oxford OX5 1GB, UK

Notices

Knowledge and best practice in this field are constantly changing. As new research and experience broaden our understanding, changes in research methods, professional practices, or medical treatment may become necessary.

Practitioners and researchers must always rely on their own experience and knowledge in evaluating and using any information, methods, compounds, or experiments described herein. In using such information or methods they should be mindful of their own safety and the safety of others, including parties for whom they have a professional responsibility.

To the fullest extent of the law, neither the Publisher nor the authors, contributors, or editors, assume any liability for any injury and/or damage to persons or property as a matter of products liability, negligence or otherwise, or from any use or operation of any methods, products, instructions, or ideas contained in the material herein.

Library of Congress Cataloging-in-Publication Data
A catalog record for this book is available from the Library of Congress

British Library Cataloguing-in-Publication Data
A catalogue record for this book is available from the British Library

ISBN: 978-0-12-803973-1

For information on all Academic Press publications
visit our website at https://www.elsevier.com/

Publisher: Sara Tenney
Acquisition Editor: Linda Versteeg-Buschman
Editorial Project Manager: Halima Williams
Production Project Manager: Julia Haynes
Designer: Greg Harris

Typeset by Thomson Digital

Contents

Preface to the Third Edition

In my first post as a graduate bacteriologist, my manager impressed upon me the need for proper use of statistical analysis in any work in applied microbiology. So insistent was he that I attended a part-time course in statistics for use in medical research at University College, London – unfortunately the course was presented by a mathematician who failed to recognise that detailed mathematical concepts often cause non-mathematicians to 'switch off'! In those days before the ready availability of personal computers and electronic calculators, statistical calculations were done manually, sometimes aided by mechanical calculators. Even simple calculations that nowadays take only a few minutes would often take days and nights to complete. Nonetheless, my interest in the use of statistics has stayed with me throughout my subsequent career.

In the early 1970s, I was fortunate to work with the late Dr Eric Steiner, a chemist with considerable knowledge and experience of the use of statistics who opened my eyes to a wider appreciation of statistical methods. I was also privileged for a time to have guidance from Ms Stella Cunliffe, the first woman President of the Royal Statistical Society. Over several years I lectured on statistical aspects of applied microbiology at various courses, including the WHO-sponsored course in Food Microbiology that I set up at the University of Surrey, UK. During that time I became aware of a lack of understanding of basic statistical concepts by many microbiologists and the total lack of any suitable publication to provide assistance. In the early 1990s I prepared a report on statistical aspects of microbiology for the then UK Ministry of Agriculture, Food and Fisheries (MAFF). It is from that background that the first edition of this book arose.

This book is intended as an aid to practicing microbiologists and others in the food, beverage and associated industries, although it is relevant also to other aspects of applied microbiology. It is also addressed to students reading food microbiology at undergraduate and postgraduate levels. With greater emphasis now being placed on quantitative microbiology, the need to understand relevant statistical matters assumes even greater importance in food microbiology.

This book is written not by a professional statistician but by a microbiologist, in the sincere hope that it will help future applied microbiologists. Over the past 2 or 3 years, discussions with colleagues from various organisations identified a number of developments in microbiological methods and in statistical analyses of microbiological data. Some are related to the growing importance of risk analysis and particularly the need to understand the statistical implications of the distribution of over-dispersed populations of pathogenic microbes in foods, others to changes in Europe and the USA in the procedures for validation of microbiological methods; others were concerned with methods for estimation of microbiological measurement uncertainty.

I was considering whether there was a need to revise the book when the publisher asked me if such revision would be timely. In editing and revising the book, I have received helpful comment and advice from colleagues in many countries. I would particularly acknowledge numerous helpful discussions on diverse topics with members of the AOAC Presidential Taskforce on 'Best Practices in Microbiological Methodology' and members of the ISO Statistics Working Group (TC34/SC9/WG2), especially Professor Peter Wilrich (Free University of Berlin) and Dr Bertrand Lombard (ANSES, Paris, France). I am especially indebted to Dr Alan Hedges (University of Bristol Medical School) and Dr Janet E.L. Corry (University of Bristol Veterinary School) for their critiques of the second edition and on drafts of this manuscript; to Dr Andreas Kiermeier (Australia) who kindly provided a copy of the FAO/WHO spreadsheet system for setting up sampling plans; and to Dr Sharon Brunelle of AOAC for again providing the chapter on method validation. But all errors of omission or commission are mine alone. I acknowledge the help received from the editors at Academic Press, especially Halima Williams. I would also wish to thank the many authors and publishers who have kindly granted me the rights to re-publish tables and figures previously published elsewhere. Details of these permissions are quoted in the text.

Finally, I thank my wife, my family and our friends, for their continuing support and for putting up with me over this period of rewriting.

Basil Jarvis
Upton Bishop, Ross-on-Wye
Herefordshire, UK
December 2015

List of tables

List of figures

Fig

List of examples

CHAPTER

Introduction

1

One morning a professor of statistics sat alone in a bar at a conference. When his colleagues joined him at the lunch break, they asked why he had not attended the lecture sessions. He replied, saying, 'If I attend one session I miss nine others; and if I stay in the bar I miss all ten sessions. The probability is that there will be no statistically significant difference in the benefit that I obtain!' Possibly a trite example, but statistics are relevant in most areas of life.

The word 'statistics' means different things to different people. According to Mark Twain, Benjamin Disraeli was the originator of the statement 'There are lies, damned lies and statistics!' from which one is supposed to conclude that the objective of much statistical work is to put a positive 'spin' onto 'bad news'. Whilst there may be some political truth in the statement, it is generally not true in science provided that correct statistical procedures have been used. Herein lies the rub! To many people, the term 'statistics' implies the manipulation of data to draw conclusions that may not be immediately obvious. To others, especially many biologists, the need to use statistics implies a need to apply numerical concepts that they hoped they had left behind at school. But to a few, use of statistics offers a real opportunity to extend their understanding of bioscience data in order to make better use of the information available.

Much statistical analysis is used to decide whether one set of data differs from another and it is essential to recognise that sometimes a difference that is statistically significant may not be of practical significance, and vice versa. Always bear in mind that statistical methods serve merely as a tool to aid interpretation of data and to enable inferences to be drawn. It is essential to understand that data should be tested for goodness of fit by seeking to fit an appropriate statistical model to the experimental data.

Microbiological testing is used in industrial process verification and sometimes to provide an index of quality for 'payment by quality' schemes. Examination of food, water, process plant swabs, etc., for microorganisms is used frequently in the retrospective verification of the microbiological 'safety' of foods and food process operations. Such examinations include assessments for levels and types of microorganisms, including tests for the presence of specific bacteria of public health significance, including pathogens, index and indicator organisms.

During recent years, increased attention has focused, both nationally and internationally, on the establishment of numerical microbiological criteria for foods. All too often such criteria have been devised on the misguided belief that testing of foods

Statistical Aspects of the Microbiological Examination of Foods. http://dx.doi.org/10.1016/B978-0-12-803973-1.00001-2

for compliance with numerical, or other, microbiological criteria will enhance consumer protection by improving food quality and safety. I say 'misguided' because no amount of testing of finished products will improve the quality or safety of a product once it has been manufactured. Different forms of microbiological criteria have been devised for particular purposes; it is not the purpose of this book to review the advantages and disadvantages of microbiological criteria – although some statistical matters relevant to criteria will be discussed in Chapter 14.

Rather, the objective is to provide an introduction to statistical matters that are important in assessing and understanding the quality of microbiological data generated in practical situations. Examples, chosen from appropriate areas of food microbiology, are used to illustrate factors that affect the overall variability of microbiological data and to offer guidance on the selection of statistical procedures for specific purposes. In the area of microbiological methodology it is essential to recognise the diverse factors that affect the results obtained by both traditional methods and modern developments in rapid and automated methods.

The book considers the distribution of microbes in foods and other matrices; statistical aspects of sampling; factors that affect results obtained by both quantitative (eg, colony count) and quantal methods [eg, presence/absence and most probable number (MPN) methods]; the meaning of, and ways to estimate, microbiological uncertainty; the validation of microbiological methods; and the implications of statistical variation in relation to food safety and use of microbiological criteria for foods. Consideration is given also to quality monitoring of microbiological practices and the use of statistical process control for trend analysis of data both in the laboratory and in manufacturing industry.

The book is intended as an aid for practising food microbiologists. It assumes a minimal knowledge of statistics and references to standard works on statistics are cited whenever appropriate.

CHAPTER

Some basic statistical concepts

2

POPULATIONS

The true population of a particular ecosystem can be determined only by carrying out a census of all living organisms within that ecosystem. This applies equally whether one is concerned with numbers of people in a town, state or country or with numbers of microbes in a batch of a food commodity or a product. Whilst it is possible, at least theoretically, to determine the human population in a non-destructive manner by undertaking a population census, the same does not apply to estimates of microbial populations.

When a survey is carried out on people living, for instance, in a single town or village, it would not be unexpected that the number of residents differs between different houses, nor that there are differences in ethnicity, age, gender, health and well-being, personal likes and dislikes, etc. Similarly, there will be both quantitative and qualitative differences in population statistics between different towns and villages, different parts of a country and different countries.

A similar situation pertains when one looks at the microbial populations of a food. The microbial association of foodstuffs differs according to diverse intrinsic and extrinsic factors, especially the acidity and water activity, and the extent of any processing effects. Thus the primary microbial population of acid foods will generally consist of yeasts, moulds and acidophilic bacteria, whereas the primary population of raw meat and other protein-rich foodstuffs will consist largely of Gram-negative non-fermentative bacteria, with smaller populations of other organisms (Mossel, 1982). In enumerating microbes, it is essential first to define the population to be counted. For instance, does one need to obtain an estimate of the total population, that is, living and dead organisms, or only the viable population; if the latter, is one concerned only with specific groups of organisms, for example, aerobes, anaerobes, psychrotrophs and psychrophiles, mesophiles or thermophiles? Even when such questions have been answered, it would still be impossible to determine the true ecological population of a particular 'lot' or 'batch' of food, since to do so would require testing of all the food. Such a task would be technically and economically impossible.

LOTS AND SAMPLES

An individual 'lot' or 'batch' of product consists of a quantity of food that has been processed under essentially identical conditions on a single occasion. The food may be stored and distributed in bulk or as pre-packaged units each containing one or

Statistical Aspects of the Microbiological Examination of Foods. http://dx.doi.org/10.1016/B978-0-12-803973-1.00002-4

more individual units of product, for example, a single meat pie or a pack of frozen peas. Assuming that the processing has been carried out under uniform conditions, theoretically, the microbial population of each unit should be typical of the population of the whole lot. In practice, this will not always be the case. For instance, high levels of microbial contamination may be associated only with specific parts of a lot due to some processing defect or the incomplete mixing of ingredients. In addition, estimates of microbial populations will be affected by the choice of test protocol that is used.

It is not feasible to determine the levels and types of aerobic and anaerobic organisms, or of acidophilic and non-acidophilic organisms, or other distinct classes of microorganism using a single test. Thus when a microbiological examination is carried out, the types of microorganisms that are detected will be defined in part by the test protocol. All such constraints therefore provide a biased estimate of the microbial population of the 'lot'. Hence, sampling of either bulk or pre-packaged units of a 'lot' merely provides an indication of the types and numbers of microorganisms that make up the population of the 'lot' and such population samples will themselves be further sampled by the choice of examination protocol. In order to ensure that a series of samples drawn from a 'lot' properly reflect the diversity of types and numbers of organisms associated with the product it is essential that the primary samples should be drawn in a random manner, either from a bulk or as individual packaged units of the foodstuff.

Analytical chemists frequently draw large primary samples that are blended and resampled before taking one or more analytical samples – the purpose is to minimise the between-sample variation in order to determine an 'average' analytical estimate for a particular analyte. It is not uncommon for several kilograms of material to be taken as a number of discrete samples that are then combined, blended and resampled to provide the series of analytical samples. The sampling of foods for microbiological examination cannot generally be done in this way because of the risks of cross-contamination during the mixing procedure, although examples of techniques for producing composite samples have been published (see, eg, Corry et al., 2010).

A 'population sample' (eg, a unit of product) may itself be sub-divided for analytical purposes and it is important, therefore, to consider the implications of determining microbial populations in terms of the number, size and nature of the samples taken. In only a few instances is it possible for the analytical sample to be truly representative of the 'lot' sampled. Liquids, such as milk, can be sufficiently well mixed that the number of organisms in the analytical sample is representative of the milk in a bulk storage tank. However, because of problems of mixing, samples withdrawn from a grain silo, or even from individual sacks of grain, may not necessarily be truly representative. In such circumstances, deliberate stratification (*qv*) may be the only practical way of taking samples. Similar situations obtain when one considers complex raw material such as animal carcases, or composite food products such as ready-to-cook frozen meals containing slices of cooked meat, Yorkshire pudding, peas, potato and gravy. It is necessary to consider also the actual sampling protocol to be used: for instance, in sampling from a meat or poultry carcase, is the sample to be taken by swabbing, rinsing or excision of skin? Where on the carcase should the sample be taken? For instance, one area (eg, chicken neck skin) may be more likely to carry higher numbers and types of organism than other areas. Hence, standardisation of sampling protocols is essential. In situations where a composite food consists of discrete components, a sampling protocol needs to be used that reflects the purpose of the test–is a composite analytical sample required (ie, one made up from the various ingredients in appropriate proportions) or should each ingredient be tested separately? These matters are considered in more detail in Chapter 5.

AVERAGE SAMPLE POPULATIONS

If a single sample is analysed, the result provides a method-dependent single-point estimate of the population numbers in that sample. Replicate tests on a single sample provide an improved estimate of population numbers within that sample based on the average of the results, together with a measure of variability of the estimate for that sample. Similarly, if replicate samples are tested, the average result provides a better estimate of the number of organisms in the population based on the between-sample average and an estimate of the variability between samples. Thus, we can have greater confidence that the 'average sample population' will reflect more closely the population in the 'lot'. The standard error of the mean (SE_M; *qv*) provides an estimate of the extent to which the average value is reliable. If a sufficient number of replicate samples is tested, then we can derive a frequency distribution for the counts, such as that shown in Fig. 2.1 (data from Blood, 1974). Note that this distribution curve has a long left-hand tail and that the curve is not symmetrical, possibly because the data were compiled from results obtained in two different production plants. The statistical aspects of common frequency distributions are discussed in Chapter 3.

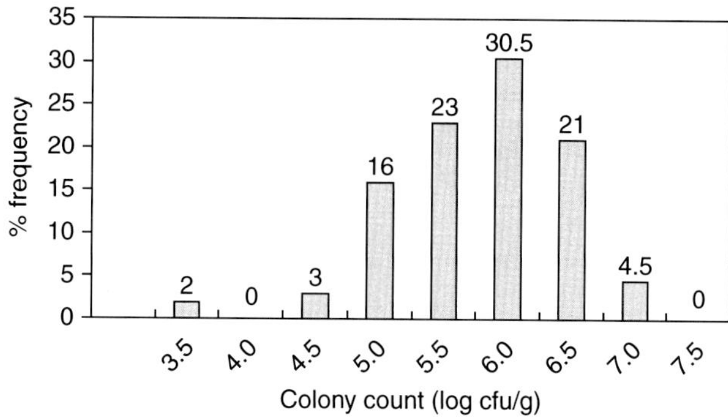

FIGURE 2.1 Frequency Distribution of Colony Count Data Determined at 30°C on Beef Sausages Manufactured in Two Factories

Modified from Blood (1974). Reproduced by permission of Leatherhead Food International

Adding the individual values and dividing by the number of replicate tests provides a simple arithmetic mean of the values $[\bar{x} = (x_1 + x_2 + x_3 + \cdots + x_n)/n = \sum_{i=1}^{n} x_i/n]$, where x_i is the value of the ith test and n is the number of tests done. However, it is possible to derive other forms of average value. For instance, multiplying the individual counts on n samples and then taking the nth root of the product provides the geometric mean value $(\bar{x}')$:

$$\bar{x}' = \sqrt[n]{x_1 \cdot x_2 \cdot x_3 \cdots x_n}$$

It is simpler to determine the approximate geometric mean by taking logarithms of the original values $(y = \log_{10} x)$, adding the log-transformed values and dividing the sum by n to obtain the mean log value $(\bar{y})$, which equals $\log \bar{x}'$. This value can be back-transformed by taking the antilog to obtain an estimate of the geometric mean value:

$$\bar{y} = \frac{\sum_{i=1}^{n} y_i}{n} = \frac{\sum_{i=1}^{n} \log_{10} x_i}{n} = \log_{10} \bar{x}'$$

The geometric mean is appropriate for data that conform to a lognormal distribution and for titres obtained from n-fold dilution series. It is important to understand the difference between the geometric and the arithmetic mean values since both are used in handling microbiological data. In terms of microbial colony counts, the log mean count is the $\log_{10}$ of the arithmetic mean; by contrast, the mean log count is the arithmetic average of the $\log_{10}$-transformed counts that, on back-transformation, gives the geometric mean count. The methods are illustrated in Example 2.1.

EXAMPLE 2.1 DERIVATION OF SOME BASIC STATISTICS THAT DESCRIBE A DATA SET

Assume that we wish to determine the statistics that describes a series of replicate colony counts on a number (n) of samples, represented by x_1, x_2, x_3, ..., x_n, for which the actual values are 1540, 1360, 1620, 1970 and 1420 as colony-forming units (cfu) per gram.

Range

The range of colony counts provides a measure of the extent of overall deviation between the largest and smallest data values and is determined by subtracting the lowest value from the highest value; for the example data, the range is 1970 − 1360 = ***610***.

Median

The median colony count is the middle value in an odd-numbered set of values or the average of the two middle values in an even-numbered set of values; for this sequence of counts the median value of 1360, 1420, *1540*, 1620, 1970 = ***1540***.

The interquartile range

This is the range covered by the middle 50% of data values, which is often more useful than the absolute range of values. We have determined the median value as 1540 – this is the *second quartile* or *Q2* value. We now determine the *first quartile* value (*Q1*) as the median of the values below and including *Q2*, and the *third quartile* value (*Q3*) as the median of the values including and greater than *Q2*. In this case because we have only 5 values we can say that *Q1* = 1420 and *Q3* = 1620, so the inter-quartile range (*IQR*) is given by 1620 − 1420 = 200.

Arithmetic mean

The arithmetic average (mean) colony count is the sum of the individual values divided by the number of values:

$$\bar{x} = \frac{x_1 + x_2 + x_3 + \cdots + x_n}{n} = \left(\sum_{i=1}^{n} x_i\right) / n$$

where $\bar{x}$ is the mean value and Σ means 'sum of'; for these data the mean count = $\Sigma x/n$ = (1540 + 1360 + 1620 + 1970 + 1420)/5 = ***1582***.

Geometric mean

The geometric mean colony count is the *n*th root of the product obtained by multiplying together each value of *x*. Hence, the geometric mean count = $\sqrt[n]{x_1 \times x_2 \times x_3 \times \cdots \times x_n}$.

Alternately, we can transform the *x* values by deriving their logarithms so that $y = \log_{10} x$; the geometric mean is the antilog of the sum of *y* divided by *n*, that is:

$$\text{antilog}\left[\frac{\Sigma_{i=1}^{n} \log x_i}{n}\right] = \text{antilog}\left[\frac{\Sigma_{i=1}^{n} y_i}{n}\right]$$

For these data the geometric mean colony count = antilog($\Sigma \log_{10} x/n$) = antilog[(log 1540 + log 1360 + log 1620 + log 1970 + log 1420)/5] = antilog[(3.1875 + 3.1335 + 3.2095 + 3.2945 + 3.1523)/5] = antilog[15.9773/5] = antilog[3.19456] = $10^{3.19456}$ = *1568*.

Note that this differs from the arithmetic mean value!

Sample variance

The sample variance (s^2) is the sum of the squares of the deviances between the values of *x* and the mean value ($\bar{x}$), divided by the degrees of freedom (df) of the data set (ie, *n* – 1). (One value of *n* was used in determining the mean value, so there are only *n* – 1 *df* available.)

The first form of the equation is as follows: $s^2 = \Sigma_{i=1}^{n} (x_i - \bar{x})^2 / (n-1)$.

Hence, with mean $(\bar{x}) = 1582$, the variance is given by

$$s^2 = \frac{(1540-1582)^2 + (1360-1582)^2 + (1620-1582)^2 + (1970-1582)^2 + (1420-1582)^2}{5-1}$$

$$= \frac{(-42)^2 + (-222)^2 + (38)^2 + (388)^2 + (-162)^2}{4} = \frac{1764 + 49{,}284 + 1444 + 150{,}544 + 26{,}244}{4}$$

$$= \frac{229{,}280}{4} = \mathit{57{,}320}$$

An alternative form of the equation is as follows:

$$s^2 = \frac{\Sigma x^2 - \left[(\Sigma x)^2 / n\right]}{n-1}$$

Hence, for our data:

$$s^2 = \frac{[(1540)^2 + (1360)^2 + (1620)^2 + (1970)^2 + (1420)^2] - [(1540 + 1360 + 1620 + 1970 + 1420)^2 / 5]}{5-1}$$

$$= \frac{12{,}742{,}900 - 12{,}513{,}620}{4} = \frac{229{,}280}{4} = \mathit{57{,}320}$$

Note that in this example, where the mean value was finite, both methods gave the same result for the variance. However, if the mean value were not finite, rounding errors could cause serious inaccuracies in the variance calculation using the first form of the equation.

Standard deviation

The standard deviation (*s*) of the mean is the square root of the variance and is given by

$$\sqrt{s^2} = \sqrt{57{,}320} = \mathit{239.4}$$

Thence the coefficient of variation (CV), sometimes called the relative standard deviation (RSD), is the ratio between the standard deviation and the mean value and is given by $(100 \times 239.4)/1582 =$ *15.1%*.

The variance of transformed values is derived similarly using the transformed values, that is, $y_i = \log_{10} x_i$.

Using the first method with a mean log count of 3.1946 gives the variance of y as follows:

$$s_2 = \frac{(3.1875-3.1946)^2+(3.1335-3.1946)^2+(3.2095-3.1946)^2+(3.2945-3.1946)^2+(3.1523-3.1946)^2}{4}$$

$$= \frac{0.01577493}{4} = 0.0039438 \approx \mathbf{0.0039}$$

The alternative equation gives

$$s^2 = \frac{5[(3.1875)^2+(3.1335)^2+(3.2095)^2+(3.2945)^2+(3.1523)^2]-(3.1875+3.1335+3.2095+3.2945+3.1523)^2}{5\times 4}$$

$$= \frac{(5\times 51.07077)-255.2741153}{20} = \frac{255.35385-255.274115}{20} = 0.0039869 \approx \mathbf{0.0040}$$

Note the small difference in the variance estimates determined by the two alternative methods.

The SD of the mean log count is $\sqrt{0.0040} = 0.063246 \approx$ *0.0632* and the CV of the mean log count is $(0.0632 \times 100)/3.1946 =$ *1.98%*.

Reverse transformations and confidence limits

The reverse transformation of the mean log count is done by taking the antilog of $\bar{y}$: $\bar{x} = 10^{\bar{y}} = 10^{3.1946} \approx 1565$. This gives an approximate estimation of the geometric mean, which requires correction. The relationship between the log mean count ($\log \bar{x}'$) and the mean log count ($\bar{y}$) is given by the following formula:

$$\log \bar{x}' = \bar{y} + \frac{\ln(10)\times s^2}{10} = \bar{y} + \frac{2.3025\times s^2}{10}$$

where s^2 is the variance of the log count.

Hence, for these data where $\bar{y} = 3.1946$ and $s^2 = 0.0040$, the log mean colony count is given by $\log \bar{x}' = 3.1946 + [(2.3025 \times 0.0040)/10] = 3.1955$.

Hence, the geometric mean count is as follows:

$$\bar{x}' = 10^{3.1955} = 1568.6 \approx 1569.$$

Note that standard deviations of the mean log count should *not* be directly back-transformed since the value obtained ($10^{0.0635} = 1.1574$) would be misleading. Rather, the approximate upper and lower 95% confidence intervals (see also Chapter 3 and Fig. 3.5) around the geometric mean would be determined as $10^{3.1946+(2\times 0.0635)}$ and $10^{3.1946-(2\times 0.0635)}$, that is, $10^{3.33216} = 2097$ and $10^{3.0676} = 1168$. Note that these confidence limits (CLs) are asymmetrical around the mean value.

For these data the geometric mean is 1569 and the 95% upper and lower CLs are 2097 and 1168, respectively. A comparison with the arithmetic mean and its 95% CLs is shown as follows:

Method	Mean	Median	95% Confidence Limits	
			Lower	Upper
Arithmetic	1582	1540	1104	2060
Geometric	1569		1168	2097

For these data the difference between the arithmetic and geometric mean values is small since the individual counts are reasonably evenly distributed about the mean value and are not heavily skewed, although the median value is smaller than both mean values indicating some skewness in the data examined. The standard deviation of the arithmetic mean value is reflected in the even dispersion of the upper and lower 95% CLs (for definition see text) around the arithmetic mean value (1582 ± 478). However, the CLs are distributed unevenly around the geometric mean value (1168 = 1569 − 401, and 2097 = 1569 + 528).

STATISTICS AND PARAMETERS

A population is described by its population parameters of which the mean (μ) and the variance (σ^2) are the most commonly used population parameters. We can determine the true values of these parameters only for a finite population all of which is analysed. However, we can estimate the parameters from the *statistics* that describe the sample population in terms of its analytical mean value ($\bar{x}$) and its variance (s^2). Such estimated statistics can be used as estimates of the true population parameters. We can also derive other *descriptive statistics* including the range, the interquartile range (IQR), the median value and the mode. The median is the mid-range value that divides the data into two exactly equal parts (see Example 2.1), whilst the mode is the value that occurs most frequently; in Fig. 2.1 the value of the mode is 6.0 $\log_{10}$ cfu/g.

VARIANCE AND ERROR

Results from replicate analyses of a single sample, and/or analyses of replicate samples, will always show some variation that reflects the distribution of microbes in the sample portions tested, inadequacies of the sampling technique and technical inaccuracies of the method and the analyst. The variation can be expressed in several ways.

The *statistical range* is the simplest way to estimate the dispersion of values by deriving the differences between the lowest and the highest estimates, for example, in Example 2.1, the colony counts range from 1360 to 1970 so the range is 610. The *statistical range* is often used in statistical process control (Chapter 12) but since it depends solely on the values for the extreme counts, its usefulness is severely limited because it takes no account of the distribution of values between the two extremes. However, the IQR, sometimes called the 'mid-spread', is a robust measure of statistical dispersion between the upper and lower quartiles of the data values that is a trimmed estimate of the data range and includes only the middle 50% of values; its estimation is shown in Example 2.1.

The estimate of *population variance* is derived from the mean of the squares of the deviations of individual results from the mean result, namely $\sigma^2 = \sum_{i=1}^{n}(x_i - \mu)^2/n$, where x_i is an individual result on the ith sample, μ is the population mean value, n is the number of samples from the population that are tested and Σ indicates 'sum of'. The individual result (x_i) differs from the population mean μ by a value ($x_i - \mu$), which is referred to statistically as the *deviation*. Since the value of μ is generally unknown, the sample mean ($\bar{x}$) is used as an estimate of the population mean. The 'sample variance' (s^2) provides an estimate of the population variance (σ^2) and is determined as the weighted mean of the squares of the deviations, weighting being introduced through the concept of *degrees of freedom* (df). In calculating the mean (or the total), you have fixed that value, so all of the individual values can each be chosen at random *except* the 'last' one – since it must give the mean (or total) that has already been fixed. There are, therefore, only (n − 1) df. The unbiased estimate (s^2) of the population variance (σ^2) is thus derived from

$$s^2 = \frac{\sum_{i=1}^{n}(x-\bar{x})^2}{n-1} = \frac{n\sum_{i=1}^{n}x_i^2 - \left(\sum_{i=1}^{n}x_i\right)^2}{n(n-1)}$$

The first form of this equation [$s^2 = \sum_{i=1}^{n}(x-\bar{x})^2/(n-1)$] is simpler and is widely used but should be used only if the estimate of the mean is finite because the deviations from the mean value provide only an approximation for the absolute infinite decimal value; since the sum of the deviations from the mean value is squared, such discrepancies are additive and the derived variance may be inaccurate. This also raises a practical issue: in calculating statistics *it is essential not to round down the decimal places until the calculation is completed.* Nowadays with computerised calculations this is easily achieved, although it was not always so.

The *standard deviation* (s) of the sample mean is determined as the square root of the variance ($s = \sqrt{s^2}$). The *relative standard deviation* (RSD) is the standard deviation expressed as a fraction of the mean. It is often referred to as the *coefficient of variation* (CV), which is a percentage. The relationship between them is as follows:

$$\text{CV\%} = \text{RSD} \times 100 = \frac{s}{\bar{x}} \times 100$$

The term 'standard error' means the 'standard deviation' of a statistic such as the mean, a ratio or some other deviance. But whereas the standard deviation provides a measure of dispersion of data around a mean value, the SE_M is a measure of the precision of the mean value. The SE_M is estimated from the square root of the variance divided by the number of observations, that is, $\text{SE}_\text{M} = \sqrt{s^2/n} = s/\sqrt{n}$.

THE CENTRAL LIMIT THEOREM

This is an important statistical concept that underlies many statistical procedures. The central limit theorem is a statement about the sampling distribution of the mean values from a defined population. It describes the characteristics of the distribution of mean values that would be obtained from tests on an infinite number of independent random samples drawn from that population. The theorem states, '*for a distribution with population mean μ and a variance σ^2, the distribution of the average (sic mean value) tends to be Normal, even when the distribution from which the average is computed is non-Normal*'. The limiting Normal distribution has the same mean as the parent distribution and its variance is equal to the variance (σ^2) of the parent distribution divided by the number of independent trials (N), that is, σ^2/N.

Individual results from a finite number of independent, randomly drawn samples from the same population are distributed around the average (mean) value so that the sum of the values greater than the average will equal the sum of the values lower than the average value. If sufficient independent random samples are tested, then the statistical distribution describes the character of the population (Chapter 3). No matter what form the actual distribution takes, the distribution of the average (mean) result in repeated tests always approaches a Normal distribution when sufficient trials are undertaken. In this situation, the number of trials relates not to the number of samples *per se* but to the number of replicate trials.

Other important parameters used to describe results from replicate analyses include the confidence interval (CI), the confidence limits (CLs) and the confidence level. A CI describes an estimated range of values that is likely to include an unknown population parameter, such as a mean, with a given statistical probability. Repeat analysis of independent samples from the same population enables calculation of a CI within which a mean value would be expected to occur with a defined probability. A two-sided 95% CI is the most widely used and refers to the likelihood that repeat tests would give an estimate of the mean that falls within the probability interval of 0.25–0.975 around the estimated value. For a Normal distribution, the bounds of the 95% CI are determined approximately as $\pm 2 \times SE_M$ and those of the 99% CI as $\pm 3 \times SE_M$ around the mean value. A wide CI indicates a high level of uncertainty about the estimated population parameter and possibly suggests that more data need to be obtained before making definitive judgements. A CL is a value that defines the lower or upper bounds of a CI.

A confidence level describes the statistical probability for which a result was calculated. Since there is an expectation (α) that the result could fall outside the calculated limits of the CI, the bounds of the Cl are determined for the probability value ($1 - \alpha$). So if $\alpha = 0.05$ (ie, 5%), then the bounds of the CI are determined for a probability of 0.95 or 95%. For instance, a result might be cited with 95% confidence as $5.4 \pm 0.4 \log_{10}$ cfu/g (*sic* 5.0–5.8 $\log_{10}$ cfu/g), values which define the 95% CLs (or bounds) around the mean estimate. Such result indicates that on 1 occasion out of 20 (ie, 5%), results of repeat tests on the *same* sample would be expected to fall outside the reported CLs of 5.0–5.8 $\log_{10}$ cfu/g.

The concept of CIs is important not only in describing a single set of data values but also in the comparison of data sets. Statistical comparison of two, or more, data sets is generally based on assessment of the statistical significance of the difference between measured parameters of each data set. Such comparisons of data require the definition of a null hypothesis and an alternative hypothesis *before* doing an analysis. The null hypothesis, usually shown as H_0, might be that the two sets of data (A and B) are from the same population, that is, they do not differ (A = B); the alternative hypothesis (H_1) might be that they do differ (A ≠ B; a two-sided comparison), or that one is greater than the other (eg, A > B; a one-sided comparison). However, statistical opinion increasingly recommends that CIs should always be used to describe data sets rather than just relying on a probability value obtained in a statistical comparison. The reasons, discussed by Aitken et al. (2010) and Hector (2015), relate to considerations of a need to place greater emphasis on the sizes of estimates and intervals for statistical inference, rather than relying purely on statistical significance. Not least amongst the arguments is the importance of recognising that although an effect may be statistically significant, the relevance of the size of the effect may be of limited practical significance. This is of particular consequence, for example, when examining the effect of treatments on the survival or growth of microorganisms in foods and other matrices. Examples will be considered in later chapters.

REFERENCES

Aitken, M.R.F., Broadhurst, R.W., Hladky, S.B., 2010. Mathematics for Biological Scientists. Garland Science, Abingdon, UK.

Blood, R.M., 1974. The Clearing House Scheme. Tech Circular No. 558. Leatherhead Food Research Association.

Corry, J.E.L., Jarvis, B., Hedges, A.J., 2010. Minimizing the between-sample variance in colony counts on foods. Food Microbiol. 27, 598–603.

Hector, A, 2015. The New Statistics With R: An Introduction for Biologists. Oxford University Press, Oxford, UK.

Mossel, D.A.A., 1982. Microbiology of Foods: The Ecological Essentials of Assurance and Assessment of Safety and Quality, third ed. University of Utrecht, the Netherlands.

GENERAL READING

Glantz, S.A., 1981. Primer of Biostatistics, fourth ed. McGraw-Hill, New York.

Hawkins, D.M., 2014. Biomeasurement – A Student's Guide to Biological Statistics, third ed. Oxford University Press, Oxford, UK.

Hoffman, H.S., 2003. Statistics Explained: Internet Glossary of Statistical Terms. <http://www.animatedsoftware.com/statglos/statglos.htm> (accessed 19.03.15).

CHAPTER

Frequency distributions

3

Replicate analyses on a single sample, or analyses of replicate samples, will always show a variation between results. A variable quantity can be either *continuous* (ie, it can assume any value within a given range) or *discrete* [ie, it assumes only whole number (integer) values]. Continuous variables are normally measurements (eg, height and weight, pH values, chemical composition data, time to obtain a particular change during incubation), although an effective discontinuity is introduced by the limitations of measurement (eg, length to the nearest 0.1 mm, acidity to the nearest 0.01 pH unit). Discrete variables are typified by whole number counts, for example, the number of bacteria in a sample unit and the number of insects in a sack of corn.

When a large number of measurements or counts has been done, the observations can be organised into classes, or groups, to derive a frequency distribution. Since counts are discontinuous, each class in the frequency distribution will be an integer or a range of integers and the number of observations falling into that class will be the *class frequency*. When more than one integer is combined, the classes must not overlap and, although not essential, it is usual to take equal class intervals in simple frequency analyses.

Although the form of a frequency distribution can be seen from a tabulation of numerical data, it is more readily recognised in a histogram (ie, a bar chart). If the class intervals are equal, then the height of each column in a bar chart is proportional to the frequency. An example of the derivation of frequency distributions is given in Example 3.1. The histograms shown in Figs 3.1 and 3.2 illustrate the effect of changing the relative positions of the class boundaries on the apparent shapes of the frequency distributions.

For a frequency distribution the arithmetic mean value ($\bar{x}$) can be derived using the following formula:

$$\bar{x} = \frac{\sum fX}{n}$$

where X is the midpoint value of the frequency class and f is the frequency of values within that class. Hence, for the distribution shown in Fig. 3.1C, the mean value is

$$\bar{x} = \frac{\sum(0.950\times 3)+(0.960\times 10)+(0.970\times 9)+\cdots+(1.090\times 1)}{50}$$

$$= \frac{49.21}{50} = 0.9842$$

Statistical Aspects of the Microbiological Examination of Foods. http://dx.doi.org/10.1016/B978-0-12-803973-1.00003-6

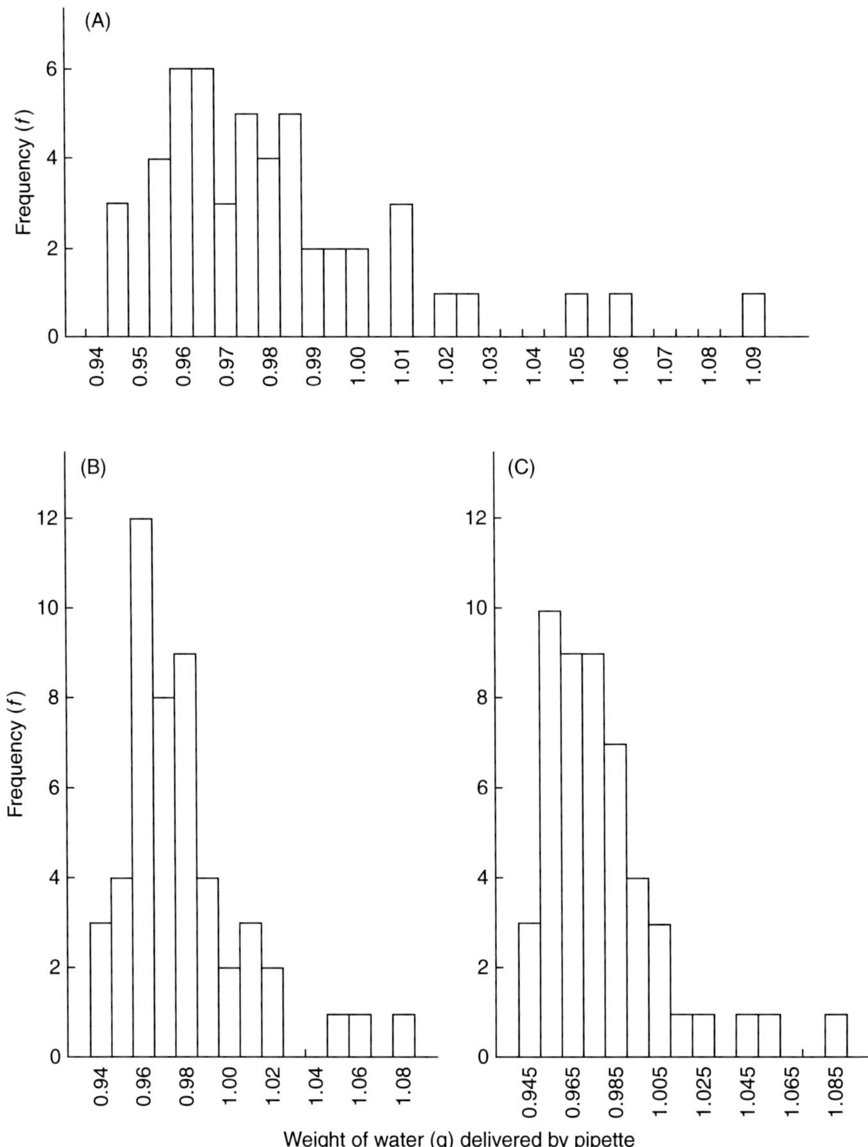

FIGURE 3.1 Frequency Distribution Histograms for Data (From the Section 'Variability in Delivery From a 1-mL (1-cm^3) Pipette' in Example 3.1) for the Delivery of Distilled Water From a 1-cm^3 Pipette

The mean value is 0.984 g and the standard deviation is 0.028 g. (A) Class intervals 0.005 g; (B and C) class intervals 0.010 g

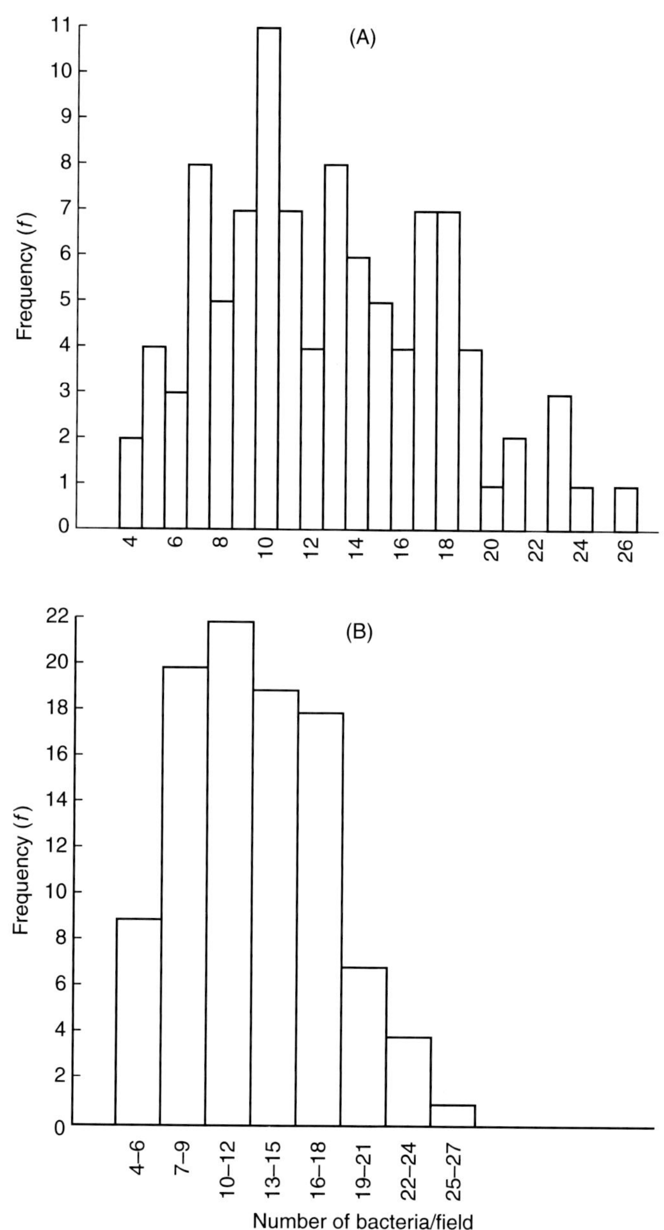

FIGURE 3.2 Frequency Distribution of Microscopic Counts of Bacteria Having a Mean Count of 12.77 Bacteria/Field and a Variance of 24.02 (From the Section 'Variability in Numbers of Bacterial Cells Counted Microscopically' in Example 3.1)

(A) Class interval = 1; (B) class interval = 3

The unbiased estimate of the population variance (s^2) is given by

$$s^2 = \frac{\sum fX^2 - \bar{x}\sum fX}{n-1}$$

with

$$\begin{aligned}\bar{x}\sum fX &= 0.9842 \times 49.21 \\ &= 48.432\end{aligned}$$

and

$$\begin{aligned}\sum fX^2 &= 3(0.95)^2 + 10(0.96)^2 + 9(0.97)^2 + \cdots + 1(1.09)^2 \\ &= 48.472\end{aligned}$$

Hence, the estimate of variance is given by

$$\begin{aligned}s^2 &= \frac{48.472 - 48.432}{49} \\ &= \frac{0.04}{49} = 0.0008163\end{aligned}$$

and the standard deviation is given by $s = \sqrt{0.0008163} = \pm 0.0286$.

Hence, for these data, the mean weight of water delivered from a 1-cm^3 pipette was 0.984 g, with a standard deviation of 0.0286 g. (Using an alternative method illustrated in Example 2.1, the calculated mean value was 0.983 g and the standard deviation was ±0.0287 g.)

TYPES OF FREQUENCY DISTRIBUTION

Mathematically defined frequency distributions can be used as models for experimental data obtained from any population. Assuming that the experimental data fit approximates to one of the models, then, amongst other things:

1. The spatial (ecological) dispersion of the population can be described in mathematical terms.
2. The variance of population parameters can be estimated.
3. Temporal and spatial changes in density can be compared.
4. The effect of changes in environmental factors can be assessed.

The mathematical models used most commonly for analysis of microbiological data include the Normal (Gaussian), the lognormal, binomial, Poisson and negative binomial distributions, which are described later. The parameters of the distributions are summarised in Table 3.1.

Table 3.1 Some Continuous and Discrete Distribution Functions

Name	Domain	Probability Density Function, f_x	Restriction on Parameters	Mean	Variance
Normal[a] (Gaussian)	$-\infty < x < \infty$	$\frac{1}{\sigma\sqrt{2\pi}}\exp\left[-\frac{1}{2}\left(\frac{x-m}{\sigma}\right)^2\right]$	$-\infty < m < \infty$ $0 < \sigma < \infty$	m	σ^2
Binomial	$x_s = s$, for $s = 0, 1, 2, \ldots, n$	$\binom{n}{s} p^s(1-p)^{n-s}$	$0 < p < 1$ $q = 1 - p$	np	npq
Poisson[b]	$x_s = s$, for $s = 0, 1, 2, \ldots, \infty$	$\frac{e^{-m} m^s}{s!}$	$0 < m < \infty$	m	m
Negative binomial	$x_s = s$, for $s = 0, 1, 2, \ldots, \infty$	$\binom{n+s-1}{s} p^n(1-p)^s$	$n \geq 0$ and $0 < p < 1$ ($p = 1/Q$) and $1 - p = P/Q$	nP	nPQ

[a]*The Normal distribution is the limiting form of the binomial distribution when* $n \to \infty$ *and* $p \to 0$.
[b]*Limiting form of binomial, as* $p \to 0$, $q \to \infty$.

EXAMPLE 3.1 DERIVATION OF THE MEAN, VARIANCE AND FREQUENCY DISTRIBUTION

Variability in delivery from a 1-mL (1-cm^3) Pipette

A pipette was used to transfer 1 cm^3 of distilled water into a tared weighing boat. The weight of water delivered was determined on an analytical balance. The experiment was performed 50 times and gave the following results (g water):

0.948, 1.012, 1.085, 1.063, 1.010, 1.000, 0.994, 0.986, 0.995, 0.999, 0.969, 0.965, 0.945, 0.977, 0.957, 0.946, 0.960, 0.955, 1.010, 0.965, 0.975, 0.972, 0.957, 0.961, 0.975, 0.988, 0.989, 0.974, 0.980, 0.980, 1.001, 0.977, 1.021, 1.051, 0.965, 0.963, 0.971, 0.983, 0.962, 0.984, 0.978, 0.968, 0.960, 1.027, 0.959, 0.985, 0.985, 0.967, 0.960, 0.992

The data are organised in distribution frequency classes as shown in Table 3.2 and Fig. 3.1.

Variability in numbers of bacterial cells counted microscopically

Data of Ziegler and Halvorson (1935): their Appendix I, slide 1.

One hundred microscope fields on a single slide were examined and the number of bacteria counted per field was determined. The frequency distributions of the data are derived from the following:

No. of Bacteria/Field (x)	Frequency (f)		No. of Bacteria/Field (x)	Frequency (f)	
4	2		16	4	
5[a]	4	9	**17**	7	18
6	3		18	7	
7	8		19	4	
8	5	20	**20**	1	7
9	7		21	2	
10	11		22	0	
11	7	22	**23**	3	4
12	4		24	1	
13	8		25	0	
14	6	19	**26**	1	1
15	5		27	0	

[a]Values in bold text are the midfield values (X).

$N = 100$

$$\bar{x} = \frac{\Sigma fX}{N}$$

$$= \frac{(5\times 9)+(8\times 20)+(11\times 22)+(14\times 19)+(17\times 18)+(20\times 7)+(23\times 4)+(26\times 1)}{100}$$

$= 12.77$

where $\bar{x}$ is the mean value, X is the mid-value of the frequency classes (ie, 5, 8, 11, …) and f is the class frequency value.

$$s^2 = \frac{\Sigma f\left(x^2\right) - \bar{x}\Sigma fX}{N-1} = \frac{18,685 - 12.77(1277)}{99} = 24.0173$$

$s = \sqrt{24.0173} = \pm 4.9007$

The frequency distributions are shown in Fig. 3.2.

Table 3.2 Arrangement of Data in Frequency Classes

Frequency Class Boundaries (g)	Frequency Class Midpoint (X)	Frequency (f) With Interval 0.005 g (Fig. 3.1A)	Frequency Class Boundaries (g)	Frequency Class Midpoint (X)	Frequency (f) With Interval 0.010 g (Fig. 3.1B)	Frequency Class Boundaries (g)	Frequency Class Midpoint (X)	Frequency (f) With Interval 0.010 g (Fig. 3.1C)
0.943–0.947	0.945	1	0.9400–0.9499	0.945	3	0.9450–0.9549	0.950	4
0.948–0.952	0.950	2	0.9500–0.9599	0.955	4	0.9550–0.9649	0.960	11
0.953–0.957	0.955	3	0.9600–0.9699	0.965	12	0.9650–0.9749	0.970	7
0.958–0.962	0.960	6	0.9700–0.9799	0.975	8	0.9750–0.9849	0.980	11
0.963–0.967	0.965	5	0.9800–0.9899	0.985	9	0.9850–0.9949	0.990	6
0.968–0.972	0.970	4	0.9900–0.9999	0.995	4	0.9950–1.0049	1.000	3
0.973–0.977	0.975	5	1.0000–1.0099	1.005	2	1.0050–1.0149	1.010	3
0.978–0.982	0.980	3	1.0100–1.0199	1.015	3	1.0150–1.0249	1.020	2
0.983–0.987	0.985	5	1.0200–1.0299	1.025	2	⋮	⋮	⋮
0.988–0.992	0.990	3	⋮	⋮	⋮	1.0450–1.0549	1.050	1
0.993–0.997	0.995	2	⋮	⋮	⋮	1.0550–1.0649	1.060	1
0.998–1.002	1.000	3	1.0500–1.0599	1.055	1	⋮	⋮	⋮
1.003–1.007	1.005	0	1.0600–1.0699	1.065	1	1.0850–1.0949	1.090	1
1.008–1.012	1.010	3	⋮	⋮	⋮			
1.013–1.017	1.015	0	1.0800–1.0899	0.085	1			
1.018–1.022	1.020	1						
1.023–1.027	1.025	0						
1.028–1.032	1.030	1						
⋮	⋮	⋮						
1.048–1.052	1.050	1						
⋮	⋮	⋮						
1.063–1.068	1.065	1						
⋮	⋮	⋮						
1.083–1.087	1.085	1						

STATISTICAL PROBABILITY

Probability is about chance and the likelihood that an event will, or will not, occur in any specific situation. For instance, the probability that a specific person will win the jackpot in the National Lottery is very low (about 1 in 14 million) because the odds against winning are enormous – yet people do win the Lottery, which shows that no matter how improbable an event there is always the possibility that it will occur. By contrast, the chance of getting snow in winter if you live in Norway, Russia or Canada is very high – one might say it is certain to occur – but if you live in England the chance is not very high.

If a standard coin is tossed once, it will fall to show either a 'head' or a 'tail', so there is a 50% probability for obtaining a head (or a tail) in a single throw. This is written as $p = 0.50$, $q = 0.50$. Note that $(p + q) = 1$, where p is the probability that an event will occur (ie, of obtaining a 'head') and q is the probability of failure (ie, a head will not occur but a tail will be obtained).

Hence, there are two mutually independent ways of obtaining either a head (H) or a tail (T) with a single coin and, for two or more coins, we could get either H or T for each coin. Hence, if the first coin falls as H, the second or subsequent coins could also fall as either H or T, that is, the first and second trials are independent. Such trials are governed by the 'multiplication rule': the probabilities for a specific event to occur in the first trial (eg, $p = 0.5$) and for a similar specific event to occur in the second (and subsequent trials) are multiplied to obtain the probability for the two (or more) independent events. So for two coins if the first outcome is H, the subsequent outcome could be either H or T so that we get HH or HT in two trials; likewise, if the first coin falls as T, then we could get TH or TT in two trials. So by tossing 2 coins there are 4 possible outcomes: HH (1 in 4 chance for which $p = 0.5 \times 0.5 = 0.25$), HT ($p = 0.25$), TH ($p = 0.25$) or TT ($p = 0.25$). These events are mutually exclusive because a 'standard' coin has both a head and a tail either, but not both, of which can occur when a spun coin falls. Although the occurrence of either HT or TH is essentially the same in that we have one of each, statistically they are mutually exclusive, as there are two ways in which the events could occur. So the probable outcome for *either* HT or TH is governed by the 'addition rule' and the combined probability for either HT or TH is $p = 0.25 + 0.25 = 0.50$. If we toss three coins, there are eight possible outcomes: HHH (0.125), HHT (including HTH and THH; 0.375), HTT (including THT and TTH; 0.375) or TTT (0.125).

We can determine the number of outcomes quite simply because for a single toss of the coin the probability for obtaining either H or T is 0.5; with 2 coins the probability of obtaining HH $= 0.5 \times 0.5 = (0.5)^2 = 0.25$ (1 in 4); with 3 coins the probability of HHH is $(0.5)^3$, that is, 0.125 or 1 in 8; and so on. We can generalise these probabilities by saying that the relative probability of a specified outcome for any given number of trials (n) is determined by the expansion of the binominal probability distribution function where $(p + q)^n = 1$. Pascal's triangle (Fig. 3.3) provides a simple 'ready reckoner' for any number of independent trials where the event is either positive or negative.

No. of trials	Possible outcomes
	1
1	1 1
2	1 2 1
3	1 3 3 1
4	1 4 6 4 1

FIGURE 3.3 Pascal's Triangle

A visual illustration of binomial outcomes from a series of trials, for instance, the toss of a coin. In a single trial there are two possible outcomes: a head or a tail, with equal probability ($p = 0.5$). In two trials there are four possible outcomes: two heads ($p = 0.25$), one head and one tail ($p = 0.5$) or two tails ($p = 0.25$). In 3 trials there are 8 possible outcomes, in 4 trials there are 16 possible outcomes, etc. Note that each value in a line is the sum of the values immediately above it. The figure can be expanded by adding data for additional trials

Provided that the outcome of one event (A) does not affect the outcome of a second event (B), the events are totally independent; however, if event A can affect the possible outcome of event B, then the events are not independent. Suppose that we have a bag containing 20 balls: 4 red (R), 6 blue (B) and 10 green (G). The probabilities of randomly drawing 1 R, 1 B or 1 G are 4/20 (0.2), 6/20 (0.3) or 10/20 (0.5), respectively. Assuming that after any ball is drawn it is returned to the bag, subsequent events are totally independent and the overall probability is described by the 'multiplication rule' (see earlier text). So the probability that we can draw sequentially 1 R, 1 B and 1 G balls, in three draws when the drawn ball is replaced before the next draw, is as follows: $P(R)P(B)P(G) = 0.2 \times 0.3 \times 0.5 = 0.03$, where $P(R)$ means the probability of drawing a red ball (R), etc. Note that this is 'sampling with replacement' because each event is totally independent. But if the first red ball is not replaced (ie, 'sampling without replacement'), the probabilities for drawing 1 R, 1 B or 1 G on a second draw are 3/19, 6/19 and 10/19, respectively. Hence, the probability for drawing 1 R, 1 B and 1 G in that order would be $P(R)P(B)P(G) = (4/20) \times (6/19) \times (10/18) = 0.2 \times 0.32 \times 0.56 = 0.036$.

If half of the balls in each colour are marked with an odd number (O) and the remainder with an even (E) number, the probability of drawing an even ball is $P(E) = 10/20 = 0.5$, as also is the probability of drawing an odd ball, $P(O) = 0.5$. The probability of drawing a ball that is either blue or green is $P(\text{not R}) = 16/20 = 0.80$ and the probability of drawing a ball that is odd but not red is $P(\text{not R})P(O) = 0.8 \times 0.5 = 0.4$. The probability of drawing an even red ball is $P(R)P(E) = (4/20) \times (10/20) = 0.2 \times 0.5 = 0.1$. These are independent events so the multiplication rule applies.

What would be the probability of drawing a ball that is either red *or* odd *but not* both? The probability of drawing an even red ball is $P(\text{R})P(\text{E}) = 0.1$ or a non–red odd ball is $P(\text{not R})P(\text{O}) = 0.4$. These are mutually exclusive ways of achieving the desired result so the *addition rule* applies: the combined probability for drawing either a red even ball or an odd ball that is not red is $P(\text{R})P(\text{E}) + P(\text{not R})P(\text{O}) = 0.1 + 0.4 = 0.50$.

Let us extend this concept. A pack of playing cards consists of 52 cards divided into 4 suits, each of which contains cards numbered from 1 to 10 plus a jack, a queen and a king. If we shuffle the pack and then draw the top card, the independent chance of drawing an ace is 1 in 13 (because there are 4 aces in the 52 cards) but the chance of drawing the ace of spades is only 1 in 52. If we replace that card and, after shuffling the pack, again draw 1 card, there is still only a 1 in 13 chance of drawing an ace or 1 in 52 chance of drawing the ace of spades – this is 'sampling with replacement'. Suppose now that we shuffle the cards and discard the top card if it is not the ace of spades, then the chance that the next card is the ace of spade is 1 in 51. Likewise, if we continue the sequence, there is now a 1 in 50 chance that the next card will be the ace of spades, and so on.

However, if we were to lay down the top four cards, the chance that any one of those four cards will be an ace is governed by the hypergeometric distribution that describes the concept of 'sampling without replacement'. The equation to calculate the probability (P) for drawing 2 aces (ie, positive events) when 4 cards are drawn from a pack of 52 cards is given by

$$P(X = x) = h(x, n, N, k) = \frac{\left({}^{k}C_{x}\right)\cdot\left({}^{N-k}C_{n-x}\right)}{{}^{N}C_{n}}$$

where C means 'choose' (ie, ${}^{k}C_{x}$ means choose x from k), N is the number of items in the population ($N = 52$), k is the possible number of successes (number of aces = 4), n is the number of items in the sample (ie, the number of cards drawn = 4) and x is the number of items in the sample that are deemed to be successes (ie, the number of aces sought = 2).

Putting these numbers into the equation the probability for picking two aces is given by

$$h(x, n, N, k) = h(2, 4, 52, 4) = \frac{\left({}^{4}C_{2}\right)\cdot\left({}^{52-4}C_{4-2}\right)}{{}^{52}C_{4}}$$

$$= \frac{(4!/\{2![(4-2)!]\})\cdot(48!/\{2![(48-2)!]\})}{52!/\{4![(52-4)!]\}} = 0.0250$$

Note that the symbol '!' means factorial, so $4! = 4 \times 3 \times 2 \times 1 = 24$, etc.; also note, for example, that $4!/(2! \times 2!) = (4 \times 3 \times 2 \times 1)/[(2 \times 1)(2 \times 1)] = 3 \times 2 = 6$.

The probabilities of drawing 0, 1, 2, 3 or 4 aces in any 4 cards drawn are approximately 0.72, 0.26. 0.025, <0.01 and $\ll 0.01$, respectively. If, as happens in a casino, 2 packs of cards are shuffled, then the chance that the first 4 cards drawn will all be aces (from the 8 aces in the 104-card deck) increases slightly from $P = 4 \times 10^{-6}$ to 1.5×10^{-5}.

In practice there is no necessity to work out probabilities manually since Excel™ and other spreadsheets provide a function for the hypergeometric calculation.

THE BINOMIAL DISTRIBUTION ($\sigma^2 < \mu$)

If we toss a number of coins, the average probability of equal numbers of heads and tails is $p = q = (1 - p) = 0.5$, but if all coins were 'double-headed', the probability of a 'head' occurring would be $p = 1.0$ (ie, there would be no chance of obtaining a 'tail' so $q = 0$). We can therefore use the concept of probability to answer the general question: 'What is the chance that a specific event will occur?' The probability scale ranges from $p = 0$ for impossible events to $p = 1$ for events that are certain to occur.

In the binomial distribution, p is the probability that an event will occur and q is the probability that the event will not occur, so $(p + q) = 1$. If p and q remain constant in each of a given number (n) of individual independent trials, then the probability series is described by the general expression $(p + q)^N$. The individual terms are given by the binomial expansion:

$$P_x = {}^N C_x (q^{N-x} p^x) = \binom{N}{x} (q^{N-x} p^x) = \frac{N!}{x!(N-x)!} (q^{N-x} p^x)$$

where P_x is the probability of finding x individuals in a sample, N is the number of times the test is repeated and $N!$ means factorial N. The factorial term

$${}^N C_x = \binom{N}{x} = \frac{N!}{x!(N-x)!}$$

counts the number of mutually exclusive ways to obtain a particular outcome. The population parameters of mean (μ) and variance (σ^2) are given by $\mu = Np$ and $\sigma^2 = Npq$.

The binomial distribution is used as a model when a specific characteristic of an individual in a sample can be recognised, for example, the prevalence of defective items in a lot. The distribution is often used as the basis for setting up sampling schemes in acceptance sampling of foods and other materials (Chapter 5). In such schemes p is defined as the probability that the sample will be defective (ie, contaminated), q is the probability that a sample will not be defective (ie, not contaminated) and N is the number of samples tested. The expected probabilities, derived from the expansion of $(p + q)^N$, can be calculated (Example 3.2) or obtained from Tables of Binomial Probability, such as those given by the National Bureau of Standards (1950), Fisher and Yates (1974) and Pearson and Hartley (1966).

Example 3.2 shows that with a mean *Clostridium botulinum* contamination level of 2 spores/fish, and a probable maximum level of 10 spores/fish, the chance that no sample would contain the organism would be 11 in 100 (approximately 1 in 10). This theme is developed further in Chapter 5 in relation to sampling schemes and the use of quantal (ie, presence/absence) tests for specific organisms.

Fig. 3.4 shows the binomial probability distribution for different values of the probability for successes (p) and different numbers of trials (N). As the values of n and p increase the shape of the binomial distribution approaches the bell-shaped curve, described as the Normal (or Gaussian) distribution for continuous variables, that is, measurements rather than counts.

EXAMPLE 3.2 CALCULATION OF EXPECTED FREQUENCIES FOR A POSITIVE BINOMIAL DISTRIBUTION

In a sample (N) of 100 farmed trout, the mean level of Clostridium botulinum spores detected was 2 spores/fish. It is widely believed that the maximum likely prevalence of contamination is 10 spores/fish (n = 10). What is the frequency distribution and what are the chances of not detecting the organism?

We can use the binomial distribution to determine the probability (P) that no contamination is detectable [$P_{(x=0)}$], or that 1, 2, ..., 10 spores will be detected [$P_{(x=1)}$, $P_{(x=2)}$, ..., $P_{(x=10)}$]. To do this, a sample estimate ($\hat{P}$) of the overall population value in relation to the maximum contamination level expected (N) is derived as follows.

The probability of contamination is given by

$$\hat{p} = \frac{x}{N} = \frac{2}{10} = 0.2$$

hence:

$$\hat{q} = 1 - \hat{p} = 0.8$$

Expected probabilities are given by the expansion of $(\hat{p}+\hat{q})^N = (0.2+0.8)^{10}$, that is:

$$P_{(x)} = \frac{N!}{x!(N-x)!}[\hat{q}^{N-x}\hat{p}^{x}]$$

For the values given, the probability of detecting no *C. botulinum* spores in a fish is

$$P_{(x=0)} = \frac{10!}{0!(10-0)!}[(0.8)^{10-0}(0.2)^0] = (0.8)^{10} = 0.1073$$

It is worthy of note that $P_{(x=0)} = q^N$: for $N = 10$ and $q = 0.80$, $q^N = 0.80^{10} = 0.1073$.

The probability of detecting 1 spore/fish is

$$P_{(x=1)} = \frac{10!}{1!(10-1)!}[(0.8)^{10-1}(0.2)^1] = 10 \times (0.8)^9 \times (0.2)^1 = 0.2684$$

The probability of detecting 2 spores/fish is

$$P_{(x=2)} = \frac{10!}{2!(10-2)!}[(0.8)^{10-2}(0.2)^2] = 45 \times (0.8)^8 \times (0.2)^2 = 0.3020$$

The probabilities for 3, 4, 5, 6, 7, 8, 9 and 10 spores/fish are derived similarly.

The total probability $\simeq$ 1.0. The expected frequencies of occurrence of *C. botulinum* spores in 100 fish are given by $f = NP_{(x)}$, where N is the number of fish examined.

x (Spores/Fish)	$P_{(x)}$	$f = NP_{(x)}$	*f* (as Integer)[a]
0	0.1074	10.74	11
1	0.2684	26.84	27
2	0.3020	30.20	30
3	0.2013	20.13	20
4	0.0881	8.81	9
5	0.0264	2.64	3
6	0.0055	0.55	1
7	0.0008	0.08	0
8	<0.0001	<0.01	0
9	<0.0001	<0.01	0
10	<0.0001	<0.01	0
Total	1.0000	100.00	101

[a]That is, to the nearest whole number.

Hence, with a mean contamination level of 2 spores/fish, and a probable maximum of 10 spores/fish, the probability of not detecting any *C. botulinum* spores would be about 11/100, or slightly more than 1 in 10. This theme is developed further in Chapter 5 with reference to sampling schemes; and Errors Associated With Quantal Response Methods in relation to the use of presence or absence tests for specific organisms.

MULTINOMIAL DISTRIBUTION

The binomial is one of a family of related distributions known collectively as the multinomial distribution. (*NB*: This is not the same as the multimodal distribution which models the occurrence of data with more than one mode.) The multinomial is an extension of the binomial distribution that applies in circumstances when there are more than two classes of data. For instance, a three-class sampling plan (Chapter 5) is a trinomial extension of a two-class binomial distribution sampling plan. The multinomial distribution could be used to model ecological changes in the proportions of populations of different organisms over time: for instance, in the curing of meats and in wine fermentation where the proportion of different species of bacteria and yeast changes as some organisms die and others take their place. Another example might be in a study of chicken caecal contents that may contain a dominant population of either *Campylobacter jejuni* or *C. coli* whilst others contain neither or both. There are therefore four classes of contamination. From knowledge of the prevalence of contamination it is possible to predict the likelihood that particular scenarios might occur.

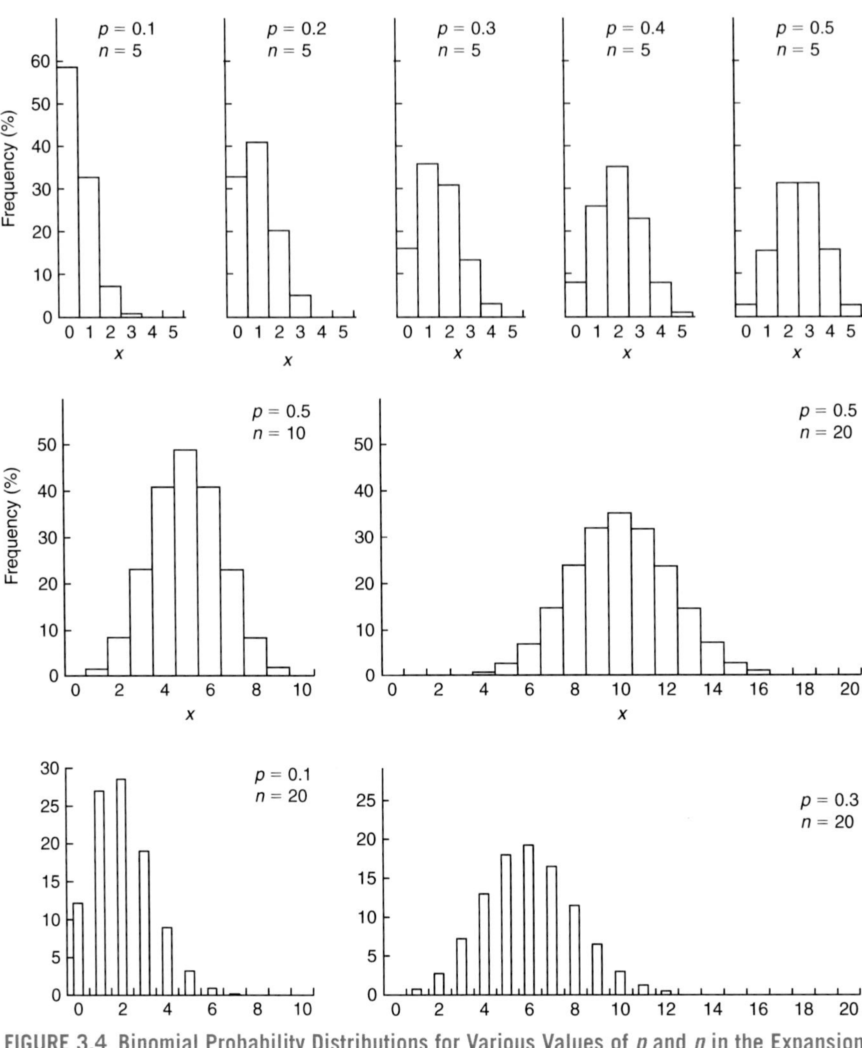

FIGURE 3.4 Binomial Probability Distributions for Various Values of *p* and *n* in the Expansion $(p + q)^n$

The probability (P) of an outcome governed by the multinomial distribution can be derived from the probability density function (pdf):

$$P = \frac{N!}{(n_1!)(n_2!)\cdots(n_k!)} p_1^{n_1} \cdot p_2^{n_2} \cdot \cdots \cdot p_k^{n_k}$$

where N is the total number of events $\left(= \Sigma_1^k n_k\right)$, with $n_1, n_2, \ldots, n_k$ the number of times outcomes 1, 2, …, k are expected to occur and $p_1, p_2, \ldots, p_k$ the probabilities that outcomes 1, 2, …, k will occur, raised to the power $n_1, n_2, \ldots, n_k$, respectively.

The pdf for the binomial is given by

$$P = \frac{N!}{n_1!n_2!}(p_1^{n_1} p_2^{n_2}) = \frac{N!}{n_1!(N-n_1)!}(p^{n_1} q^{N-n_1})$$

where $q = 1 - p_1$, and the pdf for the trinomial distribution (Chapter 5) is given by

$$P = \frac{N!}{n_1!n_2!n_3!} p^{n_1} q^{n_2} (1-p-q)^{N-n_1-n_2}$$

where $q = 1 - p$ and $N = n_1 + n_2 + n_3$.

THE NORMAL DISTRIBUTION

The *Normal distribution* refers to a family of 'continuous' probability distributions that have the same generic shape: they are symmetrical curves with more values concentrated in the centre of the curve and fewer in the tails (Fig. 3.5). They differ therefore from the probability distribution histograms shown for 'discrete data', for example, the binomial distribution (Fig. 3.4). The shape of the curve is described by the general expression:

$$f(X) = \frac{1}{\sqrt{2\pi\sigma^2}} e^{-[(x-\mu)^2/2\sigma^2]}$$

where $f(X)$ is the probable density of the derived variable X (the standard Normal deviate), with $X = (x - \mu)/\sigma$, x is the random variable (ie, observed value), μ is the population mean and σ is the standard deviation of the population.

An estimate of the value of the standard Normal deviate (X) can be determined from sample data (x) using $Z = (x - \bar{x})/s$. In other words, by subtracting the mean value ($\bar{x}$) from each observed value (x) and dividing by the estimate of the standard deviation (s) a series of standardised deviates (Z) is derived. These can be obtained from Tables of the Standardised Normal Deviate (Pearson and Hartley, 1966).

The *central limit theorem* states that if a large number of random variables (ie, sample data) are selected from (almost) any distribution, with mean μ and variance σ^2, then the means of these samples will themselves follow a Normal distribution with a mean $\mu_{\bar{x}}$ and a standard deviation $\sigma_{\bar{x}}$, which is the standard error of the mean, with $\sigma_{\bar{x}} = \sigma/\sqrt{N}$. Hence, most distributions approach the Normal distribution as the sample size increases. As will be seen later, distributions such as the Poisson and the binomial tend to approach normality when the number of samples tested tends to infinity.

A unique property of the Normal distribution is that the mean and variance are independent and the shape of the distribution is a function of the population variance parameter. From Fig. 3.5 it can be seen that 95.45% of observations occur within ±2 standard deviations (σ) from the mean (μ) and that 99.73% lie within $\pm 3\sigma$ from the mean.

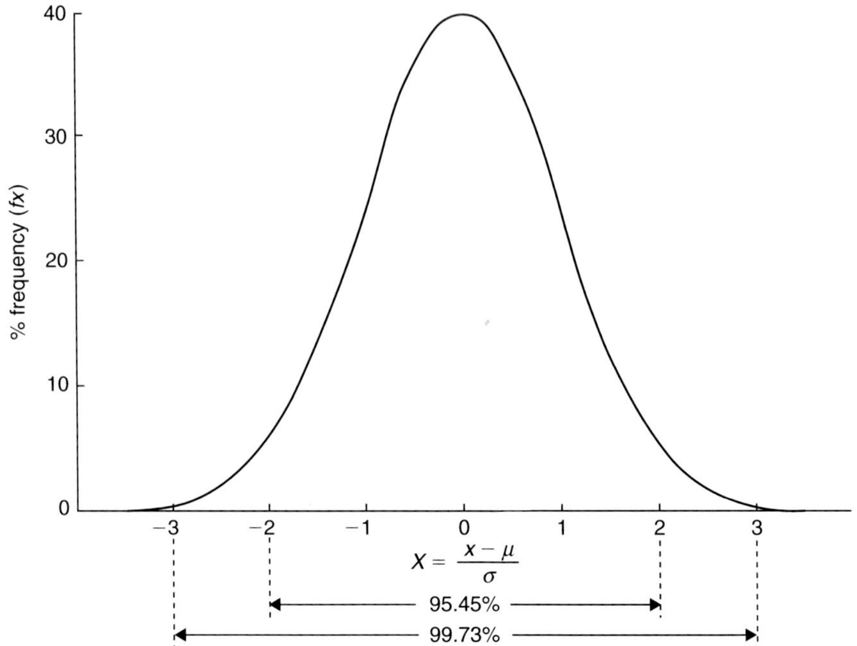

FIGURE 3.5 A Normal (Gaussian) Distribution Curve Described by the Equation $f(x) = (1/\sqrt{2\pi\sigma^2})e^{-(x-\mu)^2/\sigma^2}$, Where μ is the Population Mean With Variance σ^2

The Normal distribution is rarely a suitable model for microbiological data as measured, but it is very important because many parametric statistical tests are based on the Normal distribution. Such tests include analysis of variance (ANOVA) and tests of significance of differences. Thus, for microbiological data, steps have to be taken to transform the data such that they become 'Normally' distributed.

THE POISSON DISTRIBUTION ($\sigma^2 = \mu$)

The arithmetic mean of the discrete binomial distribution is given by $\mu = Np$ and the variance $\sigma^2 = Npq = \mu q$. Since $q = 1 - p = 1 - (\mu/N)$, $\sigma^2 = \mu q = \mu[1 - (\mu/N)] = \mu - (\mu^2/N)$. Hence, if N is finite, the variance will always be less than the mean ($\sigma^2 < \mu$).

When the probability that an event will occur is low ($p \to 0$) and n approaches infinity ($n \to \infty$) with np fixed and finite, the binomial distribution approaches the discrete Poisson distribution, since $\sigma^2 = \mu[1-(\mu/N)]$, as $N \to \infty$, $\mu/N \to 0$ and $\sigma^2 \to \mu$. The Poisson distribution is described by the equation $P_x = e^{-\lambda}(\lambda^x/x!)$, where P_x is the probability that x individuals occur in a sampling unit, λ is the Poisson parameter ($\lambda = \mu = \sigma^2$) and e is the exponential value (2.7183). Unlike the binomial distribution, which is a function of two parameters (N and p), only one parameter (λ) is needed for the Poisson series, since $\lambda = \mu = Np$. The Poisson parameter (λ) is estimated by the statistic m, where $m = s^2$; thus $P_x = \mathrm{e}^{-m}(m^x/x!)$.

The probabilities of 0, 1, 2, 3, etc., individuals per sampling unit are given by the individual terms of the expansion of this equation; thus:

$$P_{(x=0)} = \mathrm{e}^{-m}$$

$$P_{(x=1)} = \mathrm{e}^{-m}\left(\frac{m}{1!}\right) = P_{(x=0)}(m)$$

$$P_{(x=2)} = \mathrm{e}^{-m}\left(\frac{m^2}{2!}\right) = P_{(x=1)}\left(\frac{m}{2}\right)$$

$$P_{(x=3)} = \mathrm{e}^{-m}\left(\frac{m^3}{3!}\right) = P_{(x=2)}\left(\frac{m}{3}\right)$$

and so on.

This can be generalised as $P_{(x+1)} = P_{(x)} \cdot m/(x+1)$; note that this is a recurrence equation permitting each term of the series to be calculated from the previous, starting from $P_{(x=0)}$.

The individual terms can be calculated manually (Example 3.3), calculated by readily available computer programmes or obtained from standard tables; for example, Pearson and Hartley (1966) give individual terms of the Poisson distribution for different values of m from 0.1 to 15.0. The expected frequency (nP_x) is obtained by multiplying each term of the series (P_x) by the number of sampling units (n).

Since the Poisson distribution is associated with rare events which can be considered to occur randomly, for example, the distribution of industrial accidents over a long time period or the distribution of small numbers of bacteria in a large quantity of food, tests for agreement with a Poisson distribution will be tests for randomness of distribution. We have noted earlier that the Poisson distribution is a special version of the binomial distribution where $p \to 0$ as $N \to \infty$ and $\sigma^2 \to \mu$.

Certain conditions must be met if the Poisson series is to be used as a mathematical model for bacterial counts in a food sample:

1. The number of individual organisms per sampling unit (k) must be well below the maximum possible number that could occur ($k \to \infty$).
2. The probability that any given position in the sampling unit is occupied by an organism is both constant and very small (constant $p \to 0$); consequently, the probability that that position is not occupied by a particular organism is high ($q \to 1$).
3. The presence of an individual organism in any given position must neither increase nor decrease the probability that another organism occurs nearby.
4. The sizes of the samples must be small relative to the whole population; that is, the (statistical) 'sample space' is large compared to the 'event space', for example, a single bacterial cell is small compared to 1 mL.

EXAMPLE 3.3 CALCULATION OF THE EXPECTED FREQUENCIES OF A POISSON DISTRIBUTION

It is intended to inoculate 1000 bottles of meat slurry with a suspension of Clostridium botulinum spores at an average level of 10 spores/bottle. What are the expected frequencies for (a) less than 1 spore, (b) less than 5 spores and (c) more than 15 spores/bottle?

The Poisson equation is $P_x = (m^x/x!)e^{-m}$, where m is the mean count, x is the number for which the probability is needed and e is the exponential factor (=2.71828).

For the intended mean inoculum level = m = 10 with N = 1000, the probability of 0 spores/bottle (ie, <1 per bottle) is

$$P_{(x=0)} = \frac{(10)^0}{0!}e^{-10} = e^{-10} = 0.0000454$$

Hence, the expected frequency per 1000 bottles = $NP_{(x)} = 0.0454$.

An alternative way to express this is that only 1 in 22,000 bottles would be expected not to contain at least 1 spore.

The succeeding terms of the Poisson series are used to calculate the remaining probabilities. For simplicity, the generalised equation $P_{(x+1)} = (P_x \cdot m)/x$ is used:

$$P_{(x=0)} = e^{-m} = e^{-10} = 0.0000454$$

$$P_{(x=1)} = \frac{P_{(x=0)} \cdot m}{x} = \frac{0.0000454 \times 10}{1} = 0.000454$$

$$P_{(x=2)} = \frac{P_{(x=1)} \cdot m}{x} = \frac{0.000454 \times 10}{2} = 0.00227$$

$$P_{(x=3)} = \frac{0.00227 \times 10}{3} = 0.00757$$

$$P_{(x=4)} = \frac{0.00757 \times 10}{4} = 0.018917$$

and so on.

From the data in Table 3.3, the cumulative probability of less than 5, and more than 15, spores/bottle would be P = 0.291 and 0.0484, respectively. Hence, in 1000 replicate inoculated bottles the expected frequencies would be 29 and 49, respectively. Thus 922 of the 1000 bottles would be expected to contain between 5 and 15 spores/bottle and all bottles would be expected to contain at least 1 spore.

Table 3.3 Individual Terms ($x = 0$–24) of a Poisson Distribution for $m = 10$ and $N = 1000$

x	P_x	$f = NP_x$	*f* (as Integer)	
0	$P_{(x=0)} = e^{-10} = 0.0000454$	0.0454	0	29
1	$P_{(x=1)} = \frac{P_{(x=0)} \times 10}{1} = 0.000454$	0.454	0	
2	$P_{(x=2)} = \frac{P_{(x=1)} \times 10}{2} = 0.002270$	2.270	2	
3	$P_{(x=3)} = \frac{P_{(x=2)} \times 10}{3} = 0.00757$	7.567	8	
4	$P_{(x=4)} = \frac{P_{(x=3)} \times 10}{4} = 0.01892$	18.92	19	
5	$P_{(x=5)} = \frac{P_{(x=4)} \times 10}{5} = 0.03783$	37.83	38	
6	$P_{(x=6)} = \frac{P_{(x=5)} \times 10}{6} = 0.06306$	63.06	63	
7	$P_{(x=7)} = \frac{P_{(x=6)} \times 10}{7} = 0.09008$	90.08	90	
8	$P_{(x=8)} = \frac{P_{(x=7)} \times 10}{8} = 0.11260$	112.6	113	
9	And so on = 0.12511	125.1	125	
10	= 0.12511	125.1	125	
11	= 0.11374	113.7	114	
12	= 0.09478	94.78	95	
13	= 0.07291	72.91	73	
14	= 0.05208	52.08	52	
15	= 0.03472	34.72	35	
16	= 0.02170	21.70	22	49
17	= 0.01276	12.76	13	
18	= 0.007091	7.091	7	
19	= 0.003732	3.732	4	
20	= 0.001866	1.866	2	
21	= 0.000889	0.889	1	
22	= 0.000404	0.404	0	
23	= 0.000176	0.176	0	
24	= 0.000073	0.073	0	
Total		$\Sigma NP_{(x)} = 999.90717$		

f *is the expected frequency of occurrence of a spore inoculum (for details see Example 3.3).*

Table 3.4 'Exact' 95% Confidence Limits for a Poisson Distribution

Poisson Mean (λ)	Lower Limit	Upper Limit	Poisson Mean (λ)	Lower Limit	Upper Limit
0	0.00	3.69	10	4.80	18.39
1	0.03	5.57	15	8.40	24.74
2	0.24	7.22	20	12.22	30.89
3	0.62	8.77	25	16.18	36.90
4	1.09	10.44	30	20.24	42.83
5	1.62	11.77	35	24.38	48.68
6	2.20	13.06	40	28.58	54.47
7	2.81	14.42	45	32.82	60.21
8	3.45	15.76	50	37.11	65.92
9	4.12	17.08	100	81.36	121.63

Based on Garwood (1936); determined using an Excel 'add-in' downloaded from http://statpages.org/confint.htmls (accessed 22 June 2015).

The first condition implies that food samples showing high levels of contamination, or culture plates with large numbers of colonies, might be expected not to conform to a Poisson distribution but the Poisson distribution tends to the Normal distribution when λ is at least 20. However, derivation of exact confidence limits (CLs) for Poisson (using the method of Garwood, 1936) shows upper and lower limits that are not symmetrical around λ (Table 3.4), although it is generally accepted that $\pm 2\sqrt{\lambda}$ provides a reliable estimate of the 95% CLs.

The second condition implies that there must be an equal chance that any one organism will occur at any one point in the food sample or the culture. This condition is fulfilled only if the individuals are distributed randomly. The third criterion implies that if bacterial cells have replicated, then more than one organism will occur within a given location; hence, randomness is unlikely once replication occurs.

In a broth culture or in a well-mixed suspension of a liquid food, such as milk, the total volume occupied by 1,000,000 (ie, 10^6) bacterial cells is only about 1 part in 10^6 of the total volume of the liquid; hence, it is reasonable to assume that the cells will generally occur randomly. In a solid food sample (eg, a minced meat) contamination occurs randomly throughout and the condition would be met, but if the surface of a piece of meat is contaminated more than the deep tissues, randomness might apply only after total maceration of the sample in diluent. However, if contamination were relatively light, the distribution of organisms could be random as judged by use of some suitable surface sampling technique. When individual organisms are not well separated, the variance will be less than the mean ($s^2 < \bar{x}$)

and the binomial might be a more suitable model for the distribution. Tests to determine whether the Poisson distribution provides a good description of a set of data are given in Chapter 4. If clumping of organisms occurs, the third condition will not be met and it is probable that variance will be greater than the mean ($s^2 > \bar{x}$). In theory, the removal of a sample from a finite population will affect the value for p in the next sample unit (cf definition of sampling without replacement on p 21). However, if the sample forms only a minute proportion of the total 'lot' size, then this effect will be minimal and the value of p would not alter significantly from one examination to the next.

THE NEGATIVE BINOMIAL DISTRIBUTION ($\sigma^2 > \mu$)

If the second and third conditions for use of the Poisson distribution are not fulfilled, the variance of the population will usually be greater than the mean ($\sigma^2 > \mu$). This is particularly the case in microbiology where aggregates and cell clumps occur both in natural samples and in dilutions, slide preparations, etc. Of the various mathematical models available, the negative binomial is frequently the best model to describe the distribution frequencies obtained (Jones et al., 1948; Bliss and Fisher, 1953; Gurland 1959; Takahashi et al., 1964; Dodd, 1969) but other complex distributions may be appropriate.

The negative binomial, which describes the number of failures before the xth success when n is the integer, is the mathematical counterpart of the (positive) binomial, and is described by the expansion of $(q - p)^{-k}$, where $q = 1 + p$ and $p = \mu/k$. The parameters of the equation are the mean μ and the exponent k. Unlike exponent n in the binomial series, the exponent k of the negative binomial is neither an integer nor the maximum possible number of individuals that could occur in a sample population. Instead it is related to the spatial or temporal distribution of the organisms in the sample and takes into account the effects of cell clumps and aggregates.

The variance of the population is given by $\sigma^2 = kpq = \mu q = \mu[1+(\mu/k)] = \mu + (\mu^2/k)$. Therefore, the reciprocal of the constant k (ie, $1/k$) is a measure of the excess variance or clumping of individuals in the population. As $1/k \to 0$ and $k \to \infty$, the distribution converges to Poisson, with $\sigma^2 = \mu$. Conversely, if clumping is dominant, $k \to 0$ and therefore $1/k \to \infty$; the distribution converges on the logarithmic distribution (Fisher et al., 1953).

Applications of the negative binomial distribution within biological sciences are numerous. One such was the demonstration that the number of microbial colonies in soil follows a Poisson distribution and the number of organisms within colonies follows a logarithmic distribution, so that the distribution of all organisms in soil conforms to a negative binomial (Jones et al., 1948).

The individual terms of the expansion of $(q - p)^{-k}$ are given by

$$P_x = \left(1+\frac{\mu}{k}\right)^{-k}\left(\frac{(k+x-1)!}{x!(k-1)!}\right)\left(\frac{\mu}{\mu+k}\right)^x$$

where P_x is the probability that x organisms occur in a sample unit. As in other distributions, the expected frequency of a particular count is NP_x, where N is the number of sample units.

A very simple method of deriving an approximate value for k can be obtained by rearranging the equation for variance of a negative binomial:

$$\sigma^2 = \mu + \frac{\mu^2}{k}$$

hence:

$$k = \frac{\mu^2}{\sigma^2 - \mu}$$

Since the statistical estimates of the parameters μ and σ^2 are $\bar{x}$ and s^2, respectively, an estimate $(\hat{k})$ of the value of k can be derived from $\hat{k} = \bar{x}^2/(s^2 - \bar{x})$. Anscombe (1950) showed, however, that this method is inefficient because it does not give a reliable estimate for values of $\hat{k}$ below 4 unless $\bar{x}$ is also less than 4.

Estimations of the population parameters μ and k are obtained from the frequency distribution statistics $\bar{x}$ and $\hat{k}$; the arithmetic mean $(\bar{x})$ is derived in the standard manner (Example 2.1). Determination of $\hat{k}$ is much more complex. Several methods have been proposed (Anscombe, 1949, 1950; Bliss and Fisher, 1953; Debauche, 1962; Dodd, 1969) to obtain an approximate estimate, which can then be used in a *maximum likelihood* method to obtain an accurate estimate for $\hat{k}$ (Example 3.4).

The approximate value for $\hat{k}$ can be substituted in a maximum likelihood equation:

$$N \log_e\left(1+\frac{\bar{x}}{\hat{k}}\right) = \sum \frac{A_x}{\hat{k}+x}$$

where N is the total number of sample units, $\log_e$ is the natural logarithm and A_x is the cumulative frequency (ie, the total number of counts) exceeding x. Different approximations for $\hat{k}$ are tried and the equation is balanced by iteration. The method is illustrated in Example 3.4.

Tables of expected probabilities for 1480 negative binomial distributions, covering values of k from 0.1 to 200, have been published (Williamson and Bretherton, 1963). The tables are arranged in order of increasing size of a parameter p' that is equivalent to $1/q$ (not p as used in the present discussion). The discrepancy arises because they used an alternative form $[p'^k(1-q')^{-k}]$ of the negative binomial equation where $p' + q' = 1$ (whereas in this text the term $q - p = 1$ is used). An estimate of Williamson and Bretherton's p' is given by $p' = 1/[1+(\bar{x}/k)]$.

The example of a frequency distribution for a negative binomial in Fig. 3.6 shows a comparison of the observed and calculated frequency distributions.

EXAMPLE 3.4 CALCULATION OF $\hat{k}$ AND DERIVATION OF THE EXPECTED FREQUENCY DISTRIBUTION OF A NEGATIVE BINOMIAL FOR DATA FROM MICROSCOPIC COUNTS OF BACTERIAL CELLS IN MILK

The observed frequency distribution of bacterial cells per field in a milk smear was:

	Number of Cells (x)												Total
	0	1	2	3	4	5	6	7	8	9	10	>10	
f	56	104	80	62	42	27	9	9	5	3	2	1	400
fx	0	104	160	186	168	135	54	63	40	27	20	10	967
A_x	344	240	160	98	56	29	20	11	6	3	1	0	

where x is the cell count, f is the observed frequency of that count and A_x is the cumulative frequency of counts exceeding x. The total number (N) of sample units (ie, fields counted) is given by $N = \Sigma f = 400$ and the total of $A_x = \Sigma fx = 967$.

The arithmetic mean count is given by $\bar{x} = [\Sigma fx / \Sigma f] = 967/400 = 2.4175$, *and the variance by* $s^2 = \left[\Sigma fx^2 - \bar{x}\,\Sigma fx\right]/(n-1) = [3957 - 2.4175(967)]/399 = 4.0583$.

Estimation of $\hat{k}$ (method 1)

This simple method is based on the equation for variance of a negative binomial. The population variance is given by $\sigma^2 = \mu + (\mu^2/k)$ and therefore $k = \mu^2/(\sigma^2 - \mu)$. Substituting the sample statistics for the population parameters we get $\hat{k} = \bar{x}^2/(s^2 - \bar{x})$. The method is not very efficient for values of $k < 4$ but provides an approximation that can be used in other methods (*qv*).

For our data, $\hat{k}_1 = \bar{x}^2/(s^2 - \bar{x}) = (2.4175)^2/(4.0583 - 2.4175) = 3.5619 \approx 3.6$.

Test for efficiency = $\hat{k}/\bar{x} = 3.5619/2.4175 = 1.47$, which is less than 6, and

$[(\hat{k}+\bar{x})(\hat{k}+2)]/\bar{x} = [(3.5619 + 2.4175)(3.5619 + 2)]/2.4175 \approx 13.76$, which is less than 15.

Hence, the value for $\hat{k}$ is not efficient.

Estimation of $\hat{k}$ by the maximum likelihood method

The maximum likelihood equation is $N \log_e[1 + (\bar{x}/\hat{k})] = \Sigma A_x/(\hat{k} + x)$, where N is the total number of sampling units (ie, microscopic fields examined), A_x is the total number of counts exceeding x and $\log_e$ is the natural logarithm. Different values of $\hat{k}$ are tried until the equation is balanced by iteration.

We solve each side of the equation, initially using the approximation of $\hat{k}_1 = 3.6$ (derived by method 1). Solving first the left-hand side of the equation, we get

$$N \log_e\left(1 + \frac{\bar{x}}{\hat{k}_1}\right) = 400 \log_e\left(1 + \frac{2.4175}{3.6}\right) = 205.496$$

Solving the right-hand side of the equation, we get

$$\sum \frac{A_x}{\hat{k} + x} = \frac{A_{x=0}}{\hat{k}} + \frac{A_{x=1}}{\hat{k}+1} + \frac{A_{x=2}}{\hat{k}+2} + \frac{A_{x=3}}{\hat{k}+3} + \cdots + \frac{A_{x=10}}{\hat{k}+10}$$

$$= \frac{344}{3.6} + \frac{240}{4.6} + \frac{160}{5.6} + \frac{98}{6.6} + \cdots + \frac{1}{13.6} = 205.840$$

Using $\hat{k}_1 = 3.6$, the two equations differ by -0.344 (ie, the left-hand side is lower than the right-hand side).

We now select a larger value (eg, 5.0) for $\hat{k}_3$ and again solve the two sides of the equation:

$$N\log_e\left(1+\frac{\bar{x}}{\hat{k}_3}\right)=400\log_e\left(1+\frac{2.4175}{5.0}\right)=157.76$$

$$\sum\frac{A_x}{\hat{k}+x}=\frac{344}{5}+\frac{240}{6}+\frac{160}{7}+\frac{98}{8}+\cdots+\frac{1}{15}=156.51$$

For this trial value of $\hat{k}_3$, the difference between the two sides is +1.65.
The data are then arranged as follows:

	k	Difference
$\hat{k}_1$	3.6	−0.34
$\hat{k}_2$	?	0.00
$\hat{k}_3$	5.0	+1.65

Then, $\dfrac{\hat{k}_2-3.6}{5.0-3.6}=\dfrac{[0-(-0.34)]}{[1.65-(-0.34)]}=\dfrac{0.34}{1.99}\approx 0.171$.

Hence, $\hat{k}_2-3.6=1.4\times 0.171=0.239$, so $\hat{k}_2=3.6+0.239=3.84$.

The distribution of these counts can therefore be described by the statistics:

$$\bar{x}=2.1475;\quad s^2=4.0583;\quad \text{and}\quad \hat{k}=3.84$$

The negative binomial distribution curve

The distribution curve can be derived from the probability density function equation:

$$P_{(x)}=\left(1+\frac{\bar{x}}{\hat{k}}\right)^{-\hat{k}}\left[\frac{(\hat{k}+x-1)!}{x!(\hat{k}-1)!}\right]\left(\frac{\bar{x}}{\bar{x}+\hat{k}}\right)^x$$

The probability of 0 bacteria/field is given by

$$P_{(x=0)}=\left(1+\frac{2.4175}{3.84}\right)^{-3.84}\left[\frac{(3.84+0-1)!}{0!(3.84-1)!}\right]\left(\frac{2.4175}{2.4175+3.84}\right)^0$$

But since $\left[\dfrac{(3.84+0-1)!}{0!(3.84-1)!}\right]=\dfrac{2.84!}{2.84!}=1$ and $\left(\dfrac{2.4175}{2.4175+3.84}\right)^0=1$, this equation simplifies to $P_{(x=0)}=[1+(2.4175/3.84)]^{-3.84}$ and the calculation is done by taking log values of both sides:
$\log P_{x=0}=-3.84\times\log[1+(2.4175/3.84)]=-3.84\times\log(1.62955)=-0.8143$.

Therefore:

$$P_{x=0}=\text{antilog}(-0.8144)=0.1528$$

The expected frequency of zero cell counts is: $NP_{(x=0)}=400\times 0.1528=61.12$.
The probability of 1 bacterial cell/field is given by

$$P_{(x=1)}=\left(1+\frac{\bar{x}}{k}\right)^{-k}\left[\frac{(k+1-1)!}{1!(k-1)!}\right]\left(\frac{\bar{x}}{\bar{x}+k}\right)^1$$

Since

$$P_{(x=0)}=\left(1+\frac{\bar{x}}{k}\right)^{-k}$$

we have

$$P_{(x=1)} = P_{x=0} \cdot \left[\frac{(k+1-1)!}{1!(k-1)!}\right]\left(\frac{\bar{x}}{\bar{x}+k}\right)^1 = P_{x=0} \cdot \frac{k!}{(k-1)!}\left(\frac{\bar{x}}{\bar{x}+k}\right)^1 = P_{x=0} \cdot k\left(\frac{\bar{x}}{\bar{x}+k}\right)^1$$

Hence:

$$P_{x=1} = (0.1528)(3.84)(0.3866)^1 = 0.2267$$

The expected frequency of a count of 1 bacterial cell/field is: $NP_{(x=1)} = 400 \times 0.227 = 90.67$.

Similarly, the probability of 2 bacterial cells/field is given by

$$P_{(x=2)} = P_{(x=0)} \cdot \left[\frac{(k+2-1)!}{2!(k-1)!}\right]\left(\frac{\bar{x}}{\bar{x}+k}\right)^2 = P_{(x=0)} \cdot \left[\frac{(k+1)(k)}{2}\right]\left(\frac{\bar{x}}{\bar{x}+k}\right)^2$$

$$P_{(x=2)} = 0.1533\left(\frac{4.84 \times 3.84}{2}\right)(0.3866)^2 = 0.2129$$

Hence, the expected frequency of a count of 2 bacterial cells/field is

$$P_{(X=2)} = 400 \times 0.2129 = 85.16 \approx 85.2$$

This process is continued until $\Sigma P_{(x)} = 1$ and $\Sigma f = 400$, as illustrated in Table 3.5. A plot of the observed and expected frequencies is given in Fig. 3.6. A goodness-of-fit test between the observed and expected frequencies gave $\chi^2 = 5.49$ with $\nu = 7$, for which $0.70 > P > 0.50$, so the null hypothesis that the observed counts conform to a negative binomial distribution was not rejected.

Note 1: In the equation $P_{(x)} = \left(1+\frac{\bar{x}}{\hat{k}}\right)^{-\hat{k}}\left(\frac{(\hat{k}+x-1)!}{x!(\hat{k}-1)!}\right)\left(\frac{\bar{x}}{\bar{x}+\hat{k}}\right)^x$ the factor $\left(1+\frac{\bar{x}}{\hat{k}}\right)^{-\hat{k}} = P_{(x=0)}$ and the second factor $\left(\frac{(\hat{k}+x-1)!}{x!(\hat{k}-1)!}\right)$ simplify as follows:

$$\text{For } x=0, \quad \left[\frac{(\hat{k}+0-1)!}{0!(\hat{k}-1)!}\right] = \left[\frac{(\hat{k}-1)!}{1(\hat{k}-1)!}\right] = 1$$

$$\text{For } x=1, \quad \left[\frac{(\hat{k}+1-1)!}{1!(\hat{k}-1)!}\right] = \left[\frac{(\hat{k})!}{1(\hat{k}-1)!}\right] = \hat{k}$$

$$\text{For } x=2, \quad \left[\frac{(\hat{k}+2-1)!}{2!(\hat{k}-1)!}\right] = \left[\frac{(\hat{k}+1)!}{2(\hat{k}-1)!}\right]$$

$$\text{For } x=3, \quad \left[\frac{(\hat{k}+3-1)!}{3!(\hat{k}-1)!}\right] = \left[\frac{(\hat{k}+2)!}{6(\hat{k}-1)!}\right], \text{ and so on}$$

Note 2: Computer algorithms are available for fitting experimental data to negative binomial models. The 'goodness of fit' between the observed and calculated distributions (eg, Table 3.5) can be tested using χ^2 (see Chapter 4). It is essential that in all calculations to derive k, $P_{(x)}$, etc., at least seven significant figures are retained in the calculator to avoid 'rounding' errors.

Source: Data of Morgan et al. (1951).

Table 3.5 Individual Terms for Values of x From 0 to 10, for a Negative Binomial Distribution With $\bar{x} = 2.4175$, $N = 400$ and $k = 3.84$

Number of Cells/ Field, x	Probability of Occurrence, (P_x)	Calculated Frequency (NP_x)	Frequency as Integer (NP_x)	Observed Frequency (f)
0	$P_{(x=0)} = \left(1+\frac{\bar{x}}{k}\right)^{-k} = R$ [a] $= 0.1528$	61.12	61	56
1	$P_{(x=1)} = R\left(\frac{k}{1!}\right)(Y)$ [b] $= 0.2267$	90.67	91	104
2	$P_{(x=2)} = R\left[\frac{(k+1)k}{2!}\right](Y) = 0.2119$	84.77	85	80
3	$P_{(x=3)} = R\left[\frac{(k+2)(k+1)k}{3!}\right](Y) = 0.1594$	63.76	64	62
4	And so on = 0.1053	42.12	42	42
5	And so on = 0.0638	25.51	26	27
6	= 0.0363	14.52	15	9
7	= 0.0197	7.89	8	9
8	= 0.0103	4.13	4	5
9	= 0.0052	2.10	2	3
10	= 0.0026	1.04	1	2
>10			2	1
N	$P = 0.994$	397.64	400	400

For details see Example 3.4.

[a] $R = \left(\frac{1+\bar{x}}{k}\right)^{-k}$.

[b] $Y = \left(\frac{\bar{x}}{\bar{x}+k}\right)^{x}$.

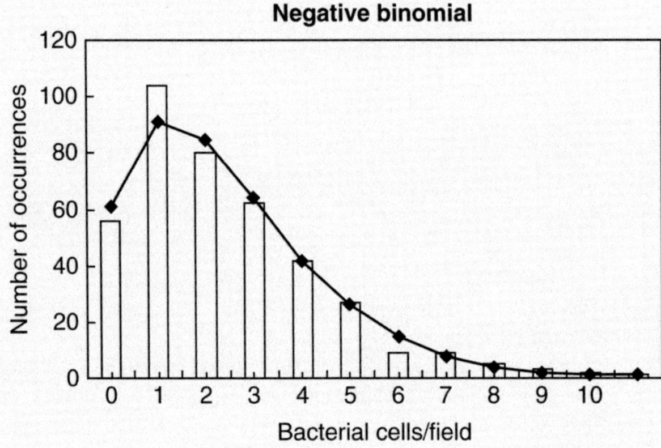

FIGURE 3.6 Negative Binomial Frequency Distributions Showing Expected (Line) and Observed (Histogram) Counts of Bacterial Cells/Field

RELATIONSHIPS BETWEEN THE FREQUENCY DISTRIBUTIONS

Elliott (1977) illustrated the general relationships amongst the binomial family distributions, whose parameters are summarised in Table 3.1, as follows:

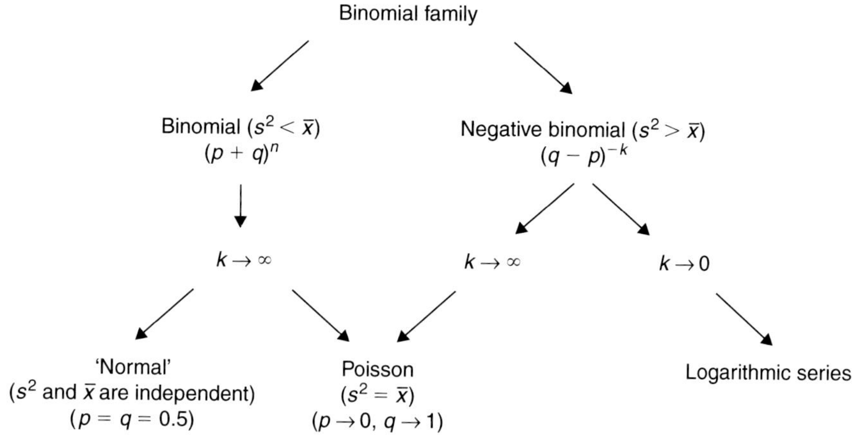

Source: *Reproduced from Elliott (1977), by permission of the Freshwater Biological Association*

This effect can be seen by comparison of the curves in Figs 3.4 and 3.6–3.8. The shapes of both the binomial (Fig. 3.4) and the negative binomial frequency distribution curves, with $\mu = 10$ and $k = 1000$ (Fig. 3.8), are very similar to that of a Poisson distribution with $\lambda = 10$ (Fig. 3.9). It can also be seen that the binomial is asymmetrical for low values of p (or q) (Fig. 3.4), that negative binomial curves are asymmetrical for low values of μ and k (Figs 3.7 and 3.8) and that Poisson curves are asymmetrical for low values of λ (Fig. 3.9).

TRANSFORMATIONS

Whenever it is required to make comparisons of data (eg, tests for the difference between mean values), the parametric test methods require that the data conform to a Normal distribution, that is, the variance of the sample should be independent of the mean and the components of the variance (ie, the variances due to actual differences between the samples and those due to random error) should be additive (Example 3.5).

The binomial distribution approximates to a Normal distribution when the number of sample units is large ($N > 20$) and the variance is greater than 3. Since $s^2 = Npq$, the Normal approximation can be used when $p = 0.4$–0.6 and N is greater than 12, or when $p = 0.1$–0.9 and $N > 33$. The Normal approximation cannot be used if $N < 12$.

The Poisson distribution is asymmetrical for low values of its parameter λ (estimated by $m = \bar{x} = s^2$) but approaches the binomial when λ is large and the binomial itself approaches the Normal distribution when N is large (Fig. 3.7). The Normal

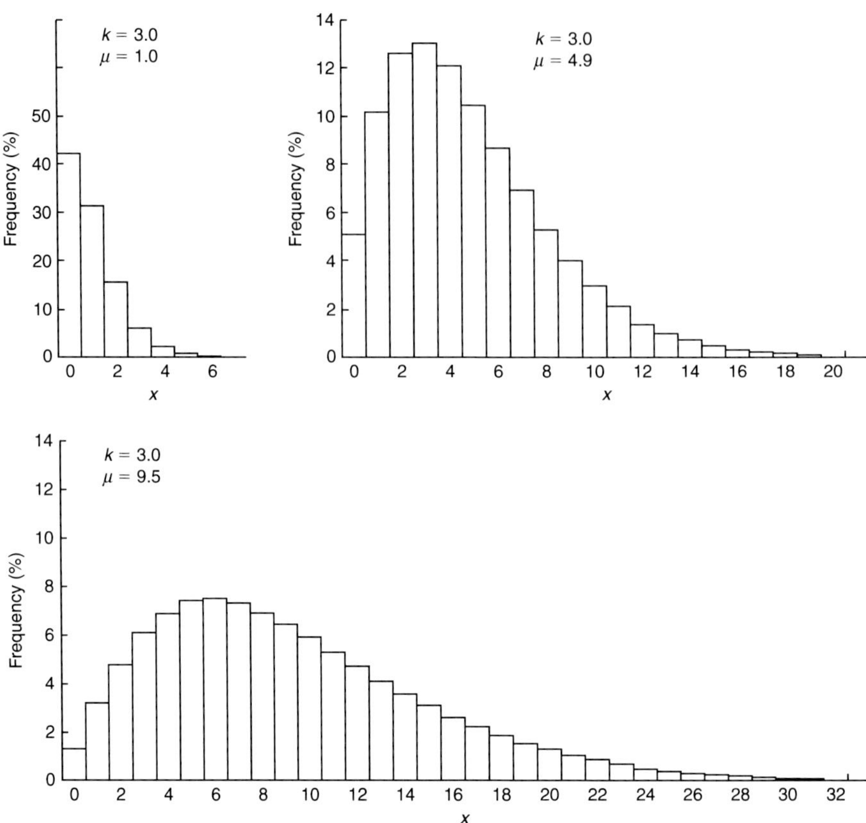

FIGURE 3.7 Negative Binomial Distributions for $k = 3.0$ and for Various Mean Values (μ) Based on Expansion of the formula $(q - p)^{-k}$

approximation to Poisson can be used when λ is generally ≥ 20. For small values of k, the negative binomial distribution is asymmetrical but it approaches Normality for large values of k when the mean (μ) is also large (eg, $\mu = 10$, $k = 1000$ in Fig. 3.8).

The first condition of Normality (ie, symmetrical distribution of values around the mean) can be attained by all three distributions in certain circumstances, so that some methods associated with the Normal distribution (eg, standard error of the mean and CLs) may then be applied. However, as mean and variance increase together in all three distributions, the condition requiring independence of mean and variance can never be fulfilled. Consequently, standard parametric methods intended to answer a question such as ‘Does the arithmetic mean value of one set of colony counts differ from that of a second set?’ and parametric ANOVA cannot be done without risk of introducing considerable errors.

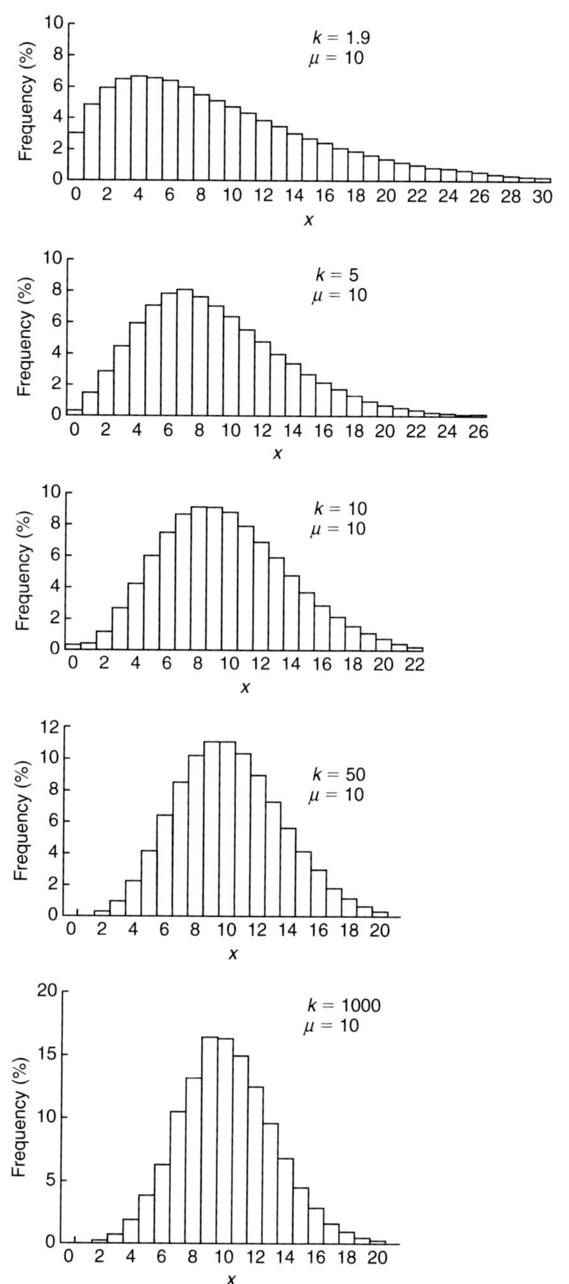

FIGURE 3.8 Negative Binomial Distributions for $\mu = 10$ and Values of k From 1.9 to 1000 Based on the Expansion of the Formula $(q - p)^{-k}$

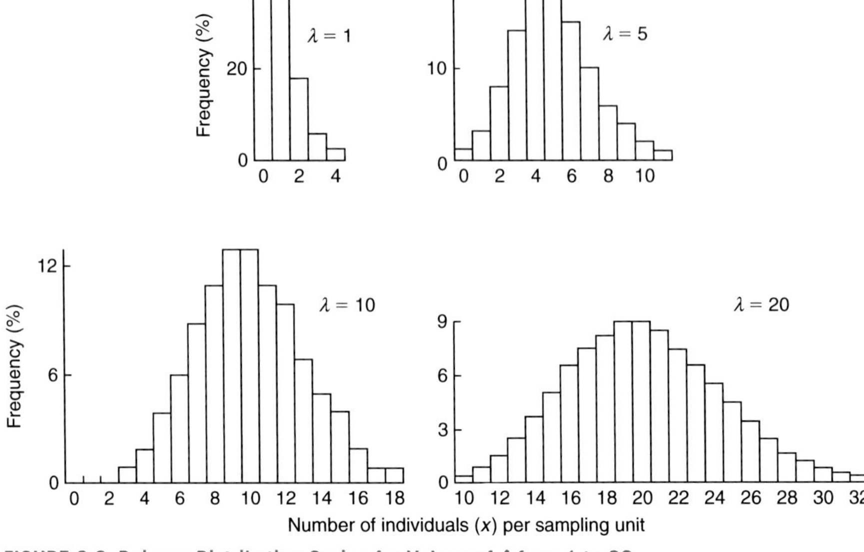

FIGURE 3.9 Poisson Distribution Series for Values of λ from 1 to 20

The frequency of each count is expressed as percentage of total number of counts

EXAMPLE 3.5 TRANSFORMATION OF A NEGATIVE BINOMIAL DISTRIBUTION

The data cited in Example 3.4 are transformed (a) as $\log[x + (k/2)]$ *using the best value calculated for* $\hat{k} = 3.915$ *and (b) as* $\log(x + 1)$*, which assumes that the distribution is unknown but that* $s^2 > \bar{x}$*. The original frequency distribution had* $\bar{x} = 2.4175$ *and* $s^2 = 4.058$*.*

The derived frequencies (a) and (b) are as follows:

Observed Frequency (f)	Cell Count (x)	$\log[x + (k/2)]$ (a)	$\log(x + 1)$ (b)
56	0	0.291	0.000
104	1	0.471	0.301
80	2	0.597	0.477
62	3	0.695	0.602
42	4	0.775	0.699
27	5	0.842	0.778
9	6	0.901	0.845
9	7	0.952	0.903
5	8	0.998	0.954
3	9	1.040	1.000
2	10	1.108	1.104
1	19	1.321	1.301

A comparison of the frequency curves is given in Fig. 3.10; note that the asymmetry is markedly reduced by transformation.

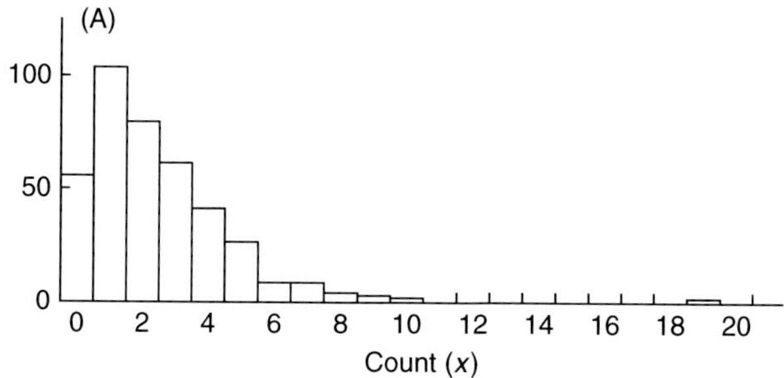

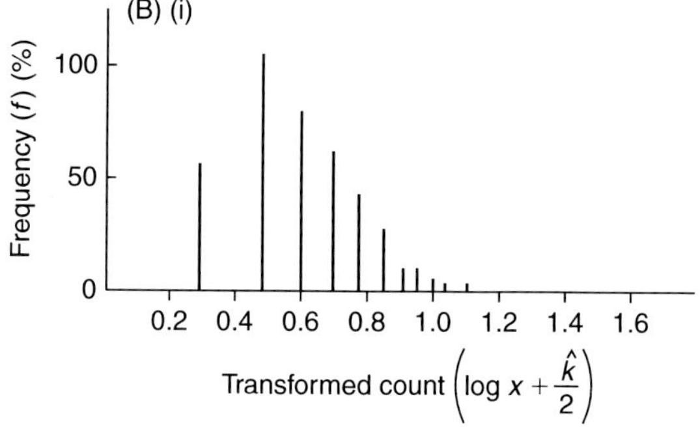

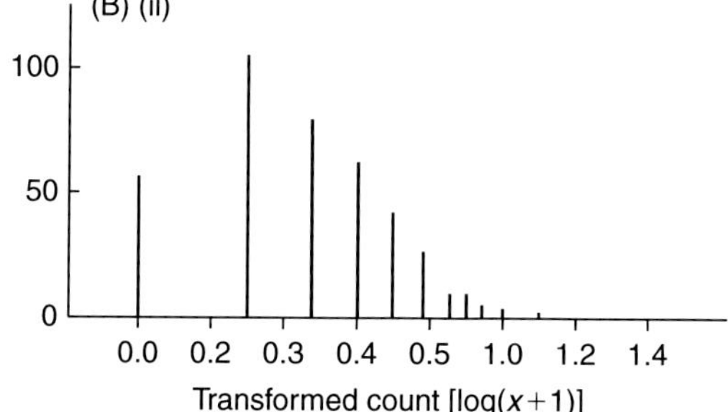

FIGURE 3.10

Frequency distributions of microscopic counts of bacterial cells/field (A) before and (B) after transformation of data. Transformation (B, i) used the formula $\log(x + k/2)$ and transformation (B, ii) used the formula $\log(x + 1)$, where x is the actual number of bacteria/field

The problem can be overcome by transforming the data, using an appropriate mathematical model such that the distribution frequency is Normalised (see Fig. 3.10) and the interdependence of mean and variance is removed. Plotting the mean against the variance on a log–log scale can provide an assessment as to whether or not the mean and variance of the original and transformed data are independent. If, as in Figs 9.1 and 9.3, the log variance increases with increasing mean values, the mean and variance are not independent. Independence of mean and variance ensures

Table 3.6 Transformation Functions

Original Distribution		Transformation – Replace *x* With	Special Conditions
Known	**Not Known**		
Poisson	$s^2 = \overline{x}$	$\sqrt{x}$	No counts <10
Poisson	$s^2 = \overline{x}$	$\sqrt{x+0.5}$	Some counts <10
Binomial (proportions)	$s^2 < \overline{x}$	$\sin^{-1}\sqrt{x}$	[a]
Negative binomial		$\sinh^{-1}\sqrt{\dfrac{x+0.375}{k-2(0.375)}}$	$k > 5$
Negative binomial		$\log\left(x+\dfrac{k}{2}\right)$	$5 > k > 2$
Lognormal	$s^2 > \overline{x}$	$\log x$	No zero counts
Unknown	$s^2 > \overline{x}$	$\log(x + 1)$	Some zero counts

[a] It is necessary first to record the proportion as a fraction of 1.0 and then to take the square root before determining the arcsin ($\sin^{-1}$) value.
Modified from Elliott (1977).

that the variance remains constant across groups compared. A general method to find a suitable transformation is described by Wardlow (1985).

Transformation also results in the components of variance becoming additive, thereby permitting application of ANOVA. The choice of transformation to be used is governed by the frequency distribution of the original data. In many routine operations, the number of sample units may be too small to permit the data to be arranged in a frequency distribution. In such circumstances, the relationship between the mean value ($\overline{X}$) and the variance (s^2) of the data can be used as a guide in the choice of a suitable transformation (Table 3.6).

Although in some microbiological situations the distribution of microorganisms conforms to Poisson, in most circumstances, additional method-based components may affect the pure Poisson sampling variance such that the distribution conforms to either a lognormal or a negative binomial distribution. For routine purposes one can be reasonably confident that a logarithmic transformation will be appropriate. That is to say the data value x is replaced by a value y, where $y = \log x$ or $y = \log(x + 1)$ depending on whether or not any zero counts are involved (Table 3.6).

Occasionally it may be necessary to back-transform the derived arithmetic mean value to the original scale (see calculation of geometric mean, Example 2.1), although transformed values are frequently cited in microbiological texts as log cfu/*g*. If back-transformation is required, it is essential that the transformation is totally reversed, that is, for the $\sqrt{x+0.5}$ transformation of Poisson data, square the transformed value and subtract 0.5; for $\log(x + 1)$, take the antilog and then subtract 1 (although this correction is usually insignificant).

Transformation of data is an essential requirement for most parametric statistical analysis of quantitative data obtained in microbiological analysis. Non-parametric procedures offer alternative means of data analysis, since such methods are by definition distribution-free, but they may be unreliable since they make certain assumptions about the shape and dispersion of the distribution, which must be the same for all the groups compared.

REFERENCES

Anscombe, F.J., 1949. The statistical analysis of insect counts based on the negative binomial distribution. Biometrika 5, 165–173.

Anscombe, F.J., 1950. Sampling theory of the negative binomial and logarithmic series distributions. Biometrika 37, 358–382.

Bliss, C.I., Fisher, R.A., 1953. Fitting the binomial distribution to biological data and a note on the efficient fitting of the negative binomial. Biometrics 9, 176–200.

Debauche, H.R., 1962. The structural analysis of animal communities in the soil. In: Murphy, P.W. (Ed.), Progress in Soil Zoology. Butterworth, London, pp. 10–25.

Dodd, A.H., 1969. The Theory of Disinfectant Testing With a Mathematical and Statistical Section, second ed. Swifts (P and D) Ltd., London.

Elliott, J.M., 1977. Some Methods for the Statistical Analysis of Samples of Benthic Invertebrates, second ed. Scientific Publication No. 25. Freshwater Biological Association, Ambleside, Cumbria, UK.

Fisher, R.A., Corbett, A.S., Williams, C.B., 1953. The relation between the number of species and the number of individuals in a random sample of an animal population. J. Anim. Ecol. 12, 42–58.

Fisher, R.A., Yates, F., 1974. Statistical Tables for Biological, Agricultural and Medical Research, sixth ed. Longman Group, London.

Garwood, F., 1936. Fiducial limits for the Poisson distribution. Biometrika 28, 437–442.

Gurland, J., 1959. Some applications of the negative binomial and other contagious distributions. Am. J. Pub. Health 49, 1388–1399.

Jones, P.C.T., Mollison, J.E., Quenouille, M.H., 1948. A technique for the quantitative estimation of soil microorganisms. Statistical note. J. Gen. Microbiol. 2, 54–69.

Morgan, M.R., MacLeod, P., Anderson, E.O., Bliss, C.I., 1951. A Sequential Procedure for Grading Milk by Microscopic Counts. Storrs Agricultural Experimental Station Bulletin No. 276.

National Bureau of Standards, 1950. Tables of binomial probability distribution. In: Applied Mathematics Series No. 6. National Bureau of Standards, Washington, DC.

Pearson, E.S., Hartley, H.O., 1966. Biometrika Tables for Statisticians, third ed. Cambridge, UK.

Takahashi, K., Ishida, S., Kurokawa, M., 1964. Statistical consideration of sampling errors in total bacteria cell count. J. Med. Sci. Biol. 17, 73–86.

Wardlow, A.C., 1985. Practical Statistics for Experimental Biologists. Wiley & Sons, Chichester.

Williamson, E., Bretherton, M.H., 1963. Tables of the Negative Binomial Probability Distribution. Wyman, New York.

Ziegler, N.R., Halvorson, H.O., 1935. Application of statistics to problems in bacteriology. IV. Experimental comparison of the dilution method, the plate count and the direct count for the determination of bacterial populations. J. Bacteriol. 29, 609–634.

CHAPTER

4 The distribution of microorganisms in foods in relation to sampling

The results of a microbiological analysis will indicate the level and/or types of microorganisms in a sample matrix and may reflect the dispersion of organisms within, or on, that matrix. If one considers a series of replicate samples taken from a stored food, changes in numbers will occur with increasing storage time. Not all organisms will grow at the same rate at any specific temperature and, indeed, some strains of organism may decrease in number, so the relative proportions of organisms in the population will also change. In a localised ecological situation, the growth of an organism will result in development of microcolonies that may then affect the growth of other organisms by removing or providing essential nutrients, by antibiosis, etc. Hence, knowledge of the temporal changes occurring over time in a population is of practical importance.

Three types of spatial distribution of organisms may occur (Elliott, 1977):

1. *random distribution*;
2. *regular, uniform or even distribution (sometimes called under-dispersion)*;
3. *contagious (aggregated or clumped) distribution, or over-dispersion.*

The three basic distribution types are illustrated diagrammatically in Fig. 4.1, but contagious distribution can also take other forms (Fig. 4.2). It is possible for two or more of the three basic types of distribution to occur simultaneously. For example, Jones et al. (1948) showed that although bacterial colonies were randomly distributed in soil (ie, distribution conformed to a *Poisson* series), individual cells within a colony conformed to a logarithmic series and the total bacterial population followed a negative binomial (contagious) distribution.

The dispersion of a population determines the relations between variance (σ^2) and mean (μ). Various mathematical distributions, such as those discussed in Chapter 3, can be used to model the relationships between variance and mean. A Poisson series ($\sigma^2 = \mu$) is a suitable model for a random distribution and the binomial ($\sigma^2 < \mu$) is an approximate model for a regular distribution. The negative binomial is one of several distributions that may be used to describe a contagious distribution ($\sigma^2 > \mu$).

Statistical Aspects of the Microbiological Examination of Foods. http://dx.doi.org/10.1016/B978-0-12-803973-1.00004-8

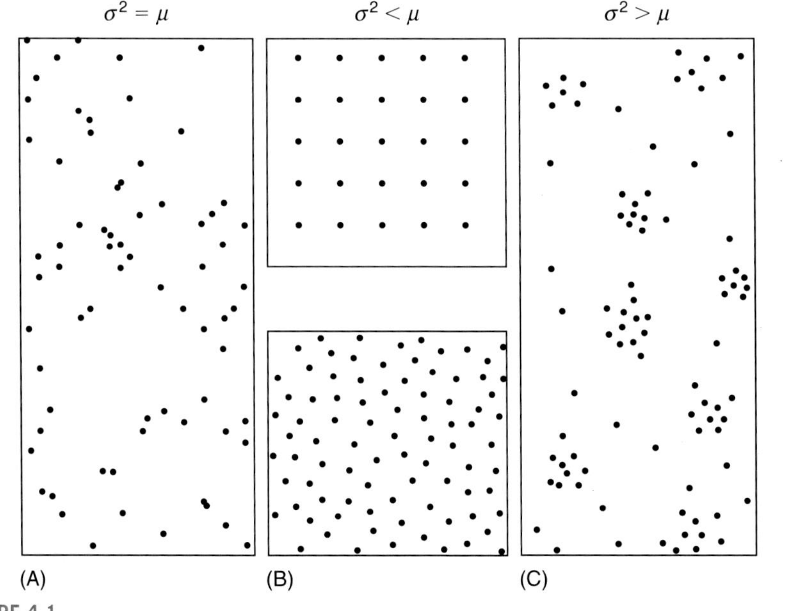

FIGURE 4.1

Three types of spatial distribution: (A) random, (B) regular (upper, ideal form; lower, normal form) and (C) contagious

Reproduced from Elliott (1977) with permission of the Freshwater Biological Association

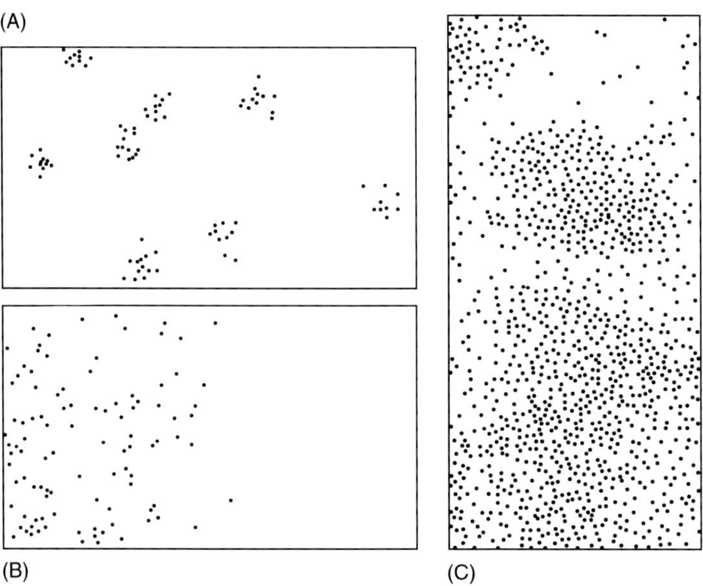

FIGURE 4.2

Different types of contagious distribution: (A) small clumps; (B) large clumps with individuals randomly distributed in each clump; (C) large clumps with individuals uniformly distributed in each clump

Reproduced from Elliott (1977) with permission of the Freshwater Biological Association

RANDOM DISTRIBUTION

In a random distribution there is an equal chance for any one individual microbial cell to occupy any specific unoccupied position in a suspension and, therefore, the presence of one organism will not affect the position of adjacent organisms. Since randomness implies the lack of any systematic distribution, some cells occur in close proximity, whereas others are more distantly spread (Fig. 4.1A).

A typical bacterial cell measures about 0.5–1.5 μm long by about 1 μm in diameter so that the relative volume occupied by such a cell will be about 1 μm^3. Since a volume of 1 cm^3 comprises 10^{12} μm^3, 1 million (10^6) bacterial cells will occupy only about 0.0001% of the available volume. It is reasonable, therefore, to assume that individual cells, clumps and microcolonies in a suspension will be distributed randomly, at least at levels up to about 10^9 colony-forming units (cfu)/mL. The accepted test for randomness is agreement to a Poisson distribution, which requires compliance with the four conditions given previously (Chapter 3).

Assuming that a hypothesis of randomness is not rejected, there remains a possibility that non-randomness occurs but that it cannot be detected. The most plausible cause of non-randomness is that of chance effects. Such effects might include environmental factors that affect microbial distribution resulting in a tendency for organisms to form microcolonies or to migrate in or on a food. On many occasions, therefore, if a test of randomness is satisfied, one must conclude that any non-randomness cannot be detected using the sampling, analytical and statistical techniques available.

It is important to take into consideration the effect of sample size when considering randomness. If the sample unit is much larger than the average size of clumps of individuals, and these clumps are randomly distributed, then the apparent population dispersion will be random and non-randomness will not be detected. Maceration and dilution of a sample to estimate numbers of colony-forming units will result in disruption of cell clumps and aggregates, such that the results obtained may indicate a random distribution of organisms.

Only for low-density populations can randomness be a true hypothesis and the implications must be considered carefully before the hypothesis is accepted. However, in practical food microbiology, the advantages associated with use of the random dispersion (ie, the Poisson series, with simple methods of calculation of confidence limits, etc.) have often been considered to outweigh the disadvantages of rejecting the hypothesis for randomness. Such expediency can lead to inaccurate conclusions since sources of variation other than simple sampling will usually inflate the overall variance and lead to a non-Poisson distribution!

TESTS FOR AGREEMENT WITH A POISSON SERIES

Confidence limits for a Poisson variable

An approximate test for agreement with a Poisson series (Fig. 4.3) is provided by confidence limits for the mean. For the data used in Example 4.1 (section 'For Large

Sample Numbers ($n > 30$)'), the mean value is 2.68 and the counts range from 0 to 6. From Fig. 4.3, the 95% confidence limits for $m = 2.68$ range from 0 to about 8. Consequently, it is not unreasonable to suppose that since all the counts lie within these limits, the counts come from a Poisson series, and therefore that the parent population could be distributed randomly. This test will not distinguish regular from random populations and the dispersion could still be random if only one or two counts lie outside the confidence limits. Although rapid, this test may be unreliable.

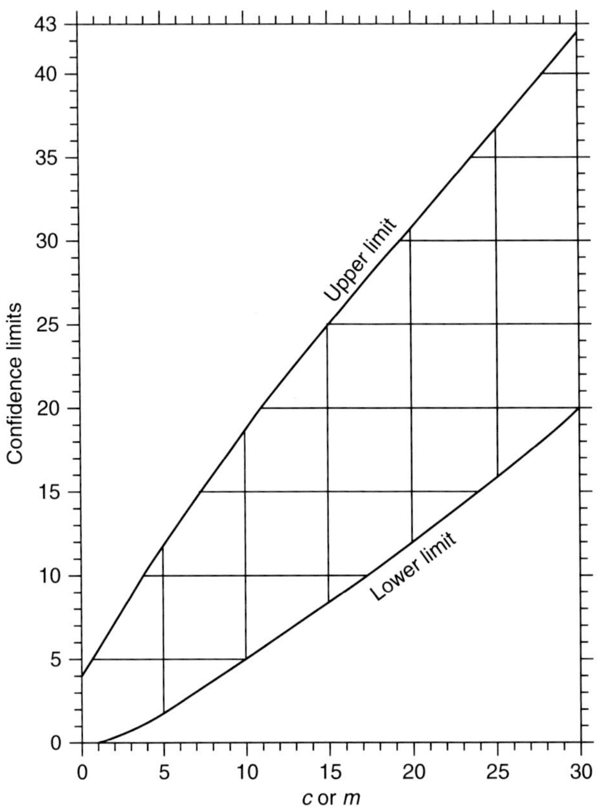

FIGURE 4.3 The 95% confidence limits for *c* from a Poisson series

c is a single count or the mean ($\bar{x} = m$) of a small sample (n) where $nm < 30$

Reproduced from Elliott (1977) with permission of the Freshwater Biological Association

EXAMPLE 4.1 INDEX OF DISPERSION AND χ^2 TESTS FOR AGREEMENT WITH A POISSON SERIES

For small sample numbers ($N < 30$)

Apparent under-dispersion

In a series of replicate plate counts the following numbers of colonies were counted on the 10^{-8} dilution plates (data of Ziegler and Halvorson, 1935, their Experiment 4a):

69, 69, 68, 79, 81, 90, 81, 80, 72, 78

The average count of the n = 10 tests is $\bar{x} = 76.7$, with variance $(s^2) = 49.79$ and $\upsilon = 10 - 1 = 9$ degrees of freedom.

First we determine the index of dispersion (I) value for the data using the following equation:

$$I = \frac{s^2}{\bar{x}} = \frac{49.79}{76.7} = 0.65$$

A value of $I < 1$ suggests the occurrence of under-dispersion and therefore the possibility of a 'regular' distribution. We use χ^2 to test for compliance with a random distribution:

$$\chi^2 = \frac{s^2(n-1)}{\bar{x}} = I(n-1) = 0.65 \times 9 = 5.85$$

From tables (eg, Pearson and Hartley, 1976) the observed probability value for $\chi^2 = 5.85$ is slightly greater than $P = 0.75$ with 9 degrees of freedom ($\upsilon = 9$). Hence, although the index of dispersion is less than 1, agreement with a Poisson series is not rejected at the 5% significance level and the hypothesis of randomness cannot be rejected.

Apparent over-dispersion

Numbers of colonies per plate from Ziegler and Halvorson (1935), their Experiment 4b at 10^{-7} dilution: 287, 307, 340, 332, 421, 309, 327, 310, 320, 358, 302, 304.

The sample statistics are as follows: $\bar{x} = 326.4$; $s^2 = 1254.8$; n = 10; $\upsilon = n - 1 = 9$.

The index of dispersion (I) is given by $s^2/\bar{x} = 1254.8/326.4 = 4.84 \gg 1$, which is indicative of over-distribution (ie, contagious distribution).

The χ^2 test is applied, as before:

$$\chi^2 = \frac{s^2(n-1)}{\bar{x}} = I(n-1) = 4.84 \times 9 = 43.60$$

From tables, the observed value for χ^2 is $P \ll 0.001$ with $\upsilon = 9$; therefore, we must reject the hypothesis that the distribution of colonies is random and accept the alternative hypothesis that distribution is contagious.

For large sample numbers ($n > 30$)

The frequency distribution of bacterial colonies in soil determined by microscopy (Jones et al., 1948; Table 4.1) had a mean count ($\bar{x}$) of 2.68 and a variance (s^2) of 2.074 with n = 80.

The index of dispersion test

The null hypothesis (H_0) is that the numbers of colonies are distributed randomly; the alternative hypothesis (H_1) is non-randomness. The index of dispersion $(I) = s^2/\bar{x} = 2.074/2.68 = 0.774$. for which $\chi^2 = I(n-1) = 0.774 \times 79 = 61.15$, with $\nu = 79$. The probability associated with this value of χ^2 is $0.90 < P < 0.95$, so the null hypothesis cannot be rejected.

However, since the sample is large ($n = 80$), with $\nu = 79$ degrees of freedom, the absolute value of the standardised Normal deviate (d) is calculated:

$$d = \left|\sqrt{2\chi^2} - \sqrt{2\upsilon - 1}\right| = \left|\sqrt{122.3} - \sqrt{157}\right| = |11.059 - 12.530| = 1.471$$

Table 4.1 χ^2 Test for 'Goodness of Fit' to a Poisson Distribution

No. of Colonies	Frequency: Observed (O)	Frequency: Expected[a] (E)	$\chi^2 = (O - E)^2/E$
0	7	5.5	0.409
1	8	14.7	3.054
2	24	19.7	0.939
3	20	17.6	0.327
4	11	11.8	0.054
5 and over	10	10.7	0.046
Total	80	80.0	4.829

[a]*Derived using:* $P_{(x)} = e^{-m}(m^x/x!)$ *and* m = 2.68.

Since the absolute value (1.47) of d is less than 1.96 (ie, 1.47 < 1.96), the null hypothesis for agreement with a Poisson series is not rejected at the 5% probability level.

The goodness-of-fit test

The index of dispersion is checked by a 'goodness-of-fit' test. The observed and expected colony numbers in the frequency distribution shown in Table 4.1 are used to determine a χ^2 value for each frequency class based on the ratio (observed − expected)^2/expected. These values are summed to give a cumulative χ^2 value of 4.83, for which $0.50 > P > 0.30$, with 5 degrees of freedom (ie, $\upsilon = 6 - 1 = 5$). Hence, the null hypothesis of randomness is again not rejected.

Fisher index of dispersion (I)

This index is a measure of the equality of variance and mean in Poisson series, namely:

$$I = \frac{\text{sample variance}}{\text{theoretical variance}} = \frac{s^2}{\bar{x}} = \frac{\sum(x-\bar{x})^2}{\bar{x}(n-1)}$$

where s^2 is the sample variance, $\bar{x}$ is the sample mean and n is the number of sample units. The significance of the extent to which this ratio departs from unity can be determined from a table of χ^2, since $I(n - 1)$ is approximated by χ^2 with $\nu = n - 1$ degrees of freedom, under the null hypothesis:

$$\chi^2 = I(n-1) = \frac{s^2(n-1)}{\bar{x}} = \frac{\sum(x-\bar{x})^2(n-1)}{(n-1)\bar{x}} = \frac{\sum(x-\bar{x})^2}{\bar{x}}$$

From tables of χ^2, or more approximately from Fig. 4.4, agreement with a Poisson series is not rejected at the 5% probability level if the χ^2 value lies between the upper and lower significance levels ($0.975 > P > 0.025$) for $\nu = n - 1$ degrees of freedom. When agreement is perfect, $I = 1$ and $\chi^2 = (n - 1)$.

If the sample is large ($n \geq 30$), it can be assumed that the absolute value of $\sqrt{2\chi^2}$ is distributed normally about $\sqrt{2\nu - 1}$ with unit variance, where ν is the degrees of freedom. Agreement is then accepted ($P > 0.05$) if the absolute value of d is less than

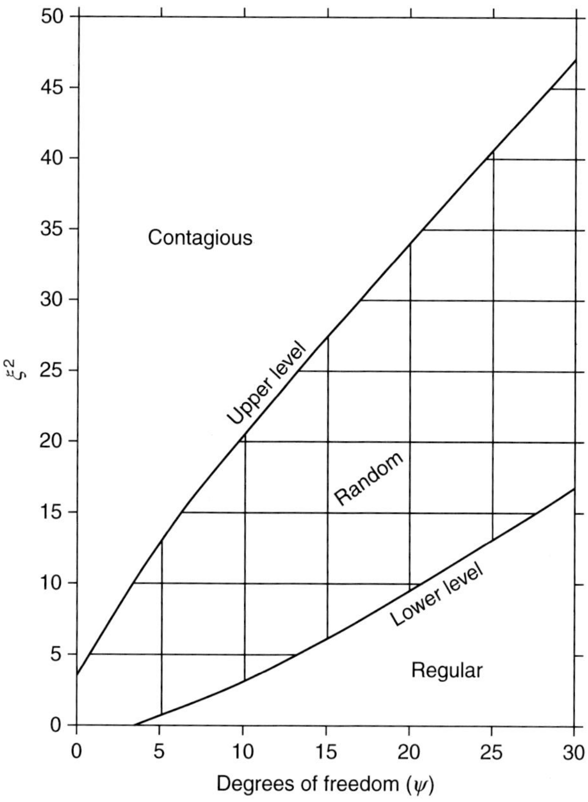

FIGURE 4.4 The 5% significance levels of χ^2

If χ^2 value lies between the upper and lower significance levels, then agreement with a Poisson series is accepted at 95% probability level ($P > 0.05$)

Reproduced from Elliott (1977) with permission of the Freshwater Biological Association

1.96, where d is the Normal variate with zero mean and v degrees of freedom. The value of d is derived from the following:

$$d = \sqrt{2\chi^2} - \sqrt{2v-1}$$

Departure from a Poisson series (ie, non-randomness) is shown by the following: (1) χ^2 less than expected (ie, $d > 1.96$ with a negative sign) – suspect a regular distribution with $s^2 < \bar{x}$; (2) χ^2 greater than expected ($d > 1.96$ with a positive sign) – suspect contagious distribution. When the sample number is large, the result of the index of dispersion test should be checked by the χ^2 test for 'goodness of fit' (see Examples 4.1 and 4.3) since this is a more robust method of testing for compliance with a specific distribution.

The likelihood ratio index (G^2 test)

The likelihood ratio index (G^2 test) as described by Anon (2005) provides an alternative method, which is slightly more complicated but more accurate. First, it is necessary to carry out an evaluation of colony numbers on replicate culture plates for evidence of under-dispersion or over-dispersion.

To test the homogeneity of colony counts, the G^2 test uses the following equation:

$$G_{n-1}^2 = 2\sum_{i=1}^{n}\left[O_i \ln\left(\frac{O_i}{E_i}\right)\right]$$

where O_i is the ith observed colony count, E_i is the ith expected colony count, i is the number of the colony count (from 1, 2, ... n), ln is the natural log and n is the total number of colony counts. The assumption is made that the observed counts should follow the volumes of sample tested. The expected value (E_i) is calculated as the expected fraction of the sum of all colonies that each of the test volumes should contain at each dilution level:

$$E_i = \frac{V_i}{\sum V_i}\sum C_i$$

where V_i is the volume of dilution i tested in each replicate plate, ΣV_i is the sum of volumes tested at that dilution and ΣC_i is the sum of colonies counted at that dilution level. In practice it is not necessary to calculate E_i since the formula can be inserted in the general equation:

$$G_{n-1}^2 = 2\left[\sum_{i=1}^{n} C_i \cdot \ln\frac{C_i}{V_i} - \sum_{i=1}^{n} C_i \cdot \ln\frac{\sum_{i=1}^{n} C_i}{\sum_{i=1}^{n} V_i}\right]$$

where C_i is the number of colonies counted on the ith plate, V_i is the volume tested on the ith plate and i is the number of plates counted ($i = 1, 2, \ldots, n$) at that dilution (n is the total number of plates used). The value determined for G^2 can be compared with values of the χ^2 distribution for $\nu = n - 1$ degrees of freedom (see Example 4.2A). The ISO standard method (Anon, 2005) includes a 'BASIC' computer programme for calculating the G^2 index.

The G^2 index can be used also to provide an overall index of homogeneity of parallel plating. Since all the volumes tested in a parallel series at a given dilution are nominally identical, the formula can be simplified and rewritten as follows:

$$G_{n-1}^2 = 2\left[\sum C_i \cdot \ln C_i - \sum C_i \cdot \ln\frac{\sum C_i}{n}\right]$$

Although a solitary value is of limited use, if there are m sets of plates with equal numbers of parallel tests, then the additive property of G^2 can be used to provide an overall measure of homogeneity across the series. This value, G_{p}^2, can be derived as follows:

$$G_{\mathrm{p}}^2 = G_{m(n-1)}^2 = \sum_{j=1}^{m} G_{(n-1)}^2$$

The derivation of G^2 and G_{p}^2 is illustrated in the section 'General Test of Homogeneity Followed by an Analysis of Deviance' in Example 4.2.

The χ^2 test for goodness of fit

This test can be undertaken when sufficient samples have been analyzed to permit comparison of the observed frequency distribution with the expected frequency derived mathematically (see the section 'The Goodness-of-Fit Test' in Example 4.1). The experimental data (Table 4.1) are a good fit to the model (Fig. 4.4) if the observed and estimated frequencies agree, and are tested by χ^2, where

$$\chi^2 = \sum \frac{(\text{observed} - \text{expected})^2}{\text{expected}}$$

The χ^2 value is calculated for each frequency class and the cumulative χ^2 value is derived, with υ degrees of freedom, where

$$v = (\text{number of frequency classes}) - (\text{number of estimated parameters}) - 1$$

Since a Poisson series has only one parameter (λ), we have

$$v = (\text{number of frequency classes}) - 2$$

It has been recommended that the number of expected values should not be less than 5, and that frequencies should be combined when necessary. Cochran (1954) considers that this weakens the test too far and recommends combinations of expected values such that none is less than 1.

Agreement with a Poisson series is not rejected if the χ^2 value is less than the value for $\alpha = 0.025$ and greater than the value for $\beta = 0.975$, with υ degrees of freedom, for instance, if the derived value for χ^2 with $\upsilon = 5$ lies between 0.831 and 12.832 (see Fig. 4.4). If $\upsilon \geq 30$, it is assumed that $\sqrt{2\chi^2}$ is distributed normally about $\sqrt{2v-1}$ with unit variance. Agreement with Poisson is accepted ($P > 0.05$) if the absolute value of d is less than 1.96, where

$$d = |\sqrt{2\chi^2} - \sqrt{2v-1}|$$

The 'goodness-of-fit' test is an all-purpose test which can be applied to test agreement of experimental data with any specific statistical distribution model. By contrast, the index of dispersion test (see earlier text) is aimed directly at a property of the Poisson distribution, namely the expected equivalence of variance and mean. Consequently, one would normally expect that if the hypothesis of randomness is not rejected by the index of dispersion test, then the 'goodness-of-fit' test should also not reject the hypothesis.

EXAMPLE 4.2 DETERMINATION OF THE OVERALL AGREEMENT BETWEEN COLONY COUNTS ON PARALLEL PLATES USING THE GENERAL HOMOGENEITY TEST METHOD (ANON, 2005)

The general homogeneity test

Assume that duplicate colony counts are made on each of two 10-fold dilutions of a bacterial suspension; the numbers of colonies counted, together with the relative volumes of suspension tested, are as follows:

Dilution	Colony Counts		Total Count	Relative Volumes		Total Volume
10^{-3}	211	226	437	10	10	20
10^{-4}	26	24	50	1	1	2
Total			487			22

Are these colonies randomly dispersed? The null hypothesis H_0 is that the colonies are derived from a randomly dispersed population of organisms; the alternative hypothesis H_1 is that they are not derived from a random population.

Randomness of the whole set is measured by the likelihood ratio index (G^2) using the following equation:

$$G^2_{n=1} = 2\left[\sum\left(C_i \cdot \ln\frac{C_i}{V_i}\right) - \sum C_i \cdot \ln\frac{\sum C_i}{\sum V_i}\right]$$

Hence:

$$G^2_{n-1} = 2\left\{\left[211 \cdot \ln\left(\frac{211}{10}\right) + 226 \cdot \ln\left(\frac{226}{10}\right) + 26 \cdot \ln\left(\frac{26}{1}\right) + 24 \cdot \ln\left(\frac{24}{1}\right)\right] - \left[487 \cdot \ln\left(\frac{487}{22}\right)\right]\right\}$$
$$= 2[(643.40 + 704.66 + 84.71 + 76.27) - 1508.35] = 2(1509.04 - 1508.35) = 2 \times 0.69 = 1.38$$

Since there are four terms in the equation, there are 4 − 1 = 3 degrees of freedom. The calculated value for G^2 (=1.38) is compared with the χ^2 value for 3 degrees of freedom for which $0.75 > P > 0.70$. Hence, the null hypothesis of randomness cannot be rejected.

The ratio of counts between the two dilution levels is given by 437/50 = 8.74:1. Whilst this is lower than the ideal ratio of 10:1, the difference could have arisen purely by chance and there is no reason to consider the data to lack homogeneity.

We can check this using the Fisher index of dispersion (Example 4.1): the mean colony count calculated at the 10^{-3} level of dilution is [211 + 226 + (26 × 10) + (24 × 10)]/4 = 234.25 and the variance of the counts with 3 degrees of freedom, $s^2 = 435.92$. So the Fisher index of dispersion (I) = 234.5/435.9 = 0.54 and the Fisher χ^2 value = $I(n - 1) = 0.54 \times 3 = 1.62$. For χ^2 of 1.62 and $v = 3$, the probability is $0.70 > P > 0.50$. Again, the hypothesis of randomness is not rejected – possibly because of the small number of values tested.

General test of homogeneity followed by an analysis of deviance

This is a similar data set but with three parallel tests at each dilution:

Dilution	Colony Count			Total	Relative Volume			Total
10^{-4}	190	220	165	575	10	10	10	30
10^{-5}	10	21	9	40	1	1	1	3
Total				615				33

$$G^2 = 2\left[190 \cdot \ln\left(\frac{190}{10}\right) + 220 \cdot \ln\left(\frac{220}{10}\right) + \cdots + 21 \cdot \ln\left(\frac{21}{1}\right) + 9 \cdot \ln\left(\frac{9}{1}\right) - 615 \cdot \ln\left(\frac{615}{33}\right)\right] = 19.64$$

For $v = 6 - 1 = 5$ degrees of freedom, the χ^2 value of 19.64 shows with a probability of $0.005 > P > 0.001$ that the hypothesis of randomness should be rejected and suggests over-dispersion of the counts.

The G^2 equation is then used repeatedly to evaluate the causes of the over-dispersion between the three components of the data set, that is, the replicate plates at dilution 10^{-4}, the replicate plates at dilution 10^{-5} and the differences between the dilution levels.

For the counts at 10^{-4}:

$$G^2 = 2\left[190 \cdot \ln\left(\frac{190}{10}\right) + 220 \cdot \ln\left(\frac{220}{10}\right) + 165 \cdot \ln\left(\frac{165}{10}\right) - 575 \cdot \ln\left(\frac{575}{30}\right)\right] = 7.91, \quad \text{with } v = 2$$

For the counts at 10^{-5}:

$$G^2 = 2\left[10 \cdot \ln(10) + 21 \cdot \ln(21) + 9 \cdot \ln(9) - 40 \cdot \ln\left(\frac{40}{3}\right)\right] = 6.25, \quad \text{with } v = 2$$

For the dilution ratio:

$$G^2 = 2\left[575 \cdot \ln\left(\frac{575}{30}\right) + 40 \cdot \ln\left(\frac{40}{3}\right) - 615 \cdot \ln\left(\frac{615}{33}\right)\right] = 5.48, \quad \text{with } v = 1$$

The derived values of G^2 can be entered into an analysis of deviance table, where P is the probability based on the value of χ^2 for the appropriate degrees of freedom (v):

Source of Variance	G^2	v	P
Between replicate counts at 10^{-4}	7.91	2	$0.02 > P > 0.01$
Between replicate counts at 10^{-5}	6.25	2	$0.05 > P > 0.025$
Between dilutions	5.48	1	$0.02 > P > 0.01$
Total	19.64	5	$0.005 > P > 0.001$

This analysis of deviance demonstrates that the statistically significant causes of over-dispersion are associated with all stages in the test, and especially with the poor replication in counts at the 10^{-4} dilution and the non-linear dilution from 10^{-4} to 10^{-5}. Such differences in counts at successive dilutions are not unusual in microbiological practice and indicate a need to improve the technical practices used in the laboratory.

REGULAR DISTRIBUTION

If high numbers of individual cells in a population are crowded together, yet are neither clumped nor aggregated, the dispersion of the population tends towards a regular (ie, homogeneous) distribution. In such circumstances, the number of individuals per sampling unit approaches the maximum possible and the variance is less than the mean ($\sigma^2 < \mu$). The binomial distribution provides an approximate mathematical model. The characteristic features of the distribution are illustrated diagrammatically in Fig. 4.1B.

In terms of microbiological analysis, such a situation might be expected when considerable growth has occurred, for example, on the cut surface of a piece of meat or when colonies are crowded on a culture plate. Such distributions are rarely seen in microbiological analysis, because normal laboratory homogenisation procedures destroy any regular distribution that may have occurred; they may, however, occur as experimental artefacts.

THE BINOMIAL DISTRIBUTION AS A MODEL FOR A REGULAR DISPERSION

The expected frequency distribution of a positive binomial is given by $n(p + q)^k$, where n is the number of sampling units, k is the maximum possible number of individuals in a sampling unit, p is the probability for occurrence of a specific organism in a sample unit and $q = 1 - p$ is the probability that any one place in a sampling unit will not be occupied. Estimates of the parameters p, q and k are obtained from the sample. The highest count on the sample provides a rough estimate for k but is frequently low. A more accurate estimate can be derived from the mean and variance, where the estimate of $k = \hat{k} = \bar{x}^2/(\bar{x} - s^2)$, to the nearest integer (whole number). Expected probabilities are given by the expansion of $(p + q)^k$. Estimates of p and q are given by $\hat{p} = \bar{x}/\hat{k}$ and $\hat{q} = 1 - \hat{p}$.

The binomial distribution is used extensively for analysis of qualitative data such as that derived from proportions of positive and negative results in a tube dilution assay (Chapter 8) and for data on prevalence of pathogenic organisms and/or disease.

CONTAGIOUS (HETEROGENEOUS) DISTRIBUTIONS

The spatial distribution of a population of organisms is rarely regular or truly random, but is often contagious and has a variance significantly greater than the mean ($\sigma^2 > \mu$). In a contagious distribution, clumps and aggregates of cells always occur, but the overall pattern can vary considerably (Figs 4.1C and 4.2). The detection of contagious distributions in microbiology may reflect the growth of microbial colonies, and indirectly reflects the consequence of environmental factors on growth, but it may be due also to artefacts associated with the counting procedure or, in certain circumstances, by localised contamination of a food product during manufacture (eg, contamination of dried milk or infant feeds with small numbers of salmonellae or other pathogens).

The overall dispersion pattern is dependent on the sizes of the clumps and aggregates; the spatial distribution of, and distance between, the clumps themselves; and the spatial distribution of organisms within a clump. A pattern found frequently is that shown in Fig. 4.1C, where patches of high density (clumps, microcolonies) are dispersed on a background of low-density random contamination.

Several mathematical models have been proposed to describe contagious distributions. Of these, the negative binomial (a gamma–Poisson distribution) is probably the most useful but other models including Poisson–lognormal, beta–Poisson and zero-inflated Poisson distributions may be appropriate in special circumstances. Other models, such as Taylor's Power Law (Taylor, 1971) and, for asymmetrical (ie, skew) distributions, the Thomas (1949), Neyman Type A (Neyman, 1939) and Pôlya–Aeppli (Pôlya, 1931) distributions, may also be of value. Elliott (1977) gives examples of the application of some of these models.

THE NEGATIVE BINOMIAL AS A MODEL FOR CONTAGIOUS DISTRIBUTION

This distribution, described fully in Chapter 3, has two parameters (the mean μ and an exponent k) and one mode (ie, most frequent count). The negative binomial can be applied as a model to a wide variety of contagious distributions:

1. *True contagion*: The presence of one individual increases the chances that another will occur in the same place. Since bacterial and yeast cells frequently adhere to one another, contamination may be by a single cell or by more than one cell. Furthermore, growth of a microcolony from a single cell will occur frequently.
2. *Constant birth–death–immigration rates*: This is unlikely to be of significance other than in the context of growth and senescence, and/or in chemostat situations, which are outside the scope of this book.
3. *Randomly distributed clumps*: If clumps of cells are distributed randomly and the numbers of individuals within a clump are distributed in a logarithmic fashion, then a negative binomial distribution will result (Jones et al., 1948; Quenouille, 1949). Hence, growth of microcolonies within a product may give rise to this situation.

   The mean number of clumps (m_1) per sampling unit and the mean number of individuals per clump (m_2) are given by

$$m_1 = k\ln\left(1+\frac{\mu}{k}\right)$$

   and

$$m_2 = \frac{\mu}{m_1} = \frac{\mu}{k\ln[1+(\mu/k)]} = \frac{\mu/k}{\ln[1+(\mu/k)]}$$

   where $\boldsymbol{m}_1 \times \boldsymbol{m}_2 = \mu$ and $\boldsymbol{k}$ is the exponent. The size of the sample unit will affect the values of μ and $\boldsymbol{m}_1$, but not of $\boldsymbol{m}_2$. If $\boldsymbol{m}_2$ is constant, the ratio $\mu/\boldsymbol{k}$ will also be constant. Hence, $\boldsymbol{k}$ is directly proportional to μ and the sample size will affect both $\boldsymbol{k}$ and μ.
4. *Heterogeneous Poisson distributions*: Compound Poisson distributions, where the Poisson parameter (λ) varies randomly, have a χ^2 distribution with 2 degrees

of freedom (Arbous and Kerrich, 1951). Southwood (1966) considers that in some respects this is a special variant of (3) given earlier, where

$$m_1 = \frac{\mu}{m_2}\frac{2k}{\chi^2}$$

Since the value of m_1 would be close to unity and would not be influenced by sample size, and m_2 must always be slightly less than μ, it is difficult to see its relevance in microbiological distributions.

The compound Poisson distributions include the gamma–Poisson distribution (where the Poisson parameter is itself a variable distributed according to the gamma distribution) and the beta–Poisson (where the Poisson parameter is a variable distributed according to the beta distribution). Work by Jongenburger (2012) and Mussida et al. (2013a,b) has shown that Poisson–lognormal, gamma–Poisson (negative binomial), beta–Poisson and zero-inflated Poison distributions are the most appropriate models for very low-level contamination of infant feeds by *Cronobacter sakazakii* where many negative test results occur together with a very small number of positive test results.

Test for agreement of large samples (n > 50) with a negative binomial

The simplest and most useful test is that for goodness of fit. The maximum likelihood method (Chapter 3, Example 3.4) provides the most accurate estimate of k and should always be used with large samples. The arithmetic mean ($\bar{x}$) of the sample provides an estimate of μ. As the two parameters are estimated from the sample data, the number of degrees of freedom (υ) = number of frequency classes after combination − 3.

Test for agreement of small samples (n < 50) with a negative binomial

If the data can be arranged in a frequency distribution, the χ^2 test can be applied as described earlier for Poisson distributions. When this cannot be done, other tests are used based on the observed and expected moments [the *first moment about the origin* = the arithmetic mean ($\bar{x}$); the *second moment about the mean* = the variance (s^2); the *third moment* is a measure of skewness; and the *fourth moment* is a measure of kurtosis or flatness].

The statistic U is the difference between the sample estimate of variance (s^2) and the expected variance of a negative binomial:

$$U = s^2 - \left(\bar{x} + \frac{\bar{x}^2}{\hat{k}}\right)$$

where $\hat{k}$ is estimated from the frequency of zero counts (see following text).

The statistic T is the difference between the sample estimate of the third moment and the expected third moment:

$$T = \left(\frac{\sum x^3 - 3\bar{x}\sum x^2 + 2\bar{x}^2\sum x}{n}\right) - s^2\left(\frac{2s^2}{\bar{x}} - 1\right)$$

The expected values of U and T are zero for perfect agreement with a negative binomial, but agreement is accepted if the values of U and T differ from zero by less than their standard errors. The standard errors of U and T can be calculated from the formulae of Anscombe (1950) and Evans (1953) or derived from the nomogram in Fig. 4.6. A large positive value for U or T indicates greater skewness than the negative binomial and suggests that the lognormal distribution might be a more suitable model. A large negative value indicates that other distributions (eg, Neyman Type A) might be more appropriate because of the reduced skewness:

1. *Moment estimate of k*: The estimate of k is given by $\hat{k} = \bar{x}^2/(s^2 - \bar{x})$.

   This can be derived, as described previously (Chapter 3), from the theoretical variance or from the statistics X and Y, where

$$X = \bar{x}^2 - \frac{s^2}{n} \quad \text{and} \quad Y = s^2 - \bar{x}$$

The expectations of X and Y are given by

$$E_X = \mu^2 \quad \text{and} \quad E_Y = \frac{\mu^2}{k}$$

Hence:

$$k = \frac{\mu^2}{E_Y} = \frac{E_X}{E_Y}$$

For small samples:

$$\hat{k} = \frac{E_X}{E_Y} = \frac{(\bar{x}^2 - s^2)}{n(s^2 - \bar{x})}$$

where n is the number of sample units. As n increases, s^2/n decreases and the equation for $\hat{k}$ approaches $x^2/(s^2 - \bar{x})$.

   This method is more than 90% efficient for small values of $\bar{x}$ when $\hat{k}/\bar{x} > 6$, for large values of $\bar{x}$ when $\hat{k} > 13$ and for medium values of $\bar{x}$ when $[(\hat{k} + \bar{x})(\hat{k} + 2)/\bar{x}] \geq 15$ (Anscombe, 1949, 1950).

   Frequently, $\hat{k} < 4$, so the use of this method is limited to small mean values ($\bar{x} < 4$) but it can be used to find an approximate value for $\hat{k}$ which can then be applied in the other methods.

2. *Estimate of $\hat{k}$ from the proportion of zeros*: Equation $\hat{k}\log[1 + (\bar{x}/\hat{k})] = \log(n/f_{x=0})$ is solved by iteration for various values of $\hat{k}$, where n is the number of sample units and $f_{x=0}$ is the frequency of occurrence of samples with no organisms. The method is over 90% efficient when at least one-third of the counts is zero (ie, $f_{x=0} > n/3$) but more zero counts are needed if $\bar{x} > 10$ (Anscombe 1949, 1950). The use of this method was illustrated in Example 3.4.

3. *Transformation method for estimation of* $\hat{k}$: After deriving an approximate value for $\hat{k}$ by method (1), each count (x) is transformed (Table 3.4) to a value y, where $y = \log[x + (\hat{k}/2)]$ if the approximate value for $\hat{k} = 2-5$ and $\bar{x} \geq 15$, or to $y = \sinh^{-1}\sqrt{(x+0.375)/(\hat{k}-0.75)}$ if the approximate value for $\hat{k} < 2$ and $\bar{x} > 4$. The function $\sinh^{-1}$ is the reciprocal of the hyperbolic sine and is found in standard mathematical tables (Chambers Mathematical Tables, vol. 2: Table VIA; Comrie, 1949).

The expected variance of the transformed counts is independent of the mean and equals 0.1886 trigamma $\hat{k}$ for the first transformation and 0.25 trigamma $\hat{k}$ for the second transformation. Selected values for the function 'trigamma $\hat{k}$' are given in Table 4.2; other values can be obtained from Tables 13–16 of Davis (1963).

To estimate k, try different values of $\hat{k}$ in the appropriate transformation until the variance of the transformed counts equals the expected variance (trigamma $\hat{k}$; Table 4.2). These transformations can also be used to estimate the confidence limits for small samples from a negative binomial distribution. Because of the work involved, method (3) can rarely be justified in terms of increased efficiency of determination of $\hat{k}$ unless the number of terms involved is few. Elliott (1977) provides a key to the choice of method for deriving $\hat{k}$ (Table 4.3).

THE LOGNORMAL AND OTHER CONTAGIOUS DISTRIBUTION MODELS

The lognormal distribution is a special form of contagious distribution that has only one mode, but is more skewed than the negative binomial. When logarithms of counts follow a Normal frequency distribution, the original counts follow a discrete lognormal distribution. The logarithmic transformation of counts (Chapter 3, Table 3.4) often provides a useful approximate model to Normalise data in a negative binomial distribution. However, the distribution of bacterial cells in colonies usually follows a logarithmic distribution (Jones et al., 1948; Quenouille, 1949), the individual terms of which are given by $\alpha x, \alpha x^2/2, \alpha x^3/3, \ldots, \alpha x^n/n$, where $\alpha = -1/\ln(1-x)$ and x is the bacterial count (Fisher et al., 1943).

An intermediate distribution, the Poisson–lognormal, with skewness intermediate between negative binomial and the discrete lognormal has been shown to be a special form of the heterogeneous Poisson distribution (Cassie, 1962). Other types of contagious distribution have been described with various degrees of skewness and in some instances have more than one mode (for review see Elliott, 1977).

The absolute nature of the population distributions in microbiological analysis will be masked by the effects of sampling (see following text) and the dilution procedures used. The apparent distributions of counts will appear to follow Poisson or, more usually, lognormal or negative binomial distributions; for routine purposes both can be approximately 'Normalised' using the logarithmic transformation (see Fig. 3.9). The moments of the lognormal distribution are given by median = $\exp(\mu)$, mean = $\exp[\mu + (\sigma^2/2)]$ and variance = $\exp(2\mu + \sigma^2)\exp(\sigma^2 - 1)$, where exp is the exponential factor, μ is the mean and σ^2 is the population variance of the natural log values of the counts, as estimated by $\bar{x}$ and s^2, respectively.

Table 4.2 Expected Variance of Transformed Counts From a Negative Binomial Distribution

0.1886 Trigamma k for $\bar{x} > 15$		**0.25 Trigamma k for $\bar{x} > 4$**					
k		***k***		***k***		***k***	
2.0	0.1216	2.0	0.1612	6.5	0.0416	12.0	0.0217
2.1	0.1138	2.1	0.1517	6.6	0.0409	12.2	0.0214
2.2	0.1081	2.2	0.1432	6.7	0.0402	12.4	0.0210
2.3	0.1023	2.3	0.1356	6.8	0.0396	12.6	0.0207
2.4	0.0972	2.4	0.1288	6.9	0.0390	12.8	0.0203
2.5	0.0925	2.5	0.1226	7.0	0.0384	13.0	0.0200
2.6	0.0882	2.6	0.1170	7.1	0.0378	13.2	0.0197
2.7	0.0843	2.7	0.1118	7.2	0.0373	13.4	0.0194
2.8	0.0808	2.8	0.1071	7.3	0.0367	13.6	0.0191
2.9	0.0775	2.9	0.1028	7.4	0.0362	13.8	0.0188
3.0	0.0745	3.0	0.0987	7.5	0.0357	14.0	0.0185
3.1	0.0717	3.1	0.0950	7.6	0.0352	14.2	0.0183
3.2	0.0691	3.2	0.0916	7.7	0.0347	14.4	0.0180
3.3	0.0667	3.3	0.0884	7.8	0.0342	14.6	0.0177
3.4	0.0644	3.4	0.0854	7.9	0.0338	14.8	0.0175
3.5	0.0623	3.5	0.0826	8.0	0.0333	15.0	0.0172
3.6	0.0603	3.6	0.0800	8.1	0.0329	15.2	0.0170
3.7	0.0585	3.7	0.0775	8.2	0.0324	15.4	0.0168
3.8	0.0567	3.8	0.0752	8.3	0.0320	15.6	0.0166
3.9	0.0551	3.9	0.0730	8.4	0.0316	15.8	0.0163
4.0	0.0535	4.0	0.0710	8.5	0.0312	16.0	0.0161
4.1	0.0521	4.1	0.0690	8.6	0.0308	16.2	0.0159
4.2	0.0507	4.2	0.0672	8.7	0.0305	16.4	0.0157
4.3	0.0494	4.3	0.0654	8.8	0.0301	16.6	0.0155
4.4	0.0481	4.4	0.0638	8.9	0.0297	16.8	0.0153
4.5	0.0469	4.5	0.0622	9.0	0.0294	17.0	0.0152
4.6	0.0458	4.6	0.0607	9.1	0.0291	17.2	0.0150
4.7	0.0447	4.7	0.0593	9.2	0.0287	17.4	0.0148
4.8	0.0437	4.8	0.0579	9.3	0.0284	17.6	0.0146
4.9	0.0427	4.9	0.0566	9.4	0.0281	17.8	0.0145
5.0	0.0417	5.0	0.0553	9.5	0.0278	18.0	0.0143
5.2	0.0400	5.1	0.0542	9.6	0.0275	18.2	0.0141
5.4	0.0384	5.2	0.0530	9.7	0.0272	18.4	0.0140
5.6	0.0369	5.3	0.0519	9.8	0.0269	18.6	0.0138
5.8	0.0355	5.4	0.0509	9.9	0.0266	18.8	0.0137
6.0	0.0342	5.5	0.0498	10.0	0.0263	19.0	0.0135
6.5	0.0314	5.6	0.0489	10.2	0.0258	19.2	0.0134
7.0	0.0290	5.7	0.0479	10.4	0.0252	19.4	0.0132
7.5	0.0273	5.8	0.0470	10.6	0.0247	19.6	0.0131
8.0	0.0251	5.9	0.0462	10.8	0.0243	19.8	0.0130
8.5	0.0235	6.0	0.0453	11.0	0.0238	20.0	0.0128
9.0	0.0222	6.1	0.0445	11.2	0.0234		
9.5	0.0209	6.2	0.0438	11.4	0.0229		
10.0	0.0198	6.3	0.0430	11.6	0.0225		
		6.4	0.0423	11.8	0.0221		

Reproduced from Elliott (1977) by permission of the Freshwater Biological Association; with additional data from Davis (1963)

Table 4.3 Key to Methods for Deriving $\hat{k}$

	Approximate Method	Accurate Method	Test by
(A) Counts can be arranged in frequency distribution			
	Moment estimate (1)[a]	Maximum likelihood (see Chapter 3)	χ^2
(B) Counts cannot be arranged in frequency distribution			
(a) $\bar{x}$ and $\bar{x}/\hat{k}$ meet in *U*-half of Fig. 4.6 and $f_{x=0} > n/3$	Moment estimate (1)	Proportion of zeros (2)[a]	*U* statistic
(b, i) $\bar{x}$ and $\bar{x}/\hat{k}$ meet in *T*-half of Fig. 4.6 and $\bar{x} < 4$	Moment estimate (1)	Moment estimate (1)	*T* statistic
(b, ii) $\bar{x}$ and $\bar{x}/\hat{k}$ meet in *T*-half of Fig. 4.6 and $\bar{x} > 4$	—	Transformation (3)	*T* statistic

[a]*Figures in ( ) refer to methods described in the text.*
Modified from Elliott (1977)

EFFECTS OF SAMPLE SIZE

Reference was made earlier to the effect of sample size on apparent randomness. If it is assumed that the population distribution is essentially contagious (Fig. 4.5), then for various sample quadrat sizes (A, B, C, D) the apparent distribution of counts would vary.

If the smallest sampling quadrat (A) were moved across the figure, dispersion would appear to be random, or slightly contagious, with $s^2 \approx \bar{x}$ but for quadrat (B) it would appear contagious because each sample would contain either very few or very many organisms (hence $s^2 > \bar{x}$). With sample quadrat (C), regularly distributed cell clumps would suggest that the overall dispersion would be random ($s^2 \approx \bar{x}$) whilst for quadrat (D), dispersion might appear regular ($s^2 < \bar{x}$) since the clumps would affect only the dispersion within the sample unit and each would contain about the same number of individuals. However, if the distribution of the clumps were random or contagious, such effects would not be seen with sample quadrats (C) and (D). These are purely hypothetical illustrations but the concept is of practical importance where microbiological samples are obtained, for instance, by swabbing the surface of a piece of meat or during hygiene tests on process equipment.

The occurrence of bacterial cells in food materials sometimes appears to be random, but is usually contagious. In circumstances where the cells can be viewed in situ they are most often seen to be in a contagious association, due to replication, senescence and death of cells within the microenvironment. Increasing the sample size does not change the intrinsic distribution of the organisms but the processes of sample maceration and dilution introduce artificial changes. Only in special circumstances

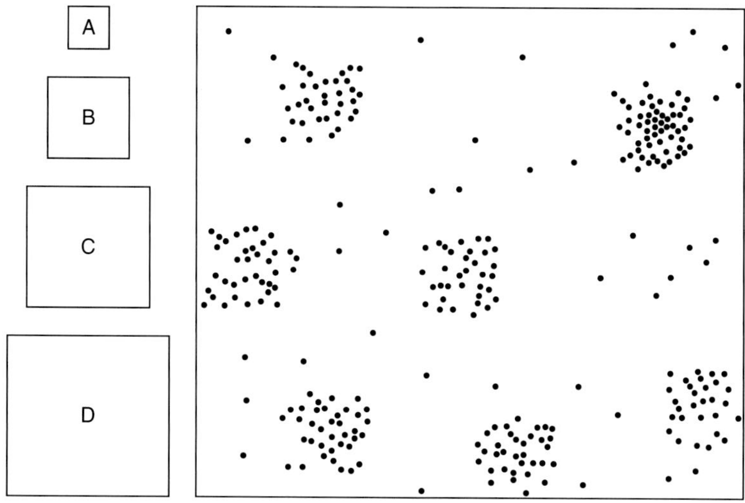

FIGURE 4.5

Four sample templates (A, B, C and D) and a contagious distribution with regularly distributed clumps. The area of sample quadrat D is 16× greater than A, C is 9× greater than A, and B is 4× greater than A

Reproduced from Elliott (1977) with permission of the Freshwater Biological Association

is it possible to examine directly very small samples drawn directly from a food, for example, the smear method for total bacterial counts on milk (Breed, 1911) and, even then, spreading the sample across a microscope slide will affect the spatial distribution. In addition, the size of the sample (20 μL) will still be large compared with the size of individual bacterial cells and clumps. It is therefore pertinent to consider that the distribution of organisms in a food will follow a contagious rather than a random distribution.

The numbers of discrete colonies on a culture plate derived from higher serial dilutions will usually appear to conform to a Poisson series (ie, random distribution of colony-forming units will be observed) partly as a consequence of maceration and dilution. But it is important to remember that it is not possible to distinguish whether the colonies grew from individual cells, from cell clumps or from both.

When organisms have been subjected to some form of sub-lethal treatment (eg, freezing, mild heating, treatment with disinfectants or food preservatives), the total number of viable microbial cells (and clumps) will often follow a Poisson distribution in the dilutions plated, but the numbers of apparently viable cells may not do so. Since some fraction of the organisms will have greater intrinsic resistance, or may have received a less severe treatment than other cells, the distribution of viable organisms detected in replicate colony counts will frequently follow a contagious distribution with $s^2 > \bar{x}$ (see Examples 4.3 and 4.4; Dodd, 1969; Abbiss and Jarvis, 1980).

EXAMPLE 4.3 USE OF METHOD (2) AND THE *U* STATISTIC TO ESTIMATE THE NEGATIVE BINOMIAL PARAMETER (*K*) AND DETERMINE THE GOODNESS OF FIT

Microscopic counts of bacterial spores in a suspension gave the following distribution:

	No. of Spores/Field (x)							**Total**
x	0	1	2	3	7	8	9	
Frequency (f)	14	6	8	4	2	4	2	40
fx	0	6	16	12	14	32	18	98
x^2	0	1	4	9	49	64	81	
$f(x^2)$	0	6	32	36	98	256	162	590

Does the negative binomial provide a good model for the distribution of these data? We test the null hypothesis that the data distribution conforms to the negative binomial; the alternative hypothesis is that the data do not conform.

The mean value ($\bar{x}$) is given by $\bar{x} = \sum fx / \sum f = 98/40 = 2.45$, where fx is determined by multiplying the number of spores counted (x) by the frequency (f) of occurrence, and $\Sigma fx = 98$ is the cumulative sum of these values.

The variance (s^2) is derived as follows:

$$s^2 = \left[\sum f(x^2) - \bar{x}\sum fx\right]/(n-1) = [590 - 2.45(98)]/39 = 8.97.$$

Then the approximate value of $\hat{k}$ [by method (1); see text] is given by

$$\hat{k} = \frac{\bar{x}^2}{s^2 - \bar{x}} = \frac{(2.45)^2}{8.97 - 2.45} = 0.9206$$

The coordinates, $\bar{x} = 2.45$ and $\bar{x}/\hat{k} = 2.45/0.9206 = 2.66$, meet in the U-half of the standard error nomogram (Fig. 4.6) close to the line for $SE_U = 2$, and the number of zero counts (f_0) = 14 which is slightly greater than $n/3$ (13.3).

Then $\hat{k}$ can be estimated by method (2) (see text), first trying $\hat{k} = 0.92$ in the equation:

$$\log\left(\frac{n}{f_0}\right) = \hat{k}\log\left(1 + \frac{\bar{x}}{\hat{k}}\right)$$

For the left-hand side of the equation:

$$\log\left(\frac{n}{f_0}\right) = \log\left(\frac{40}{14}\right) = \log 2.857 = 0.45593 \qquad [4.1]$$

Inserting the value $\hat{k} = 0.92$ on the right-hand side of the equation, we get

$$\hat{k}\log\left(1 + \frac{\bar{x}}{\hat{k}}\right) = 0.92 \cdot \log\left(1 + \frac{2.45}{0.92}\right) = 0.5187 \qquad [4.2]$$

Since the value of Eq. [4.2] is greater than that of Eq. [4.1], calculation [4.2] is repeated using a lower value ($\hat{k} = 0.60$); this gives

$$0.60\log\left(1 + \frac{2.45}{0.60}\right) = 0.4237 \qquad [4.3]$$

Since value [4.3] is lower than both values [4.1] and [4.2], the value of $\hat{k}$ lies between 0.60 and 0.92. By iteration we find that $\hat{k} = 0.696$ gives $\hat{k}\log[1 + (\bar{x}/\hat{k})] = 0.45598 \approx \log(n/f_0)$.

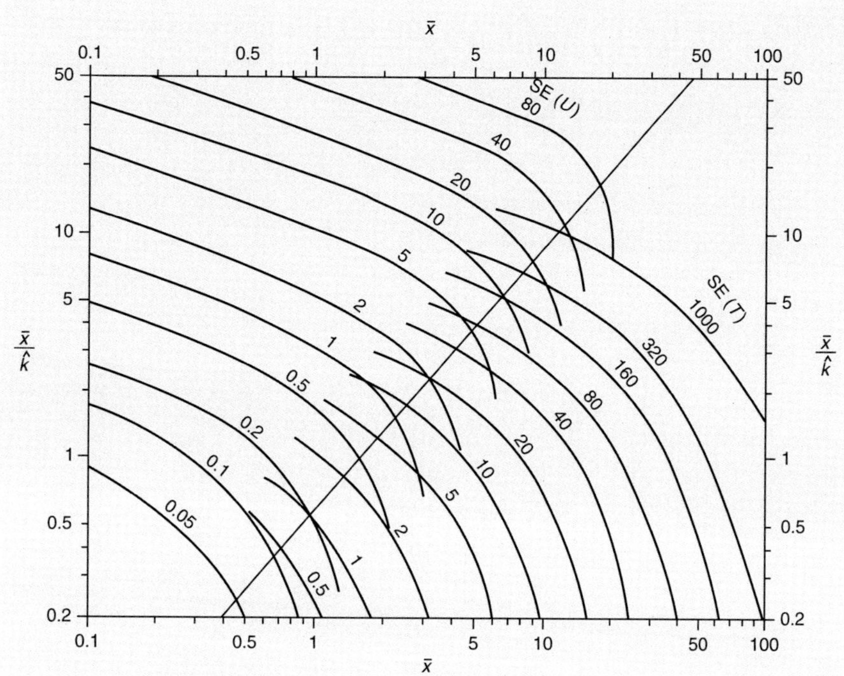

FIGURE 4.6 Standard errors of *T* and *U* for *n* = 100

For other values of *n*, multiply the standard error by $10/\sqrt{n}$

After Evans (1953); reproduced from Elliott (1977) with permission of the Freshwater Biological Association

Hence, $\hat{k} = 0.696$ provides a reliable estimate for *k* (if required, the process could be continued to further places of decimals).

Now, for $\bar{x} = 2.45$, $s^2 = 8.97$, $\hat{k} = 0.696$ and $n = 40$; the absolute value of *U* is determined:

$$U = \left| s^2 - \left(\bar{x} + \frac{\bar{x}^2}{\hat{k}} \right) \right| = \left| 8.97 - \left[2.45 + \frac{(2.45)^2}{0.696} \right] \right| = 2.104$$

From Fig. 4.6, the standard error of *U* is obtained for $\bar{x} = 2.45$ and $\bar{x}/\hat{k} = 8.97$ for which the approximate value of the standard error is given by

$$2\left(\frac{10}{\sqrt{n}}\right) = 2\left(\frac{10}{6.324}\right) = 3.16$$

Since the absolute value of *U* (2.104) is less than its standard error (3.16), the hypothesis that the data conform to the negative binomial distribution is not rejected at the 5% probability level ($P > 0.05$). It should be noted that for the *U* statistic it is not possible to state the actual probability level.

EXAMPLE 4.4 USE OF METHOD (3) AND THE *T* STATISTIC TO ESTIMATE $\hat{k}$ AND TEST AGREEMENT WITH A NEGATIVE BINOMIAL DISTRIBUTION

After treatment with detergent and disinfectant, the colony counts (cfu/mL) on rinses from replicate miniaturised Lisboa tubes (Blood et al., 1979) were as follows:

21, 19, 21, 34, 38, 30, 9, 8, 12, 16

Does the negative binomial provide a good model for the distribution of these counts?

The statistics of the count are the mean $= \bar{x} = 20.8$, variance $= s^2 = 106.84$ and $n = 10$.

Then by the method of matching moments [method (1)], the approximate value for $\hat{k}$ is given by $\hat{k} = [\bar{x}^2 - (s^2/n)]/(s^2 - \bar{x}) = 4.70$.

Determine the square and cube for each value of x and summate:

	x	x^2	x^3
	21	441	9,261
	19	361	6,859
	21	441	9,261
	34	1,156	39,304
	38	1,444	54,872
	30	900	27,000
	9	81	729
	8	64	512
	12	144	1,728
	16	256	4,096
Total	208	5,288	153,622

Calculate the value of the T statistic from the following equation:

$$T = \left(\frac{\sum x^3 - 3\bar{x}\sum x^2 + 2\bar{x}^2\sum x}{n}\right) - s^2\left[\left(2\frac{s^2}{\bar{x}}\right) - 1\right]$$

$$= \frac{153{,}622 - 329{,}971.2 + 179{,}979.24}{10} - 106.84\left[\frac{2(106.84)}{20.8} - 1\right]$$

$$= 363.004 - 990.736 = -627.73$$

For values of $\bar{x} = 20.8$, $\bar{x}/\hat{k} = 4.25$ and $n = 10$ the standard error of T is determined from Fig. 4.6 as $T \approx 320(10/\sqrt{10}) = 1012$.

The absolute value (627.73) of T is less than its standard error (1012) so the hypothesis that the data conform to a negative binomial distribution is not rejected at the 5% significance level ($P > 0.05$). The negative value of T indicates a tendency towards reduced skewness.

Since $\bar{x} > 4$ and the approximate value of $\hat{k} > 2$ and < 5, estimate $\hat{k}$ from the transformation $y = \log[x + (\hat{k}/2)]$. First, try the approximate estimate, that is, $\hat{k} = 4.7$:

x	21	19	21	34	38	30	9	8	12	16
y	1.3683	1.3294	1.3682	1.5605	1.6058	1.5099	1.0550	1.0149	1.1569	1.2636

The mean transformed value is $\bar{y} = 1.3233$ and the variance of the transformed counts is $s_y^2 = 0.0415$. The expected variance for $\hat{k} = 4.7$ is 0.0447 (Table 4.2) and since the actual variance ($s_y^2 = 0.0415$) is slightly greater than that for $\hat{k} = 5.0$, repeat the transformation using other values for $\hat{k}$ and set out the data in a table:

$\hat{k}$	$\bar{y}$	Variance (s_y^2)		Difference
		Observed (*O*)	Expected (*E*)[a]	(*O* − *E*)
4.7	1.323	0.0415	0.0447	−0.0032
4.9	1.326	0.0411	0.0427	−0.0016
5.0	1.327	0.0408	0.0417	−0.0009
5.1	1.328	0.0406	0.0408	−0.0002
5.2	1.329	0.0404	0.0400	+0.0004

[a]The expected variances are the 0.1886 trigamma values, from Table 4.2.

Hence, the best estimate of k is approximately $\hat{k} = 5.1$ since the value of the observed variance of the transformed data approximates closely to the expected variance. If the estimate of k is required to more decimal places, or for values not shown in Table 4.3, then it is necessary to refer to mathematical tables such as those of Davis (1963) for values of trigamma k, or determine the value by iteration from the data obtained.

Data of Abbiss and Jarvis (1980).

REFERENCES

Abbiss, J.S., Jarvis, B., 1980. Validity of statistical interpretations of disinfectant test results. Research Report No. 347. Leatherhead Food Research Association, Leatherhead, UK.

Anon, 2005. Milk and Milk Products – Quality Control in Microbiological Laboratories. Part 1. Analyst Performance Assessment for Colony Counts. ISO 14461-1:2005. International Standards Organization, Geneva.

Anscombe, F.J., 1949. The statistical analysis of insect counts based on the negative binomial distribution. Biometrics 5, 165–173.

Anscombe, F.J., 1950. Sampling theory of the negative binomial and logarithmic series distributions. Biometrika 37, 358–382.

Arbous, A.G., Kerrich, J.E., 1951. Accident statistics and the concept of accident-proneness. Biometrics 7, 340–432.

Blood, R.M., Williams, A.P., Abbiss, J.S., Jarvis, B., 1979. Evaluation of disinfectants and sanitizers for use in the meat processing industry. Part 1. Research Report No. 303. Leatherhead Food Research Association, Leatherhead, UK.

Breed, R.S., 1911. The determination of the number of bacteria in milk by direct microscopical examination. Zentralbl. Bakteriologie II Abt. Bd. 30, 337–340.

Cassie, R.M., 1962. Frequency distribution models in the ecology of plankton and other organisms. J. Anim. Ecol. 31, 65–92.

Cochran, W.G., 1954. Some methods for strengthening the common χ^2 tests. Biometrics 10, 417–451.

Comrie, L.J., 1949. Chambers Six-Figure Mathematical Tables. vol. 2, Chambers, Edinburgh.

Davis, H.T., 1963. Tables of the Higher Mathematical Functions. vol. 2 (revised), Principia Press, Bloomington, IN.

Dodd, A.H., 1969. The Theory of Disinfectant Testing With Mathematical and Statistical Section, second ed. Swifts (P & D) Ltd., London.

Elliott, J.M., 1977. Some Methods for the Statistical Analysis of Samples of Benthic Invertebrates, second ed. Scientific Publication No. 25. Freshwater Biological Association, Ambleside, Cumbria.

Evans, D.A., 1953. Experimental evidence concerning contagious distributions in ecology. Biometrika 40, 186–211.

Fisher, R.A., Corbett, A.S., Williams, C.B., 1943. The relation between the number of species and the number of individuals in a random sample of an animal population. J. Anim. Ecol. 2, 42–58.

Jones, P.C.T., Mollison, J.E., Quenouille, M.H., 1948. A technique for the quantitative estimation of soil microorganisms. Statistical note. J. Gen. Microbiol. 2, 54–69.

Jongenburger, I., 2012. Distributions of microorganisms in foods and their impact on food safety. PhD Thesis. Wageningen University, Netherlands. <http://edepot.wur.nl/196895> (accessed 22.06.15).

Mussida, A., Vose, D., Butler, F., 2013a. Efficiency of the sampling plan for *Cronobacter* spp. assuming a Poisson lognormal distribution of the bacteria in powder infant formula and the implications of assuming a fixed within and between-lot variability. Food Control 33, 174–185.

Mussida, A., Gonzales-Barron, U., Butler, F., 2013b. Effectiveness of sampling plan by attributes based on mixture distributions characterizing microbial clustering in food. Food Control 34, 50–60.

Neyman, J., 1939. On a new class of "contagious" distributions, applicable in entomology and bacteriology. Ann. Math. Stat. 10, 35–57.

Pearson, E.S., Hartley, H.O., 1976. Biometrika Tables for Statisticians, third ed. Biometrika Trust, London.

Pôlya, G., 1931. Sur quelques points de la théorie des probabilités. Ann. Inst. Henri Poincaré 1, 117–161.

Quenouille, M.H., 1949. A relation between the logarithmic, Poisson and negative binomial series. Biometrics 5, 162–164.

Southwood, T.R.E., 1966. Ecological Methods. Methuen, London.

Taylor, L.R., 1971. Aggregation as a species characteristic. Patil, G.P. et al., (Ed.), Statistical Ecology, vol. 1, Penn State University Press, Pennsylvania, pp. 357–372.

Thomas, M., 1949. A generalization of Poisson's binomial limit for use in ecology. Biometrika 36, 18–25.

Ziegler, N.R., Halvorson, H.O., 1935. Application of statistics to problems in bacteriology. IV. Experimental comparison of the dilution method, the plate count and the direct count for the determination of bacterial populations. J. Bacteriol. 29, 609–634.

CHAPTER

5

Statistical aspects of sampling for microbiological analysis

Whenever we wish to make a decision about the microbiological (or other) quality, safety or acceptability, of a food material, it is necessary to carry out tests on representative samples drawn from the 'batch' or 'lot' in question. Strictly, the estimates reflect only the parameters of the random samples tested but decisions concerning the food 'lot' are often based on the results of such analyses. The precision of the estimate will depend on the test(s) employed (Chapters 6–11), whether the samples are truly representative and the number of samples tested. The 'correctness' of the decision will increase with increasing sample size and sample number, but the number of samples that can be tested will be limited *inter alia* by the practicalities and economics of testing.

A large bulk of a relatively uniform material, such as milk, can be tested by drawing a representative sample after thorough mixing of the bulk. Suppose, however, that samples are taken from a total volume of 10,000 L, that 1 mL of each of those samples is diluted 10-fold for testing by a colony count procedure and colonies are counted on the 10^{-2} dilution; then the amount used for each test would be only 1 in $10^2 \times 10^4 \times 10^3$ or 1000-millionth (10^{-9}) of the 'lot'. To ensure that the final dilution used is truly representative of the 'lot' under test requires very efficient and intimate mixing at all stages of sampling, both in the production plant and in the laboratory.

Whilst such representative sampling may be theoretically feasible for a liquid, it can rarely be so for solids (eg, grain, vegetables, meat) or for solid–liquid multiphase systems (eg, meat and gravy in a pastry case). Leaving aside for the moment the technical efficiency of testing (see Chapters 6–9), the efficiency of the sampling operation *per se* will affect the results obtained (see Board and Lovelock, 1973 and the ISO 6887 series, eg, Anon, 2013a).

ATTRIBUTES AND VARIABLES SAMPLING

A product that has been produced to a set of process and product specifications is assumed to possess certain defined compositional and other characteristics, measurement of which can serve as a means to assess the 'manufactured quality' of the product through application of appropriate sampling and testing procedures. Although testing cannot of itself change the overall quality of the product, the test results can be used to draw conclusions about the overall quality of that 'lot' of the product. This then leads to the concept that a 'lot' can be accepted, or rejected, according to the

Statistical Aspects of the Microbiological Examination of Foods. http://dx.doi.org/10.1016/B978-0-12-803973-1.00005-X

results of tests on randomly drawn representative samples; this concept is known as acceptance sampling (*qv*).

Two types of sampling scheme can be used in acceptance sampling: one is for attributes and the other for variables. In this context, the term 'variables' describes those attributes of a sample that can be estimated analytically; by contrast, the term 'attributes' describes inherent characteristics that may be assessed directly by inspection or indirectly by measurement (*qv*).

A variables sampling scheme assumes that the measurements made on a series of samples from a given population follow a Normal distribution with approximately 95% of all test values within ±2 standard deviations from the mean. Unfortunately, neither the distribution of microorganisms in a food nor estimates of chemical or physical attributes of microbes (see Chapter 9) conform to a Normal distribution. However, after logarithmic transformation, the log colony count values generally approximate to a Normal distribution (Kilsby et al., 1979). Hence, variables sampling, which is the more efficient scheme, can be applied in microbiological analyses, although it has long been considered to be unsuitable (Kilsby, 1982).

By contrast, attribute sampling assumes that each unit in a 'lot' is characterised as being defective or not; in some schemes a 'marginally defective' criterion is permitted (see below). In microbiological testing, the term 'defective' implies that the sample unit contains more than a specified number of organisms (when tested by a defined test procedure) or, in the case of a test for pathogens and/or indicator organisms, that the target organism sought is detected when a sample unit of specified size is tested by an appropriate method. Note though that non-detection does not equate to absence of the target organism – merely that it has not been detected!

When colony counts are used as an estimate of viable numbers of microbes, the counts are often considered only in relation to a pre-determined limit and the essential criterion is whether or not that limit is exceeded. Such an approach permits classification of the sample units in terms of the proportion defective, and the proportion not defective, in a given number of sample units. The frequency of occurrence of defective units is generally described by the binomial distribution, although other distribution functions (eg, Poisson or negative binomial series) may be more appropriate in some instances (see Chapter 14).

In the assessment of colony count data, a marginally defective grouping is often included in order to make some allowance for variation in the distribution of organisms in the food and for the imprecision associated with colony count procedures. The marginally defective sample is defined as one that contains a number of organisms lower than a specified upper limit (M) but greater than a lower (acceptable) specified limit (m). Such a scheme, referred to as a three-class sampling plan, is described approximately by the trinomial distribution, which makes allowance not only for the proportions of defective and non-defective items (p and q, respectively) but also for the proportion (γ) of marginally defective units, so that $p + q + \gamma = 1$. The trinomial is one of a group of related distributions referred to previously as the multinomial distribution.

BINOMIAL AND TRINOMIAL DISTRIBUTIONS

THE BINOMIAL DISTRIBUTION (TWO-CLASS SAMPLE PLAN)

It has already been shown that the probability for an event to occur x times out of n tests on a large number of occasions is given by

$$P_{(x)} = \frac{n!}{(n-x)!x!}(p^x)(q)^{n-x}$$

where $q = 1 - p$. We can derive an estimate of p ($\hat{p}$) from the expected prevalence of contamination of the 'lot' and, therefore, we can estimate the probability with which 0, 1, 2, ..., n defective sample units will be found in the samples tested, provided that the total number of sample units tested (n) is small compared to the 'lot' size (N); otherwise, each time a sample unit is removed, the proportion of defectives will be changed significantly. Sample sizes <20% of the total are generally considered satisfactory. Thus, for a sample of n units from a consignment with expected 1% defectives (ie, $\hat{p} = 0.01$), we can calculate the terms of the expansion (for detail of the procedure see Example 3.2) by substituting in the equation given earlier.

Table 5.1 shows the percentage frequency with which zero, one or two defective units would occur on average in samples of various sizes taken from 'lots' of different average quality and size. For instance, if 20 sample units were tested for salmonellae, on average it is likely that, in 82 out of 100 tests, no salmonellae would be detected if the true prevalence of defectives was 1% of the sample units in the 'lot' (even assuming no technical errors in the analysis). If 20 sample units of a 'lot' containing 5% true defectives were tested, on average no salmonellae would be expected to be found in 36 out of 100 tests, but, on average, 1 or more positive tests would be

Table 5.1 Frequency of Detection of 0, 1 or 2 Defective Units in Samples of Different Size with Different Overall Prevalence of True Defectives in a Lot

	% Frequency of Detection for Samples of Size (n)								
	10			20			100		
	With Number of Defective Units of								
Defective Items in Lot (%)	**0**	**1**	**2**	**0**	**1**	**2**	**0**	**1**	**2**
0.01	99.9	<0.1	≪0.1	99.8	0.2	≪0.1	99	1	<0.1
0.1	99	1	<0.1	98	2	<0.1	90	9	0.5
1	90	9	0.4	82	17	2	37	37	18
2	82	17	1	67	27	5	13	27	27
5	60	32	7	36	38	19	0.6	3	8
10	35	39	19	12	27	29	<0.1	0.1	0.2

Modified from Steiner (1971); reproduced by permission of Leatherhead Food International.

found in 64 out of the 100 tests. However, had a sample size of 100 units been tested, the probability of not detecting salmonellae in 'lots' with 1% or 5% true defectives would be 0.37 or 0.006 (ie, on average none would be detected in 37 out of 100 tests or in 6 out of 1000 tests), respectively.

It can be seen, therefore, that testing small numbers of samples can give little or no protection against accepting a 'lot' containing salmonellae (or other pathogen) unless the prevalence of contamination is high. The same reasoning can be applied in relation to tests of processed cans for seam faults, loss of headspace vacuum or 'blowing' on incubation, or indeed to any other sampling and testing procedure where the prevalence of defectives is sought. This does not imply that sampling is useless, since, if operated continuously on a quality control basis, it provides 'cumulative' data that a process is operating satisfactorily, or that the quality has suddenly deteriorated (see also Chapter 12).

Another way of looking at this for a 'lot' with 0.1%, 1%, 5% or 10% overall true prevalence of salmonellae is that the probability [$P_{(x=0)}$] of finding no salmonellae in 10 sample units tested is 0.99, 0.90, 0.60 and 0.35, respectively (from column 2, Table 5.1). If the probability of finding no salmonellae [$P_{(x=0)}$] in n sample units from a 'lot' is α, then the probability of obtaining one or more positive tests is $P_{(x>0)} = 1-\alpha$. Example 5.1 shows the determination of the probable maximum prevalence of defective samples per 'lot' ($\hat{d}$), using the following equation:

$$\hat{d} = 1 - \sqrt[n]{1-\alpha}$$

EXAMPLE 5.1 CALCULATION OF THE PROBABLE PREVALENCE OF DEFECTIVES

Assuming 30 × 25 g samples of a product are tested and all found to be satisfactory (eg, negative for salmonellae), what is the maximum probable prevalence for salmonellae in the 'lot'?

Assuming no evidence is obtained for contamination of n test samples, the probable upper limit of prevalence of defectives ($\hat{d}$) is given by

$$\hat{d} = 1 - \sqrt[n]{1-\alpha}$$

where $1-\alpha$ is the probability that test samples may be contaminated with salmonellae. So, for $n = 30$ and $\alpha = 0.95$ (ie, the probability of uncontaminated samples is 95%) the estimate of prevalence is given by

$$\hat{d} = 1 - \sqrt[30]{1-0.95} = 1 - \sqrt[30]{0.05} = 1 - 0.905 = 0.095$$

Note that the value of $\sqrt[30]{0.05}$ can be determined as follows: antilog[(log 0.05)/30] = $10^{(\log 0.05)/30}$.

The value $\hat{d} = 0.095$ is an estimate of the 95% upper limit of the CI for a result of 0 positive results from 30 tests and can also be calculated by the Clopper–Pearson procedure for the estimation of confidence intervals for a binomial distribution.

We can conclude, therefore, that on 95% of occasions the true proportion (prevalence) of defective samples will not exceed 0.095 or, approximately, 1 in 10 × 25 g samples.

To determine the probable prevalence of salmonellae in a given weight of product multiply $\hat{d}$ by 1000/W, where W is the weight to samples tested. For these 30 × 25 g samples the probable maximum prevalence of salmonellae in the product is:

$$(0.095 \times 1000)/25 = 3.8 \approx 4 \text{ salmonella/kg}.$$

It is also possible to determine a confidence interval for the precision of $\hat{d}$ using Wilson's asymptotic method, without continuity correction (Newcombe, 1998). The CIs to the estimate are given by

$$\frac{(2np+z^2)}{2(n+z^2)} \pm \frac{z\sqrt{(z^2+4npq)}}{2(n+z^2)}$$

where p is the estimated proportion ($\hat{d}$) of contaminated 25 g samples (ie, the prevalence), $q = 1 - p$ and z is the $(1 - \alpha)/2$ point of the standard Normal distribution. For these data, with $n = 30$, $p = 0.095$, $q = 0.905$ and $z = 1.96$, the CI is given by

$$\frac{(2\times 30\times 0.095)+(1.96)^2}{2[30+(1.96)^2]} \pm \frac{1.96\sqrt{1.96^2+(4\times 30\times 0.095\times 0.905)}}{2[30+(1.96)^2]} = \frac{9.5416}{67.6832} \pm \frac{7.3751}{67.6832} = 0.1410 \pm 0.1090$$

that is, 0.032 to 0.250.

Hence, the 95% CIs, around the estimated prevalence value of $\hat{d}$ (=0.095), range from 0.032 to 0.250 salmonellae/25 g sample, which is equivalent to CIs of 1.3–40 salmonellae/kg.

Note that although no salmonellae were detected in the 30 samples, it should not be presumed or reported that salmonellae were not present, only that they were not detected.

Similarly, for a defined probability we can determine how many samples it would be necessary to test with negative results to give an assurance that a specified prevalence of defectives $\hat{p}$ is not exceeded (see Example 5.2) by rearranging the equation as follows:

$$n = \frac{\log_{10}(1-\alpha)}{\log_{10}(1-\hat{d})}$$

From the properties of the binomial distribution given previously (Chapter 3) the mean is np, where n is the number of samples tested and p is the proportion of positive results; the standard deviation of the distribution of the number of defective units is $\sqrt{np(1-p)}$. For sample sizes greater than $n = 20$ the binomial distribution approaches the Normal distribution, which can be used as an approximation when

EXAMPLE 5.2 HOW MANY SAMPLES DO WE NEED TO TEST?

Can we calculate the number of samples that is necessary to test with negative results to ensure at a probability of 95% that not more than 0.1% of the lot is contaminated with salmonellae?

The expected prevalence of defectives is as follows: $\hat{d} = 0.1/100 = 0.001$. Then, we have

$$n = \frac{\log_{10}(1-\alpha)}{\log_{10}(1-\hat{d})} = \frac{\log_{10} 0.05}{\log_{10} 0.999} = \frac{-1.3010}{-0.00043} = 2994.2$$

where n is the number of samples to be tested from a 'lot' and $(1 - \alpha)$ is the probability of detection of positive test samples from a 'lot' that has a prevalence of defective samples of $\hat{d}$.

Hence, to ensure with 95% probability that on average not more than 0.1% of a lot is contaminated by salmonellae it would be necessary to test 2995 sample units with negative results, clearly a practical impossibility!

If the expected prevalence of defectives were 5%, then $\hat{d} = 0.05$. To ensure with a 95% probability that on average salmonellae would contaminate not more than 5% of the lot would require a minimum of 59 samples to be tested. [$n = (\log_{10} 0.05)/(\log_{10} 0.95) = -1.3010/-0.02228 = 58.4$].

the number of defective units is large. When the chance of detecting defectives is very small and the number of sample units tested is large, the binomial distribution approaches the Poisson series.

THE TRINOMIAL DISTRIBUTION (THREE-CLASS SAMPLE PLANS)

The three-class sampling plan is defined by the trinomial distribution, which is an extension of the binomial distribution, and both are special cases of the multinomial distribution. Bray et al. (1973) showed that the trinomial distribution provides an acceptable approximation when the 'lot' size is large, and when $p \neq q$. The probabilities that d_0 good units, d_1 marginal units and $d_2 = n - d_0 - d_1$ defective items occur is given by the expansion of

$$\frac{n!}{d_0!d_1!d_2!}\left(p_0^{d_0} p_1^{d_1} p_2^{d_2}\right)$$

where p_0, p_1 and p_2 are the proportions of good, marginal and defective units, respectively. If the prevalence of marginally defective units $(d_1) = 0$, then this simplifies to the binomial expansion:

$$P_{(x)} = \frac{n!}{d_0!d_2!}\left(p_0^{d_0}\right)\left(p_2^{d_2}\right) = \frac{n!}{(n-x)!x!}p^x(1-p)^{n-x}$$

We can use the trinomial distribution to determine the probabilities of occurrence for various combinations of defective and marginally defective results, for instance, using a sample of 20 units (n) drawn from a 'lot' containing 1% defective ($d_2 = 0.01$) and 10% marginally defective ($d_1 = 0.10$) units. The first few terms of the expansion are shown in Table 5.2 for the probabilities of detecting different combinations of defective and marginally defective units. The cumulative frequency values for each of 0, 1 and 2 defectives, for the first 5 marginally defective terms, provides values of 80.8%, 16.5% and 1.5%, respectively. These are essentially the same as the observed values of 81.8%, 16.5% and 1.6%, respectively, for the equivalent binomial distribution for 20 sample units containing 0, 1 and 2 defectives from a population with 1% defective units.

Table 5.2 Probable Frequency of Detection of 0, 1 or 2 Defective Units with 0, 1, 2, 3, 4 or 5 Marginally Defective Units in a Sample of 20 Units

No. of Defective Units in Sample[a]	% Frequency of Detection for						Cumulative 0–5 (%)
	0	1	2	3	4	5	
	Marginally Defective Units in Sample						
0	9.7	21.9	23.3	15.7	7.5	2.7	80.8
1	2.2	4.7	4.7	3.0	1.4	0.5	16.5
2	0.2	0.5	0.4	0.3	0.1	<0.1	1.5

[a]*Assumes sample size = 20 units; prevalence of defective units in lot = 1%; prevalence of marginally defective units = 10%.*

PRECISION OF THE SAMPLE ESTIMATE

The precision of the estimate of defectives (or marginal defectives) is derived in statistical terms by allocating upper and lower confidence limits (CLs) to the sample value. These correspond to a probability that the true 'lot' value may lie outside these limits. Commonly the 95% CL is used, which implies that up to 5% of true values may lie outside this limit; of course, other CLs can also be applied. Suppose 3 defective units are found in a 2-class plan analysis of 20 sample units. Then if the true proportion of defectives in the consignment is p, the probability (P) of obtaining 3 or less defective units in the sample is given by the sum of the last 4 terms of the binomial expansion of $(p+q)^n$, where $n = 20$:

$$\frac{20!}{20!0!}q^{20}+\frac{20!}{19!1!}p^1q^{19}+\frac{20!}{18!2!}p^2q^{18}+\frac{20!}{17!3!}p^3q^{17}$$

where $q = 1 - p$.

Equating this to $P = 0.025$ (ie, 97.5% probability) and solving for p sets an upper limit to the proportion of defectives, beyond which there is only a 2.5% chance of the true value occurring. Similarly, by solving the first 17 terms of $(p + q)^{20}$ and equating to $P = 0.975$, the lower limit for p can be determined. Taken together, these calculations define the approximate upper and lower 95% CLs of $p = 0.38$ and 0.03, respectively. The procedure of Newcombe (1998) with a continuity correction gives lower and upper 95% CLs of 0.3886 and 0.0396, respectively. A simple software programme that provides both uncorrected and continuity-corrected CLs is available on-line (Lowry, 2015). It is also possible to calculate 'exact' binomial CLs using the Clopper and Pearson (1934) procedure but for most purposes the extensive calculation required is not worthwhile.

When the number of defectives is large (≥ 10), the Normal approximation may be used and the 95% CLs are given by $\pm 1.96\sqrt{np(1-p)}$ of the observed number. In this expression, the term p should ideally be the true proportion but the observed value of p (ie, $\hat{p}$) can be used as an estimate.

Table 5.3 compares the 95% CLs for 2 cases where the proportion of defectives is 15% and the number of defectives is either 3 in 20 or 30 in 200. The Poisson series provides a good approximation to the true (binomial) limits in both cases, but the Normal approximation is of use only for the larger number of sample units. It should

Table 5.3 95% Confidence Intervals (CIs) to an Estimated Proportion of Defectives Derived From Different Distributions Using the Method of Wilson (Newcombe, 1998)

Method of Calculation Assumes	Approximate 95% CI for 15% Defectives Assuming	
	3 Defectives in 20 Units	**30 Defectives in 200 Units**
Binomial distribution	4–39	10–21
Poisson series approximation	4–44	10–21
Normal approximation	0–31	10–20

not be forgotten that even if 200 units are tested, there is still a 1 in 20 chance that the true proportion of defectives will lie outside the 95% CLs.

The CLs for the trinomial distribution can be derived in a similar way to those for the binomial distribution, but the calculations are much more involved.

VARIATION IN SAMPLE SIZE

The limiting precision of the sampling procedure depends on the number of sample units tested rather than on the size of the 'lot'. In some sampling schemes the size of the sample has been related to the 'lot' size and/or the perception of risk associated with the product. For instance, many official control programmes are based on inspection of samples of fresh and processed foods for a range of defined defects (decomposition, unwholesomeness, etc.); the number of samples to be inspected increases with increasing lot size and is based on defined sampling plans. However, tests for microorganisms in such products are generally based on relatively few samples, the actual number of samples tested being dependent on the perceived risks of microbial contamination, the likely handling practices between production and consumption (eg, whether the product will be cooked before consumption) and the potential susceptibility to pathogenic microorganisms of the intended consumer group (infants, elderly, immuno-compromised persons, etc.) (ICMSF, 1986, 2002). In most instances, the number of samples to be tested for each lot is defined as, for example, 5 units/lot for colony counts and examination of up to 60 units for pathogens in high-risk products (Andrews and Hammack, 2003).

Some older references to sampling of foods recommend that the sample size should be related to the square root of the 'lot' size. There is no scientific justification for this approach other than to increase the sample size as the 'lot' size increases. In practice, sampling schemes generally relate sample size to 'lot' size since the risks of a wrong decision become more serious as the 'lot' size increases. The most commonly used standard sampling tables include the ISO 2859 series (Anon, 2006a) and other ISO standards (eg, Anon, 2001, 2002, 2003, 2004, 2006b), many of which are based on those of the Military Standard (1989) and the American National Standards Institute (ASTM, 2010). All specify plans based on the size of the 'lot' and a criterion termed the acceptance quality level (AQL) that is defined below. Examples of the relation between sample size and 'lot' size are illustrated in Tables 5.4–5.6 for various acceptance sampling plans.

ACCEPTANCE SAMPLING BY ATTRIBUTES

The risk of accepting an unsatisfactory 'lot' or of rejecting a satisfactory 'lot' leads to another aspect of sampling theory known as acceptance sampling. Any manufacturer producing food (or other goods) to an acceptable average quality will have variations between 'lots'. The purchaser decides whether to accept or reject the 'lot' on the basis of tests on a random sample, or a series of random samples. Two types of error can result from decisions based on the results from the test sample(s):

1. a 'good lot' may be rejected (a Type I or α error), or
2. a 'bad lot' may be accepted (a Type II or β error),

Table 5.4 Acceptance Sampling – Single Sampling Plans for Proportion of Defectives by Attributes

MIL STD Code	Lot Size	Sample Size n	AQL (%)	Acceptance Number c	% Defective Items in Lots With Chance of Acceptance[a] of			
					0.99	0.95 (5% PR)	0.50	0.10 (10% CR)
B	9–15	3	4	0	0.3	1.7	21	54
C	16–25	5	2.5	0	0.2	1.0	13	37
D	26–50	8	1.5	0	0.1	0.6	8.3	25
			6.5		2.0	4.6	20	41
E	51–90	13	1.0	0	0.08	0.4	5.2	16
			4.0	1	1.2	2.8	12	27
F	91–150	20	0.65	0	0.05	0.3	3.4	11
			2.5	1	0.8	1.8	8.2	18
			4.0	2	2.2	4.2	13	24
G	151–280	32	0.4	0	0.03	0.2	2.1	6.9
			1.5	1	0.5	1.1	5.2	12
			2.5	2	1.4	2.6	8.3	16
			4.0	3	2.6	4.4	11	20
H	281–500	50	0.25	0	0.02	0.1	1.4	4.5
			1.0	1	0.3	0.7	3.3	7.6
			1.5	2	0.9	1.7	5.3	10
			2.5	3	1.7	2.8	7.3	13
			4.0	5	3.7	5.3	11	18
J	501–1,200	80	0.65	1	0.2	0.4	2.1	4.8
			1.0	2	0.6	1.0	3.3	6.5
			1.5	3	1.0	1.7	4.6	8.2
			2.5	5	2.3	3.3	7.1	11
			4.0	7	3.7	5.1	9.6	14
K	1,201–3,200	125	0.4	1	0.1	0.3	1.3	3.1
			0.65	2	0.3	0.7	2.1	4.3
			1.0	3	0.7	1.1	2.9	5.3
			1.5	5	1.4	2.1	4.5	7.4
			2.5	7	2.3	3.2	6.1	9.4
L	3,201–10,000	200	0.4	2	0.2	0.4	1.3	2.7
			0.65	3	0.4	0.7	1.8	3.3
			1.0	5	0.9	1.3	2.8	4.6
			1.5	7	1.5	2.0	3.8	5.9
			2.5	10	2.4	3.1	5.3	7.7

Modified from ASTM (2010).
AQL, acceptance quality limit; CR, consumer's risk; PR, producer's risk.
[a]*Chance of accepting lots of different quality (expressed as % individual items lying outside specification) for different sample sizes and acceptance criteria. The lot is accepted if* c *or fewer defectives are found in a sample size of* n.

Table 5.5 Standard Plans for Attribute Sampling – Single Sampling Plans for Normal Inspection (Master Table)

Sample size code letter	Sample size	Acceptance Quality Levels (Normal Inspection)																									
		0.010	0.015	0.025	0.040	0.065	0.10	0.15	0.25	0.40	0.65	1.0	1.5	2.5	4.0	6.5	10	15	25	40	65	100	150	250	400	650	1000
		Ac Re	Ac Re	Ac Re	Ac Re	Ac Re	Ac Re	Ac Re	Ac Re	Ac Re	Ac Re	Ac Re	Ac Re	Ac Re	Ac Re	Ac Re	Ac Re	Ac Re	Ac Re	Ac Re	Ac Re	Ac Re	Ac Re	Ac Re	Ac Re	Ac Re	Ac Re
A	2	↓	↓	↓	↓	↓	↓	↓	↓	↓	↓	↓	↓	↓	↓	0 1	↓	↓	1 2	2 3	3 4	5 6	7 8	10 11	14 15	21 22	30 31
B	3	↓	↓	↓	↓	↓	↓	↓	↓	↓	↓	↓	↓	↓	0 1	↑	↓	1 2	2 3	3 4	5 6	7 8	10 11	14 15	21 22	30 31	44 45
C	5	↓	↓	↓	↓	↓	↓	↓	↓	↓	↓	↓	↓	0 1	↑	↓	1 2	2 3	3 4	5 6	7 8	10 11	14 15	21 22	30 31	44 45	↑
D	8	↓	↓	↓	↓	↓	↓	↓	↓	↓	↓	↓	0 1	↑	↓	1 2	2 3	3 4	5 6	7 8	10 11	14 15	21 22	30 31	44 45	↑	↑
E	13	↓	↓	↓	↓	↓	↓	↓	↓	↓	↓	0 1	↑	↓	1 2	2 3	3 4	5 6	7 8	10 11	14 15	21 22	30 31	44 45	↑	↑	↑
E	20	↓	↓	↓	↓	↓	↓	↓	↓	↓	0 1	↑	↓	1 2	2 3	3 4	5 6	7 8	10 11	14 15	21 22	↑	↑	↑	↑	↑	↑
G	32	↓	↓	↓	↓	↓	↓	↓	↓	0 1	↑	↓	1 2	2 3	3 4	5 6	7 8	10 11	14 15	21 22	↑	↑	↑	↑	↑	↑	↑
H	50	↓	↓	↓	↓	↓	↓	↓	0 1	↑	↓	1 2	2 3	3 4	5 6	7 8	10 11	14 15	21 22	↑	↑	↑	↑	↑	↑	↑	↑
J	80	↓	↓	↓	↓	↓	↓	0 1	↑	↓	1 2	2 3	3 4	5 6	7 8	10 11	14 15	21 22	↑	↑	↑	↑	↑	↑	↑	↑	↑
K	125	↓	↓	↓	↓	↓	0 1	↑	↓	1 2	2 3	3 4	5 6	7 8	10 11	14 15	21 22	↑	↑	↑	↑	↑	↑	↑	↑	↑	↑
L	200	↓	↓	↓	↓	0 1	↑	↓	1 2	2 3	3 4	5 6	7 8	10 11	14 15	21 22	↑	↑	↑	↑	↑	↑	↑	↑	↑	↑	↑
M	315	↓	↓	↓	0 1	↑	↓	1 2	2 3	3 4	5 6	7 8	10 11	14 15	21 22	↑	↑	↑	↑	↑	↑	↑	↑	↑	↑	↑	↑
N	500	↓	↓	0 1	↑	↓	1 2	2 3	3 5	5 6	7 8	10 11	14 15	21 22	↑	↑	↑	↑	↑	↑	↑	↑	↑	↑	↑	↑	↑
P	800	↓	0 1	↑	↓	1 2	2 3	3 4	5 6	7 8	10 11	14 15	21 22	↑	↑	↑	↑	↑	↑	↑	↑	↑	↑	↑	↑	↑	↑
Q	1250	0 1	↑	↓	1 2	2 3	3 4	5 6	7 8	10 11	14 15	21 22	↑	↑	↑	↑	↑	↑	↑	↑	↑	↑	↑	↑	↑	↑	↑
R	2000	↑	↑	1 2	2 3	3 4	5 6	7 8	10 11	14 15	21 22	↑	↑	↑	↑	↑	↑	↑	↑	↑	↑	↑	↑	↑	↑	↑	↑

Use first sampling plan below arrow ↓. *If sample size equals, or exceeds, lot or batch size, do 100% inspection.*

Use first sampling plan above arrow ↑.

Ac, acceptance number; Re, rejection number.

Table 5.6 Standard Plans for Attribute Sampling – Double Sampling Plans for Normal Inspection (Master Table)

Sample size code letter	Sample	Sample size	Cumulative sample size	Acceptance Quality Levels (Normal Inspection)																									
				0.010	0.015	0.025	0.040	0.065	0.10	0.15	0.25	0.40	0.65	1.0	1.5	2.5	4.0	6.5	10	15	25	40	65	100	150	250	400	650	1000
				Ac Re	Ac Re	Ac Re	Ac Re	Ac Re	Ac Re	Ac Re	Ac Re	Ac Re	Ac Re	Ac Re	Ac Re	Ac Re	Ac Re	Ac Re	Ac Re	Ac Re	Ac Re	Ac Re	Ac Re	Ac Re	Ac Re	Ac Re	Ac Re	Ac Re	Ac Re
A				↓	↓	↓	↓	↓	↓	↓	↓	↓	↓	↓	↓	↓	↓	•	↓	↓	•	•	•	•	•	•	•	•	•
B	First second	2 2	2 4	↓	↓	↓	↓	↓	↓	↓	↓	↓	↓	↓	↓	↓	•	↑	↓	0 2 1 2	0 3 3 4	1 4 4 5	2 5 6 7	3 7 8 9	5 9 12 13	7 11 18 19	11 16 26 27	17 22 37 38	25 41 56 57
C	First second	3 3	3 6	↓	↓	↓	↓	↓	↓	↓	↓	↓	↓	↓	↓	•	↑	↓	0 2 1 2	0 3 3 4	1 4 4 5	2 5 6 7	3 7 8 9	5 9 12 13	7 11 18 19	11 16 26 27	17 22 37 38	25 31 56 57	↑
D	First second	5 5	5 10	↓	↓	↓	↓	↓	↓	↓	↓	↓	↓	↓	•	↑	↓	0 2 1 2	0 3 3 4	1 4 4 5	2 5 6 7	3 7 8 9	5 9 12 13	7 11 18 19	11 16 26 27	17 22 37 38	25 31 56 57	↑	↑
E	First second	8 8	8 18	↓	↓	↓	↓	↓	↓	↓	↓	↓	↓	•	↑	↓	0 2 1 2	0 3 3 4	1 4 4 5	2 5 6 7	3 7 8 9	5 9 12 13	9 11 18 19	11 16 26 27	17 22 37 38	25 31 56 57	↑	↑	↑
F	First second	13 13	13 26	↓	↓	↓	↓	↓	↓	↓	↓	↓	•	↑	↓	0 2 1 2	0 3 3 4	1 4 4 5	2 5 6 7	3 7 8 9	5 9 12 13	7 11 18 19	11 16 26 27	↑	↑	↑	↑	↑	↑
G	First second	20 20	20 40	↓	↓	↓	↓	↓	↓	↓	↓	•	↑	↓	0 2 1 2	0 3 3 4	1 4 4 5	2 5 6 7	3 7 8 9	5 9 12 13	7 11 18 19	11 16 26 27	↑	↑	↑	↑	↑	↑	↑
H	First second	32 32	32 61	↓	↓	↓	↓	↓	↓	↓	•	↑	↓	0 2 1 2	0 3 3 4	1 4 4 5	**2** 5 **6** 7	3 7 8 9	5 9 12 13	7 11 18 19	11 16 26 27	↑	↑	↑	↑	↑	↑	↑	↑
J	First second	50 50	50 100	↓	↓	↓	↓	↓	↓	•	↑	↓	0 2 1 2	0 3 3 4	1 4 4 5	2 5 8 6	3 7 8 9	5 9 12 13	7 11 18 19	11 16 26 27	↑	↑	↑	↑	↑	↑	↑	↑	↑
K	First second	80 80	80 160	↓	↓	↓	↓	↓	•	↑	↓	0 2 1 2	0 3 3 4	1 4 4 5	2 5 6 7	3 7 8 9	5 9 12 13	7 11 18 19	11 16 26 27	↑	↑	↑	↑	↑	↑	↑	↑	↑	↑
L	First second	125 125	125 250	↓	↓	↓	↓	•	↑	↓	0 2 1 2	0 3 3 4	1 4 4 5	2 5 6 7	3 7 8 9	5 9 12 13	7 11 18 19	11 16 26 27	↑	↑	↑	↑	↑	↑	↑	↑	↑	↑	↑
M	First second	200 200	200 400	↓	↓	↓	•	↑	↓	0 2 1 2	0 3 3 4	1 4 4 5	2 5 6 7	3 7 8 9	5 9 12 13	7 11 18 19	11 16 26 27	↑	↑	↑	↑	↑	↑	↑	↑	↑	↑	↑	↑
N	First second	315 315	315 630	↓	↓	•	↑	↓	0 2 1 2	0 3 3 4	1 4 4 5	2 2 6 7	3 7 8 9	5 9 12 13	7 11 18 19	11 16 26 27	↑	↑	↑	↑	↑	↑	↑	↑	↑	↑	↑	↑	↑
P	First second	500 500	500 1000	↓	•	↑	↓	0 2 1 2	0 3 3 4	1 4 4 5	2 4 6 7	3 7 8 9	5 9 12 13	7 11 18 19	11 16 26 27	↑	↑	↑	↑	↑	↑	↑	↑	↑	↑	↑	↑	↑	↑
Q	First second	800 800	800 1600	•	↑	↓	0 2 1 2	0 3 3 4	1 4 4 5	2 5 6 7	3 7 8 9	5 9 12 13	7 11 18 19	11 16 26 27	↑	↑	↑	↑	↑	↑	↑	↑	↑	↑	↑	↑	↑	↑	↑
R	First second	1250 1250	1250 2500	↑	↑	0 2 1 2	0 3 3 4	1 4 4 5	2 5 6 7	3 7 8 9	5 9 12 13	7 11 18 19	11 16 26 27	↑	↑	↑	↑	↑	↑	↑	↑	↑	↑	↑	↑	↑	↑	↑	↑

Use first sampling plan below arrow ↓. *If sample size equals or exceeds lot or batch size, do 100% inspection.*

Use first sampling plan above arrow ↑.

Use corresponding single sampling plan (or alternatively, use double sampling plan below, where available). (•)

Ac, acceptance number; Re, rejection number.

Permission to reproduce extracts from BS 6001-1:1999 is granted by BSI (1999, 2007). British Standards can be obtained in PDF format from the BSI online shop: http://www.bsi-global.com/en/Shop, or by contacting BSI Customer Services for hardcopies: Tel +44(0)20 8996 9001, email: mailto:cservices@bsi-global.com.

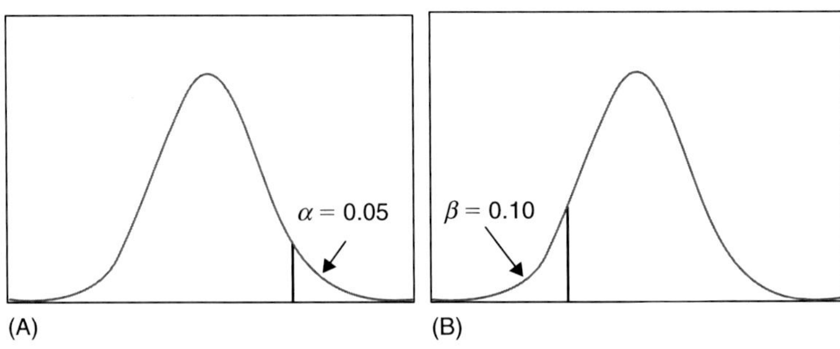

FIGURE 5.1

The Normal Distribution Curve Showing the Upper (A, $\alpha = 0.05$) and Lower (B, $\beta = 0.10$) Expected Percentiles Used in Setting Acceptance Quality and Rejectance Quality Limits Note that 95% of the cumulative distribution lies to the left of the α value and that 90% of the cumulative distribution lies to the right of the β value

where α and β represent the statistical probabilities for occurrence of the errors and are probabilities of the Normal distribution (Fig. 5.1).

Assume that the null hypothesis H_0 for acceptance of a 'lot' of product requires that it is not contaminated by specific pathogens; then, the alternative hypothesis H_1 is that the product is contaminated. If there is evidence of contamination, then we must reject H_0 and accept H_1. However, although the result that led to a decision to accept or reject a lot may be genuine, there is a risk that it might be a false-negative or a false-positive result. In setting up an attribute sampling plan it is normal to allow for a producer's risk that acceptable product may be wrongly rejected – this is a Type I error.

Conversely, there is a consumer's risk that a product is not acceptable. The null hypothesis for acceptance (H_0) is that the product is not contaminated; H_1 is that the product is contaminated. The Type II error is failure to reject H_0 when it is false (ie, when the product is contaminated). So if a lot is accepted on the basis of a false-negative test but the product is actually contaminated, the decision error is known as a Type II error.

The aim of acceptance sampling is to reduce both risks to a minimum but without the need for excessive sampling. Two expressions commonly used are the acceptance quality level (AQL) and the rejectance quality level (RQL) [sometimes called lot tolerance percent defective (LTPD)]. The AQL defines the quality of good 'lots' that the purchaser is prepared to accept most of the time (usually set at $P = 0.95$). Hence, the Type I error (producer's risk) of rejecting good-quality 'lots' is 5%. The RQL denotes the quality of poor 'lots' that the purchaser wishes to reject as often as possible – usually set at $P = 0.90$. Hence, the Type II error (consumer's risk) is 10%. Consignments of intermediate quality will be accepted at a frequency somewhere between $P = 0.10$ and 0.95. The balancing of producer's and consumer's risks is of critical importance. For the producer, if a good 'lot' is wrongly rejected, he loses both the cost of production and the potential profit from sales; however, if a 'bad' lot is accepted, the producer's risks may include the costs of product recall, including the value of the recalled product, loss of reputation and legal costs arising from consumer and customer litigation, whilst the consumer will be dissatisfied with the product received.

TWO-CLASS PLANS

Tables based on the binomial and Poisson distributions showing the size of sample and the acceptable number of defectives for different values of AQL and RQL have been published, for example, the ISO 2859 series (Anon, 2006a) and the Military Standard (1989). Examples are shown in Tables 5.5 and 5.6; note that these were derived primarily for acceptance of goods by inspection, although the term 'inspection' can include examination and testing – it is important to note the high levels of samples required for statistically valid sampling schemes. Generally, the AQL is designated in a purchase contract or some other documentation. The AQL value depends on the sample size, being higher for large samples than for small ones; the AQL does not describe the level of protection to the consumer for individual 'lots' but more directly relates to what might be expected from a series of 'lots'. It is necessary to refer to an operating characteristics (OC) curve (eg, Fig. 5.2) of the sampling plan in order to determine the relative risks attached to the plan in terms of both producer's risk and consumer's risk. Generally the sampling plan varies according to 'lot' size so that more samples are taken from larger 'lots' in order to reduce the risk of accepting a 'lot' with a high proportion of poor-quality products. For instance, it can be seen from Table 5.4, sampling plan code J, that for a 'lot' containing between 501 and 1200 units, 80 sample units should be examined. Assuming an AQL of 1.0%, the 'lot' would be accepted if not more than 2 units out of the 80 were found to be defective – this critical acceptance number is termed c. At such a level, there would still be a chance that defective items could occur in the acceptable 'lot' to the extent that 1.03% defectives would be accepted with a probability of 0.95 and that 6.52% defectives would be accepted with a probability of 0.10 (ie, the consumer's risk).

Table 5.5 shows the critical (c) levels for acceptance and rejection by normal inspection for various sample sizes and different levels of AQL. Again, taking sample size code J as an example, with 80 samples examined (n), an AQL of 10% would permit acceptance with not more than 14 defective units (c) and would require rejection with 15 defective units. However, for an AQL of 2.5%, $c = 5$ and for an AQL of 0.25%, $c = 0$. These single sampling plans could therefore be described as follows:

For AQL = 10%, $n = 80$, $c = 14$.
For AQL = 2.5%, $n = 80$, $c = 5$.
For AQL = 0.25%, $n = 80$, $c = 0$.

Because of the very large numbers of samples required, double or other multiple sequential sampling plans may be adopted. For a double sampling scheme (eg, Table 5.6) and again, taking the data for sample code J, 100 samples would be drawn from the 'lot', but initially only 50 samples would be examined. For an AQL of 10% the 'lot'

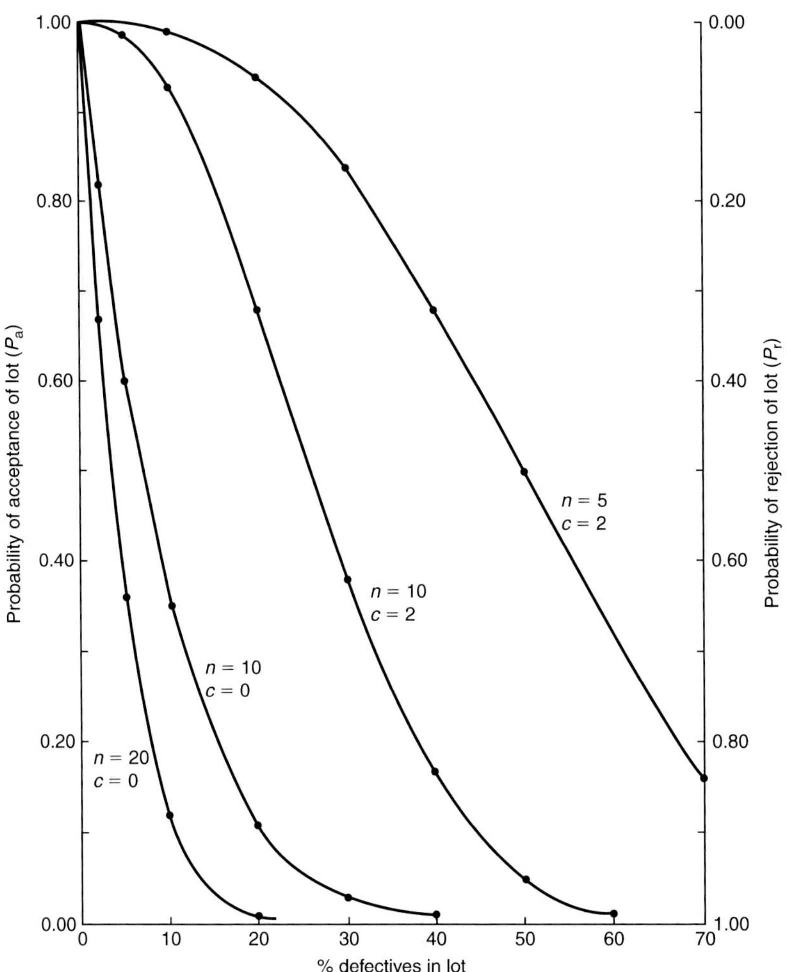

FIGURE 5.2 Operating Characteristics (OC) Curves for Typical Attribute Sample Plans Used in Microbiological Criteria for Foods

Here *n* is the number of samples tested and *c* is the number of unsatisfactory test results that can be accepted. Note that the gradient of the lines increases as *n* increases and as *c* becomes smaller

would be accepted with 7 or fewer defective units (c_1) and rejected if 11 or more were found (c_2). However, if an intermediate level of defectives (c') occurred (ie, between c_1 and c_2), then the second set of 50 samples would be tested. The cumulative critical defective level (c) would then be 18. Such a plan could be described as follows:

double sampling, AQL = 10%; $n_1 = 50$, $c_1 = 7$, $c_2 = 11$;
$n = 100$, $c = 18$.

Table 5.7 Percentage of Defectives in Consignments Which Would Have a 95% Chance of Acceptance (AQL) and a 90% Chance of Rejection (RQL) by Different Sampling Schemes

	% Defectives in Consignment When			
	$c = 0$		$c = 1$	
No. of Sample Units (*n*)	**95% Chance of Acceptance**	**90% Chance of Rejection**	**95% Chance of Acceptance**	**90% Chance of Rejection**
10	0.5	25	3.5	39
20	0.3	12	1.8	20
30	0.2	8	1.2	13
50	0.1	5	0.7	8
100	0.05	2.5	0.35	4
150	0.04	1.7	0.23	2.6
200	0.03	1.2	0.18	2.0
500	0.01	0.5	0.07	0.8

c, maximum acceptable number of defectives.
Reproduced from Steiner (1971) by permission of Leatherhead Food International.

Table 5.7 illustrates the 95% probability for acceptance (5% AQL) and 90% probability for rejection (10% RQL) for different sample numbers and 2 levels of maximum acceptable number of defectives based on Poisson distribution. The values can be used also to derive the upper boundaries describing the number of specific organisms in a given quantity of a product. For instance, if the sampling scheme required testing of 10 sample units ($n = 10$) with no salmonellae detected ($c = 0$), then the 'lot' would be accepted with a 95% probability that the average prevalence of contamination in the 'lot' is less than 0.5% and rejected with a 90% probability that the average prevalence of contamination is 25% or more. However, even if no salmonellae were detected, there would be a 5% probability that the prevalence of contamination would be between 0.5% and 25%. If we assume that a positive test requires the presence of at least 1 organism per sample and if the sample unit is 25 g, then the prevalence of contamination is ≥ 1 organism/25 g or 40 organisms/kg. Now if 25% of the 25 g sample units were positive, then on average the overall contamination level would be at least 10 organisms/kg (compare with Example 8.5, which assumes binomial distribution and determines the maximum level as 10.4 salmonellae/kg). There is also a 10% probability that a rejected 'lot' may contain more than 25% prevalence of contamination.

Internationally agreed Statistical Sampling Tables are available for attribute sampling schemes and for variables sampling schemes with either known or unknown standard deviations. They are intended for use in conjunction with continuous production and provide facility to switch from normal to tightened or reduced inspection based on previous performance. International standards exist for other criteria, for instance, Sequential Attribute Sampling Plans according to defined AQL (Anon, 2005a), Attribute Sampling Plans for skip-lot procedures (Anon, 2005b) and

Attribute Sampling Plans for limiting AQL for isolated lots (Anon, 1985). Other standard tables exist also for specific products, for example, milk and milk products (Anon, 2004). However, the schemes are basically suitable only for inspection-type tests and, therefore, whilst of value to the microbiologist in field studies (eg, in assessing the quality of sacks of stored grain with respect to overt mould or insect damage, or the incidence of bottles of a beverage showing evidence of microbial growth), the sheer quantity of testing required would totally swamp any laboratory facility. They serve, therefore, as ideals towards which we might aim if effective automated microbiological testing ever became established.

THREE-CLASS PLANS

In precisely the same way that two-class plans are constructed for various AQL and RQL levels based on binomial or Poisson distributions, three-class plans can be derived from the trinomial distribution. Examples of such plans are cited by Bray et al. (1973) and by ICMSF (1986, 2002). Since three-class plans are used only for microbiological colony count data (and not for presence/absence tests), the criteria for acceptable or defective samples are defined in terms of two colony count levels. The lower criterion (designated m) is the acceptable target level below which colony counts should fall most of the time; the upper criterion (M) is that level which must not be exceeded. Hence, marginal defectives will have counts intermediate between m and M.

Since no colony count should exceed M, the maximum acceptable level of defective units (c) will be zero, but some marginal defective units (c_1) will be permitted out of n samples tested. Hence, the acceptable proportion of defectives is $p_2 = 0$ and the proportion of marginal defectives is $p_1 = c_1/n$.
Hence, the probability for acceptance of samples (P_a) can be derived from a simplification of the trinomial expansion, for values of i from 0 to c_1:

$$P_a = \sum_{i=0}^{c_1} \binom{n}{i} p_0^{n-i} p_1^i$$

where p_0 is the proportion of good samples, p_1 is the proportion of marginal samples, n is the number of sample units tested and c_1 is the number of marginal defectives permitted.

This has the same form as that of a binomial expansion but is not a general binomial since $p_0 + p_1 < 1$. Values of P_a can be calculated from the following equation:

$$P_a = \left[\sum_{i=0}^{c} \binom{n}{i} \left(\frac{p_0}{p_0 + p_1}\right)^{n-i} \left(\frac{p_1}{p_0 + p_1}\right)^{i}\right] (p_0 + p_1)^n$$

The expression inside the brackets [ ] is a cumulative binomial term, which can be read directly from standard tables of the binomial distribution (Pearson and Hartley, 1966, Table 37). Example 5.3 illustrates the derivation of data for a three-class sampling plan.

EXAMPLE 5.3 DERIVATION OF PROBABILITIES FOR ACCEPTANCE (P_A)

Suppose that it is desirable to assess the probability for acceptance (P_a) of a 'lot' of product expected to contain up to 2% defective and up to 30% marginal defective items. The proposed sampling plan permits acceptance of no defectives ($c_2 = 0$) in 10 sample units ($n = 10$). What would be the probability of acceptance for sample plans that permit 0, 1 or 2 marginal defectives ($c_1 = 0$, 1 or 2)?

First, let us assume that the test methods can determine only acceptable or unacceptable items (ie, a two-class sample plan). The binomial probability that an event will occur x times out of n tests is given by

$$P_{(X)} = \left[\frac{n!}{(n-x)!x!}\right] p^x (1-p)^{n-x}$$

where P is the probability that the event will occur. An estimate of p ($\hat{d}_0$) can be derived from the expected 2% prevalence of defectives, so that $\hat{d}_0 = 0.02$. Then for $c_2 = 0$:

$$P_a = \left[\frac{10!}{(10-0)!0!}\right](0.02)^0(1-0.02)^{10-0} = (0.98)^{10} \approx 0.82$$

Hence, we would expect that, on average, no defective units would be found in 82 out of every 100 tests undertaken on 10 replicate sample units with an expected prevalence of 2% defective samples (cf Table 5.1, column 2).

Next, let us consider the occurrence of marginal defectives at levels up to 30% but with zero defectives. The proportion of defectives is given by $(\hat{d}_0) = 0.00$, and the proportion of marginal defectives by $(\hat{d}_1) = 0$, 0.10, 0.20 or 0.30. From the binomial expansion used earlier or from standard Tables of Binomial Frequencies (eg, Pearson and Hartley, 1976, Table 37) we can derive the following values for the probability of occurrence of 0, 1 or 2 'defectives' in a sample of 10 units, where 'defectives' are now defined as 'marginal defectives':

Probability for Occurrence of 'Defectives'	No. of 'Defectives' in Sample	Probability of Acceptance with a Proportion of Marginal Defectives ($\hat{d}_1$) of			
(P_x)	**(c_2)**	**0.0**	**0.10**	**0.20**	**0.30**
$P_{(x=0)}$	0	1.0	0.35	0.11	0.03
$P_{(x=1)}$	1	0.0	0.39	0.27	0.12
$P_{(x=2)}$	2	0.0	0.19	0.30	0.23
Cumulative probability	$P_x \leq 1$	1.0	0.74	0.38	0.15
	$P_x \leq 2$	1.0	0.93	0.68	0.38

Hence, if $n = 10$ and $c_1 = 0$, the probabilities of acceptance of a batch with 0% defectives and 0%, 10%, 20% or 30% marginal defectives would be 1.0, 0.35, 0.11 or 0.03, respectively. But if $n = 10$ and $c_1 = 2$, the probabilities of acceptance (P_a) for these levels of marginal defectives would be 1.0, 0.93, 0.68 and 0.38, respectively.

It has already been shown that if the batch contains 2% defectives and 0% marginal defectives, the probability of acceptance, with $c_2 = 0$, is $P_a = 0.82$. Hence, for a batch containing 2% defectives *and* 30% marginal defectives, the cumulative probability of acceptance is given by

$$P_a = 0.82 \times 0.03 = 0.025 \quad \text{if } c_1 = 0 \quad \text{and } c_2 = 0$$

and

$$P_a = 0.82 \times 0.38 = 0.31 \quad \text{if } c_1 = 2 \quad \text{and } c_2 = 0$$

By extending this procedure for other values we can derive the required P_a values for the three-class plans given as follows and in Table 5.8:

Proportion of Marginal Defectives (Expected Defectives = 2%)	Probability of Acceptance (P_a) with c_2 = 0, and		
	c_1 = 0	c_1 = 1	c_1 = 2
0.0	0.82	0.82	0.82
0.1	0.29	0.61	0.76
0.2	0.09	0.31	0.56
0.3	0.025	0.12	0.31

Table 5.8 shows that for $n = 10$, the probability of acceptance of a 'lot' containing 2% defectives and 20% marginal defectives is 0.55 when $c_1 = 2$, but only 0.09 when $c_1 = 0$. The sampling plan is more stringent when $c_1 = 0$, but there is an equal chance of accepting a 'lot' with 5% defective units and 0% marginal defectives with either $c_i = 0$ or $c_i = 2$.

In setting a three-class plan for a specified AQL, it is necessary to decide on an arbitrarily defined probability of acceptance (P_a) of, say, 95%, for 'lots' of specific quality (eg, 90% good, 10% marginal, 0% bad) and then to choose a provisional sample size n (eg, $n = 10$). From Table 5.9 for $n = 10$, $p_0 = 0.90$, $p_1 = 0.10$, $p_2 = 0$, we find

Table 5.8 Probability of Acceptance for a Three-Class Plan with $N = \infty$, $n = 10$, $c_1 = 0$ or 2 and $c_2 = 0$

Marginal Defectives in Lot (%)	Defectives in Lot (%)							
	0	2	5	10	20	30	40	50
c_1 = 2								
70	0.00	0.00	0.00					
60	0.01	0.01	0.01	0.00	0.00			
50	0.05	0.04	0.03	0.02	0.01			
40	0.17	0.14	0.10	0.06	0.02	0.00		
30	0.38	0.31	0.23	0.13	0.04	0.01		
20	0.68	0.55	0.41	0.24	0.07	0.02	0.00	
10	0.93	0.76	0.56	0.32	0.10	0.03	0.01	
0	1.00	0.82	0.60	0.35	0.11	0.03	0.01	0.00
c_1 = 0								
50	0.00							
40	0.01	0.00	0.00	0.00				
30	0.03	0.02	0.02	0.01	0.00			
20	0.11	0.09	0.07	0.04	0.01	0.00		
10	0.35	0.29	0.21	0.12	0.04	0.01	0.00	
0	1.00	0.82	0.60	0.35	0.11	0.03	0.01	0.00

c_2, number of defective units permitted; c_1, number of marginally defective units permitted; n, sample size; N, lot size.

Table 5.9 Probabilities of Acceptance for Three-Class Plans ($N = \infty$) Using $c_2 = 0$ and c_1 = Number of Marginally Defective Samples Acceptable

% of Lot			Number of Marginal Defectives with Sample Size = 10										
Good	**Marg**	**Bad**	**0**	**1**	**2**	**3**	**4**	**5**	**6**	**7**	**8**	**9**	**10**
99	1	0	0.90	1.00	···	···	···	···	···	···	···	···	1.00
99	0	1	0.90	···	···	···	···	···	···	···	···	···	0.90
97	3	0	0.74	0.97	1.00	···	···	···	···	···	···	···	1.00
97	1	2	0.74	0.81	0.82	···	···	···	···	···	···	···	0.82
95	5	0	0.60	0.91	0.99	1.00	···	···	···	···	···	···	1.00
95	3	2	0.60	0.79	0.81	0.82	···	···	···	···	···	···	0.82
90	10	0	0.35	0.74	0.93	0.99	1.00	···	···	···	···	···	1.00
90	8	2	0.35	0.66	0.78	0.81	0.82	···	···	···	···	···	0.82
90	5	5	0.35	0.54	0.59	0.60	···	···	···	···	···	···	0.60
80	20	0	0.11	0.38	0.68	0.88	0.97	0.99	1.00	···	···	···	1.00
80	18	2	0.11	0.35	0.59	0.74	0.80	0.81	0.82	···	···	···	0.82
80	15	5	0.11	0.31	0.48	0.56	0.59	0.60	···	···	···	···	0.60
80	10	10	0.11	0.24	0.32	0.34	0.35	···	···	···	···	···	0.35
80	5	15	0.11	0.17	0.19	0.20	···	···	···	···	···	···	0.20
70	30	0	0.03	0.15	0.38	0.65	0.85	0.95	0.99	1.00	···	···	1.00
70	25	5	0.03	0.13	0.29	0.45	0.54	0.58	0.60	···	···	···	0.60
70	20	10	0.03	0.11	0.21	0.29	0.33	0.34	0.35	···	···	···	0.35
70	10	20	0.03	0.07	0.09	0.10	0.11	···	···	···	···	···	0.11
60	40	0	0.01	0.05	0.17	0.38	0.63	0.83	0.95	0.99	1.00	···	1.00
60	35	5	0.01	0.04	0.13	0.28	0.42	0.53	0.58	0.59	0.60	···	0.60
60	30	10	0.01	0.04	0.10	0.20	0.27	0.32	0.34	0.35	···	···	0.35
60	20	20	0.01	0.03	0.06	0.08	0.10	0.11	···	···	···	···	0.11
60	10	30	0.01	0.02	0.02	0.03	···	···	···	···	···	···	0.03
50	50	0	0.00	0.01	0.05	0.17	0.38	0.62	0.83	0.95	0.99	1.00	1.00
50	45	5		0.01	0.05	0.13	0.27	0.41	0.52	0.58	0.59	0.60	0.60
50	40	10		0.01	0.04	0.10	0.18	0.26	0.32	0.34	0.35	···	0.35
50	30	20		0.01	0.02	0.05	0.07	0.09	0.10	0.11	···	···	0.11
50	20	30		0.00	0.01	0.02	0.02	0.03	···	···	···	···	0.03
50	10	40		0.00	0.00	0.01	···	···	···	···	···	···	0.01
40	60	0		0.00	0.01	0.05	0.17	0.37	0.62	0.83	0.95	0.99	1.00
40	50	10			0.01	0.03	0.09	0.17	0.25	0.31	0.34	0.35	0.35
40	40	20			0.01	0.02	0.04	0.07	0.09	0.10	0.11	···	0.11
40	30	30			0.00	0.01	0.02	0.02	0.03	···	···	···	0.03
30	70	0			0.00	0.01	0.05	0.15	0.35	0.62	0.85	0.97	1.00
30	60	10				0.01	0.03	0.07	0.15	0.24	0.31	0.34	0.35
30	50	20				0.00	0.01	0.03	0.06	0.08	0.10	0.11	0.11
30	40	30				0.00	0.01	0.01	0.02	0.02	0.03	···	0.03
20	80	0				0.00	0.01	0.03	0.12	0.32	0.62	0.89	1.00
20	70	10					0.00	0.02	0.06	0.14	0.24	0.32	0.35
20	60	20					0.00	0.01	0.02	0.05	0.08	0.10	0.11

Marg, marginal defectives.
Permission to reproduce sought from Bray et al. (1973).

Table 5.10 Probabilities of Acceptance for Three-Class Plans, Chosen So That AQL = 95% (*), RQL = 10% (**) and $c_2 = 0$

	Percent of Lot			Sample Size (*n*)			
				3	10	25	60
	Good	**Marginal**	**Bad**	**Maximum No. of Marginal Defectives (c_1) Acceptable**			
				1	2	5	9
	99	1	0	1.00	1.00	1.00	1.00
	95	5	0	0.99	0.99	1.00	1.00
AQL	90	10	0	0.97*	0.93*	0.97*	0.93*
	80	20	0	0.90	0.68	0.62	0.21
	70	30	0	0.78	0.38	0.19	0.01
	60	40	0	0.65	0.17	0.03	0.00
RQL	50	50	0	0.50	0.05*	0.00**	**
	40	60	0	0.35	0.01		
	30	70	0	0.22	0.00		
	99	0	1	0.97	0.90	0.78	0.55
	90	9	1	0.95	0.85	0.75	0.52
	80	19	1	0.88	0.63	0.52	0.14
	60	30	1	0.64	0.16	0.02	0.00
	40	59	1	0.35	0.01	0.00	
	98	0	2	0.94	0.82	0.60	0.30
	90	8	2	0.92	0.78	0.60	0.29
	80	18	2	0.86	0.59	0.42	0.09

$P_a = 0.93$ for $c_1 = 2$. Therefore, an appropriate sampling plan would be $n = 10$, $c_1 = 2$. Such a plan would also accept 'lots' with 50% marginal (or 45% marginal, and 5% defectives) with $P_a = 0.05$ (RQL = 5%) and would accept 60% marginal defectives (or 40% marginal defectives and 20% defectives) with $P_a = 0.01$. Table 5.10 provides an AQL chart for three-class sample plans based on Bray et al. (1973). Probabilities marked * indicate suitable plans for an AQL of about 95% and those marked ** indicate the equivalent cut-off point for an RQL of 10%.

Tables of the various probabilities for different proportions of good, marginal and defective sample units, and various sample plans (ie, values of *n* and *c*) are given by Bray et al. (1973) and by ICMSF (1986, 2002).

OPERATING CHARACTERISTICS (OC) CURVES

A convenient way of looking at the relative efficiencies of various sampling plans is to construct a series of OC curves. An OC curve is a graphical representation of the cumulative probability of acceptance (P_a) against the percentage of defective sample units in the 'lot'. This is illustrated in Fig. 5.2 for various two-class sampling plans. The steepness of the OC curve increases with both an increase in sample numbers (n) and a reduction in acceptance numbers (c). For instance, a sampling plan of $n = 20$, $c = 0$ would accept 5% defectives 36 times in 100 tests, whereas a plan with $n = 10$, $c = 0$ would accept the same level of defectives in 60 out of 100 tests. For these same sampling plans and percentage of 'lot' defectives the probabilities of rejection (P_r) would be 64 and 40 times out of 100 tests, respectively. However, for plans that allow 2 defective units out of 5 or 10 sample units, there is little difference in P_a at 5% 'lot' defectives. The ideal OC curve would have a vertical 'cut-off' for the acceptable percentage of defectives in a 'lot'. Such a curve would require the testing of so many samples per 'lot' that it would be both economically and technically impracticable.

OC curves for three-class plans can be considered to form a geometric surface on a three-dimensional 'plot' of P_a (or P_r) against the percentage of 'lot' defectives and the percentage of 'lot' marginal defectives. This is illustrated in Fig. 5.3 for the sampling plan $n = 10$, $c_1 = 2$, $c_2 = 0$, the acceptance probability (P_a) values of which are presented in Table 5.8.

ACCEPTANCE SAMPLING BY VARIABLES

As is the case for attribute sampling, international standards have been produced for variables sampling plans (see, eg, Anon, 2013b) but they are rarely used in assessment of data from the microbiological examination of food. Sampling plans based on such tables are analogous to those in Tables 5.4–5.7, but relate AQLs to the mean and standard deviation of populations.

A major disadvantage of attribute sampling schemes is that no assumptions are made about the distribution of measurable parameters within a 'lot'. Whilst this is not of consequence in schemes based purely on inspection for defects, the use of attributes schemes based on the acceptability, or otherwise, of estimates of variable parameters will result in excessive risk of rejection of otherwise acceptable materials. Indeed, Clark (1978) pointed out that in using an ICMSF sampling plan for $n = 5$, $c = 1$ with a 'lot' containing no defective material and only 10% marginal defective units (ie, with 10% counts $>m$ and $<M$) there is a 1 in 12 chance of rejection of the 'lot'. Clearly, such a risk is unacceptably high and could result in the wastage of much valuable foodstuff.

The primary advantage of a variables sampling scheme is that proper use of quantitative data should permit 'better decision processes' and lead to 'economic benefits' (Gascoigne and Hill, 1976). The primary disadvantages are claimed to be: (1) the need for analysis of laboratory data, rather than merely making a 'go–no go' decision; and (2) an absolute requirement that the data obtained for the parameter being estimated conform to a Normal frequency distribution. Neither of these disadvantages is real! The widespread availability of low-cost computing

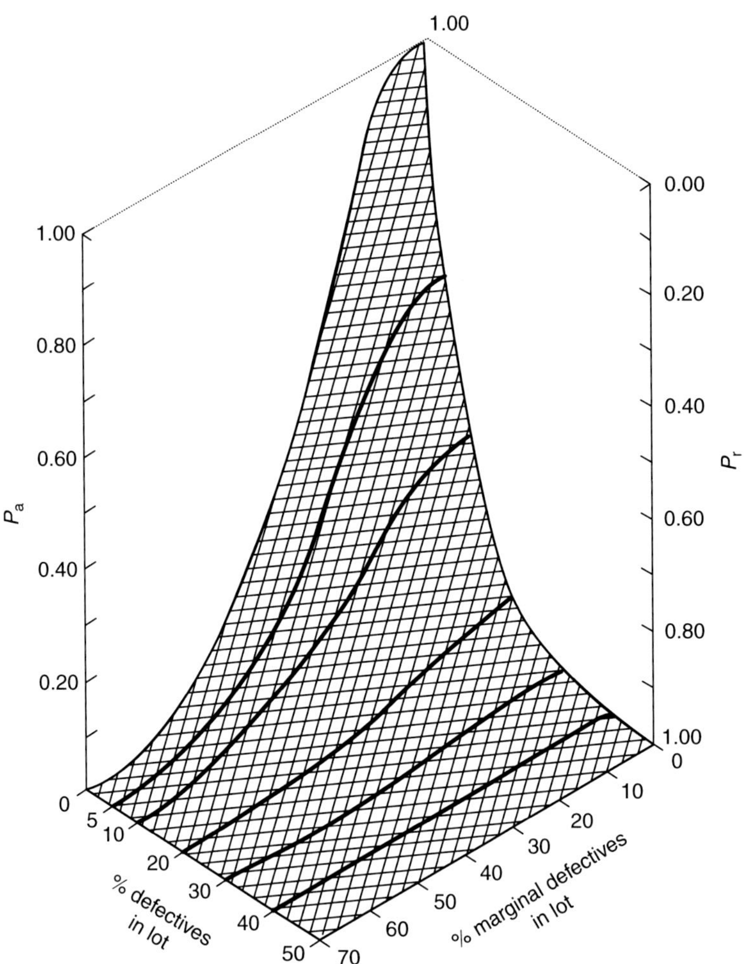

FIGURE 5.3 An Operating Characteristics (OC) Surface Plot for a 3-Class Attributes Plan with $n = 10$, $c_1 = 2$ and $c_2 = 0$ (Data From Table 5.8)

Here n is the number of samples tested, c_1 is the number of acceptable marginal defectives and c_2 is the number of defectives permitted. The right-hand OC plot gives the values for a two-class plan with $n = 10$, $c_2 = 0$ and 0% marginal defectives. The left-hand OC plot is for a two-class plan with $n = 10$, $c_1 = 2$ and 0% defectives. The other curves are for different values of prevalence of defectives plotted against the prevalence of marginal defectives. The 'surface' of the plot shows the probability of acceptance (or rejection) for various combinations of prevalence of defective and marginally defective samples

facilities provides facility for data analysis as part of laboratory quality assurance procedures. If one assumes that a set of colony counts on a 'lot' of food follows a contagious distribution with variance greater than the mean (ie, $s^2 > \bar{x}$), as is generally the case, then the data may be 'Normalised' by a logarithmic transformation (Table 3.4).

It is generally the case that microbiological colony counts conform at least approximately to a lognormal distribution (Fig. 2.1; also Kilsby et al., 1979). Hence, variables sampling schemes can be applied in microbiological control systems where colony counts are determined. However, presence–absence tests cannot be treated in this way unless sufficient replicates are tested to provide an estimate of most probable numbers (see Chapter 8). How, then, is a variables scheme formulated?

A variables scheme requires definition of a percentage, d, of units in a 'lot' that exceeds a defined critical limit in order to assess whether the 'lot' is unsatisfactory. If the critical limit is defined in terms of a log colony count (C), then, for a set of n samples, we can determine whether the percentage of values exceeding C exceeds the permissible level (d).

For a Normal distribution, the percentage (K) exceeding the value C depends solely on the population distribution and is given by $K = (C - \mu)/\sigma$, where μ is the population mean and σ is the population standard deviation. By reference to tables of the standard Normal deviate (eg, Pearson and Hartley, 1976, Table 1) the percentage beyond the value C can be determined for any value of K. For example, 16% of values greater than C would correspond to $K = 1$ and 5% of values above C correspond to $K = 1.65$. Conversely, the critical value of K (K_d) corresponding to d may be determined, for example, for $d = 5\%$, $K_d = (C - \mu)/\sigma \leq 1.65$. Low values of K imply a high percentage of values above C so that rejection may be warranted.

In most food situations the values of μ and σ will not be known, but they can be estimated from the sample data by the statistics for the mean log count (log cfu/g; $\bar{x}$) and its standard deviation (s). Hence, an estimate of K ($\hat{K}$) can be derived as follows: $\hat{K} = (C - \bar{x})/s$. Since the estimate ($\hat{K}$) of K is as likely to exceed K_d as it is to fall below it, some allowance must be made for the imprecision of $\hat{K}$. This is achieved by determining the value of $\hat{K}$, such that the probability of $\hat{K} < K$ is at least P when $K < K_d$, and P is the desired lowest probability for rejection. Values of K may be calculated for particular situations (Bowker, 1947; Kilsby et al., 1979) or may be obtained from standard tables (Bowker and Goode, 1952, Table K).

However, Malcolm (1984) showed that, for low values of n, the values of K derived by Kilsby et al. (1979) were not sufficiently precise, since they were derived by mathematical approximation. Use of the non-central t distribution for the calculation permits derivation of more precise values for K (Table 5.11). The acceptance criterion with an upper critical level C is given by $\bar{x} + Ks \leq C$ for an acceptance probability of P and the rejection criterion by $\bar{x} + Ks > C$ as illustrated in Fig. 5.4. These critical levels are essentially CLs.

Table 5.11 Values of K, Calculated Using the Non-Central t Distribution for Use in Setting Specifications for Variables Sampling

(a) Safety/quality specification ($\bar{x} + Ks > C$)

Probability of Rejection (P_r)	Proportion Exceeding C (d)	Number of Replicates (n)							
		3	4	5	6	7	8	9	10
0.99	0.1			5.4	4.4	3.9	3.5	3.2	3.0
	0.2			4.0	3.3	2.9	2.6	2.4	2.2
	0.3			3.0	2.5	2.2	2.0	1.8	1.7
	0.4			2.3	1.9	1.6	1.5	1.4	1.3
	0.5		2.3	1.7	1.4	1.2	1.1	1.0	0.9
0.95	0.05	7.7	5.1	4.2	3.7	3.4	3.2	3.0	2.9
	0.1	6.2	4.2	3.4	3.0	2.8	2.6	2.4	2.4
	0.3	3.3	2.3	1.9	1.6	1.5	1.4	1.3	1.3
	0.5	1.7	1.2	0.95	0.82	0.73	0.67	0.62	0.58
0.90	0.1	4.3	3.2	2.7	2.5	2.3	2.2	2.1	2.1
	0.25	2.6	2.0	1.7	1.5	1.4	1.4	1.3	1.3

(b) GMP limit: ($\bar{x} + Ks < Cm$)

Probability of Acceptance (P_a)	Proportion Exceeding Cm (d)	Number of Replicates (n)							
		3	4	5	6	7	8	9	10
0.95	0.10	0.33	0.44	0.52	0.57	0.62	0.66	0.69	0.71
	0.20	−0.13	0.02	0.11	0.17	0.22	0.26	0.29	0.32
	0.30	−0.58	−0.36	−0.24	−0.16	−0.10	−0.06	−0.02	0.00
0.90	0.05	0.84	0.92	0.98	1.03	1.07	1.10	1.12	1.14
	0.10	0.53	0.62	0.68	0.72	0.75	0.78	0.81	0.83
	0.20	0.11	0.21	0.27	0.32	0.35	0.38	0.41	0.43
	0.30	−0.26	−0.13	−0.05	0.01	0.04	0.07	0.10	0.12
	0.40	−0.65	−0.46	−0.36	−0.30	−0.25	−0.21	−0.17	−0.16
	0.50	−1.09	−0.82	−0.69	−0.60	−0.54	−0.50	−0.47	−0.44
0.75	0.01	1.87	1.90	1.92	1.94	1.96	1.98	2.00	2.01
	0.05	1.25	1.28	1.31	1.33	1.34	1.36	1.37	1.38
	0.10	0.91	0.94	0.97	0.99	1.01	1.02	1.03	1.04
	0.25	0.31	0.35	0.38	0.41	0.42	0.44	0.45	0.46
	0.50	−0.47	−0.38	−0.33	−0.30	−0.27	−0.25	−0.24	−0.22

GMP, good manufacturing practice.
Reproduced from Malcolm (1984) by permission of Blackwell Publishing.

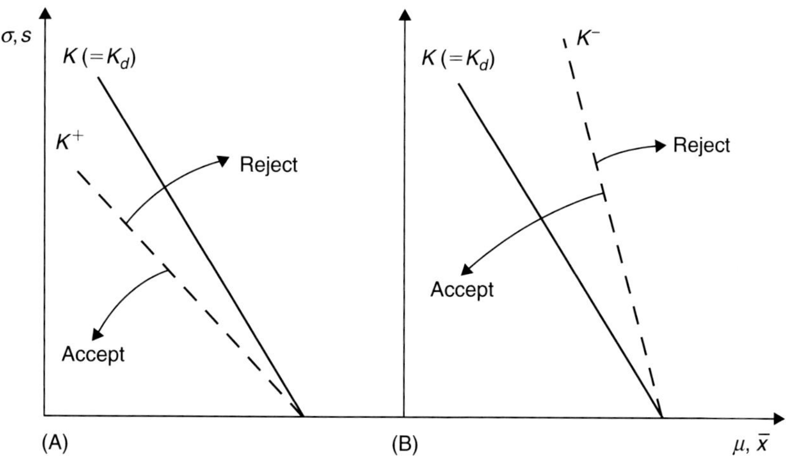

FIGURE 5.4 Variables Sampling – Rejection and Acceptance Criteria on a Standards Graph

(A) Rejection: K^+ is larger than K such that the chance of rejection (response to the right-hand side of the line K^+) is increased (>50%). (B) Acceptance: K^- is smaller than K such that the chance of acceptance (response on the left-hand side of the line K^-) is increased (>50%)

(Reproduced from Kilsby et al., 1979, by permission of Blackwell Publishing.)

Kilsby et al. (1979) recommended that the scheme be used for quality control purposes in relation to both safety and quality specifications, in a manner analogous to the ICMSF attributes scheme, but with lower producer's risk. It may be used also to ensure compliance with good manufacturing practices (GMP) by the introduction of a lower critical limit (C_m) (Table 5.11). The use of such a scheme is illustrated in Example. 5.4.

Regrettably this approach to sampling criteria has not been accepted widely, possibly since it involves replicate tests and calculation of statistical values rather

EXAMPLE 5.4 APPLICATION OF A VARIABLES SAMPLING SCHEME

Suppose: (1) that for a food product the absolute critical limit (C) that is not to be exceeded is 6.0 log_{10} cfu/g (≡M in ICMSF plans) and that in GMP a limit (C_m) of 5 log_{10} cfu/g should be achievable; (2) that 5 samples are to be tested per lot; and (3) that the producer wishes to reject with a 95% probability any lots where the proportion (d) exceeding C is 10% or greater and to accept with a probability of 0.90 only those lots where the proportion which exceeds C_m is less than 20%.

The food safety specification requires rejection if $\bar{x} + K_1 s > C$. Then from Table 5.11, for $n = 5$, $P = 0.95$ and $d = 0.10$: $K_1 = 3.4$. For the food safety criterion, reject the 'lot' if $\bar{x} + 3.4\,s > 6.0$.

The GMP specification requires acceptance if $\bar{x} + K_2 s \lessapprox C_m$ and, from Table 5.11, for $n = 5$, $P = 0.90$ and $d = 0.20$: $K_2 = 0.27$. Hence, for GMP criterion, accept if $\bar{x} + 0.27s \lessapprox 5$.

Suppose that for two lots of product the following data are obtained:

Lot	$\log_{10}$ cfu/g	Mean ($\bar{x}$)	SD (s)
A	4.52; 4.28; 4.79; 4.91; 4.50	4.60	0.25
B	4.98; 5.02; 5.28; 4.61; 5.11	5.00	0.25

Then for lot A:

$$\bar{x} + 3.4s = 4.60 + 3.4(0.25) = 5.45 < 6.0 \quad \text{(food safety criterion)}$$

and

$$\bar{x} + 0.27s = 4.60 + 0.27(0.25) = 4.67 < 5.0 \quad \text{(GMP criterion)}$$

Hence, the product from lot A would be acceptable on both food safety/quality and GMP criteria.

For lot B:

$$\bar{x} + 3.4s = 5.0 + 3.4(0.25) = 5.85 < 6.0 \quad \text{(food safety criterion)}$$

and

$$\bar{x} + 0.27s = 5.0 + 0.27(0.25) = 5.1 > 5.0 \quad \text{(GMP criterion)}$$

Hence, the product of lot B would be unacceptable in terms of the GMP specification but would not be rejected on food safety/quality criteria.

than just mindless comparison of the results of tests with defined criteria. However, Santos-Fernández et al. (2014) have proposed a robust variables sampling scheme for microbiological food safety that is claimed to be more stringent and allows for fewer samples to be tested than the current attributes schemes.

SOME STATISTICAL CONSIDERATIONS ABOUT DRAWING REPRESENTATIVE SAMPLES

It is recognised that samples should be representative of the 'lot' from which they are drawn, such that the quality of each individual sample is neither better nor worse than that of the overall population from which the samples were taken. In drawing samples to conform to the requirements of a sampling plan it is of prime importance to avoid bias. Random sampling is universally recognised as the way to avoid bias and is safer than aiming deliberately to draw samples from specific parts of a 'lot'. Whilst there is no guarantee that the quality of sample units drawn randomly will be typical of the 'lot', we can at least be confident that there was no deliberate bias in the choice of samples and statistical logic demands randomisation (Anon, 2009).

The simplest manner of doing this is to use random numbers drawn from standard tables (eg, Table 5.12) or derived using computer software (eg, Anon, 2015). Each

Table 5.12 Random Numbers Table

Row Number	Column Number 1–4	5–8	9–12	13–16	17–20	21–24	25–28	29–32	33–36	37–40
1	87 08	83 09	40 14	39 15	99 24	21 85	00 45	54 19	36 18	03 88
2	88 33	78 20	40 40	24 73	77 70	00 31	84 59	26 06	50 30	95 96
3	22 50	09 11	00 37	36 51	55 95	83 97	13 75	46 22	77 50	11 72
4	48 70	56 57	16 24	21 74	91 53	18 05	59 61	74 97	31 82	77 68
5	93 45	40 93	12 80	88 63	26 93	85 05	19 87	84 37	59 76	16 65
6	50 76	72 02	39 19	40 69	57 23	09 33	20 70	86 45	13 94	98 39
7	91 64	01 34	67 13	1100	32 09	39 76	21 64	29 85	65 14	51 74
8	33 20	63 71	95 94	13 77	12 44	12 94	91 04	41 83	79 72	44 08
9	90 59	65 46	78 82	16 45	97 85	57 75	79 96	79 08	16 83	43 99
10	05 10	93 57	80 32	86 65	26 90	27 54	34 94	46 33	65 35	56 84
11	92 85	63 26	69 69	81 54	70 56	17 62	43 17	86 78	99 62	34 15
12	08 50	36 55	82 11	26 54	76 88	85 67	82 21	65 00	83 89	06 09
13	59 36	77 09	83 87	81 77	93 77	48 44	88 30	37 21	74 02	93 10
14	05 85	86 43	25 50	76 70	36 32	26 68	54 92	84 90	02 38	77 40
15	13 46	99 31	30 29	71 70	91 10	99 84	55 31	95 20	90 28	49 78
16	56 27	09 33	66 79	32 29	50 54	76 94	27 01	45 87	29 66	23 15
17	54 15	62 11	22 33	39 39	58 30	73 43	59 32	26 43	76 12	99 10
18	83 01	86 58	89 77	68 87	29 71	49 50	46 53	56 53	41 53	52 20
19	00 28	17 33	8142	24 33	55 75	42 70	73 65	16 96	47 17	42 69
20	52 29	68 59	32 69	40 30	89 12	11 07	18 53	27 13	46 54	85 40
21	64 43	09 80	68 29	86 65	60 27	87 70	77 45	31 69	12 31	21 79
22	80 68	13 48	80 84	25 33	70 89	76 61	03 41	57 89	87 07	56 12
23	28 72	57 80	54 05	80 92	82 65	25 01	74 58	89 39	25 05	57 66
24	23 48	49 96	00 17	88 90	63 67	02 64	71 12	21 02	29 86	88 54
25	04 41	27 70	10 49	13 76	99 38	64 14	90 60	69 75	10 97	16 60

unit within a 'lot' is allocated a sequential number from 1 to N in any convenient sequence. Then, using a page of random numbers, select randomly a particular digit or series of digits (eg, by using a pencil point brought randomly onto the page). The digit, or series of digits, nearest to the pencil point is selected as the first random number. Then follow down the column(s) taking each sequential number that lies within the range 1–N until sufficient numbers have been drawn for n sample units; if numbers are drawn more than once, the second and subsequent replicates are ignored. This is illustrated in Example 5.5.

EXAMPLE 5.5 USE OF RANDOM NUMBER TABLES AND SAMPLE PLANS TO DRAW SAMPLES

Suppose that sacks of flour in a warehouse are to be examined for evidence of insect infestation and mould. The lot consists of 1400 sacks, on pallets with 20 sacks per pallet. From sampling tables (Table 5.4, assuming 'normal' inspection) it is found necessary to draw 125 samples from the lot.

Each pallet is sequentially numbered: 1–20, 21–40, 41–60, …, 1381–1400, each sack being allocated a number according to its position on the pallet. Using random number tables (eg, Table 5.12) of 4 random digits in consecutive columns and using a pencil point to identify (randomly) 1 row and column, for example, row 10 in column 1, the first random number to be taken will be 0510 (ie, row 10, columns 1–4; sample unit 510). The next value in row 11 (9285) is too large and is ignored; the third, fifth and sixth values (0850, 0585 and 1346, respectively) are retained. This process is continued as necessary until the required number (125) of sets of random digits is obtained, moving from the bottom set of digits (row 25, columns 1–4) to the first set (row 1) in columns 5–8, and so on. If a set of digits occurs more than once, it is recorded only once. The 125 sets of 4 digits obtained give the numbers of sacks to be examined (namely 510, 850, 585, 1346, 28, 441, 911, 134, 933, 980, 1348, 37, 1280, 17, 1049, etc.).

An alternative approach, which introduces some bias yet makes sampling easier, is to deliberately stratify and to draw 13 out of the 70 pallets as primary samples (eg, numbers 69, 11, 50, 29, 33, 42, 5, 17, 49, 39, 24, 36, 21, starting from row 11, columns 11–12 in Table 5.12) and then to examine 5 sacks drawn at random from each pallet [eg, for pallet no. 5 examine sack numbers 17 (row 11, columns 21–22), 11, 2, 5 and 7; for pallet no. 11 examine sacks 9 (row 16, columns 5–6), 17, 13, 20 and 11], and so on. Note that in this deliberately stratified sampling scheme the total number of sample units to be tested will be lower than if samples are drawn totally at random since a 2-tier sample plan is used [ie, for 70 pallets on normal inspection, plan E (Tables 5.4 and 5.5) requires 13 pallets to be tested, whilst for each pallet of 20 sacks plan C requires that only 5 sacks/pallet are to be tested – hence, only 65 sacks are tested]. One or more positive tests from the 5 sacks/pallet would provide evidence that that pallet is unacceptable.

If the AQL required were, say, 4%, then the following probabilities of acceptance would result:

Number of Units in Lot	Number of Units		AQL (%)	% Defectives	
	To Test (*n*)	Acceptable Defective (*c*)		Accepted at $P_a = 0.95$	Rejected at $P_a = 0.10$
1400 sacks	125	10	4	4.9	12.3
70 pallets	13	1	4	2.8	26.8
20 sacks	5	0	2.5	2.1	37.0

Note that the deliberately stratified test would be less stringent for a required AQL than would the non-stratified test. Only 1 of the 70 pallets could be accepted if found defective, and none of the 5 sacks examined per pallet could be accepted if defective; yet the overall rejection level would be higher, that is, at $P_a = 0.1$, the accepted lot might contain up to 37% defectives compared with 12% for the examination of 125 individual sacks.

STRATIFIED SAMPLING

Where subpopulations may vary considerably, or where different 'lots' are combined to make up a consignment, it is necessary to identify the relative proportion of each 'lot' or portion in a consignment. The number of samples drawn should then reflect the overall composition of the consignment. For instance, if a consignment consists of 20% of Lot A, 50% of Lot B and 30% of Lot C, then the number of samples drawn from each lot should be representative of the whole; for instance, a total of (say) 40 samples should comprise 8 samples from Lot A, 20 from Lot B and 12 from Lot C. In these circumstances, deliberate stratification is used in order to ensure that a proportionate number of sample units is taken from each of the different strata in order to provide an equal opportunity for random sample units from different strata to be included in the total sample. However, the results from the individual 'lots' can be pooled only if no evidence for heterogeneity is found in the results.

SAMPLING FRAMES

Sometimes it is not possible to sample a whole 'lot' or consignment because of problems of accessibility. Such a situation might arise in sampling stacks of cartons within a cold store, a warehouse or a ship's hold. In such situations, sample units are drawn at random from those parts of the 'lot' that are accessible (referred to as 'frames'). After analysis, the results are applicable only to the frame – not to the whole 'lot' – but if several distinct frames are tested and the results are homogeneous, then it is a reasonable presumption that the results on the frames may be considered to be indicative of those that might have been obtained on the whole 'lot'.

SINGLE OR MULTIPLE SAMPLING SCHEMES

A given number of random samples is drawn for testing from a 'lot' but sometimes the numbers required for testing may be too high. In such circumstances, a multiple sampling scheme may be used, as described previously (Table 5.6 and text). Each bulk unit (eg, a case of product) is numbered and the required number of bulk sample units is drawn using random numbers. An appropriate number of individual sample units is then drawn from each bulk unit, again using random numbers to identify the individual unit(s) to be tested. This is illustrated in Example 5.5.

ADDENDUM

As described in Chapter 14, the whole concept of setting sampling plans is to enable an appropriate risk analysis to be undertaken for any particular food type and microorganism of interest (eg, food-borne pathogens). Recent developments in this area include the availability of computer spreadsheets that permit specific sampling plans to be developed for defined limits for both producer's and consumer's risks. Links to enable download of these spreadsheets together with guidance on their use are given in Chapter 14. The most valuable of these spreadsheets is that produced as an adjunct to the draft FAO/WHO *The Statistical Aspects of Microbiological Criteria Related to Foods: A Risk Manager's Guide* (Anon, 2016).

REFERENCES

Andrews, W.H., Hammack, T.A., 2003. Food sampling and preparation of sample homogenate. In: Bacteriological Analytical Manual Online, eighth ed. US FDA, Washington, DC. <http://www.fda.gov/Food/FoodScienceResearch/LaboratoryMethods/ucm2006949.htm> (accessed 26.06.15) (Chapter 1, revision A, 1998, updated 2003).

Anon, 1985. Sampling Procedures for Inspection by Attributes – Part 2: Sampling Plans Indexed by Limiting Quality (LQ) for Isolated Lot Inspection. ISO 2859-2:1958. International Standards Organization, Geneva.

Anon, 2001. Statistical Aspects of Sampling From Bulk Materials – Part 2: Sampling of Particulate Materials. ISO 11648-2:2001. International Standards Organization, Geneva.

Anon, 2002. Sampling Procedures for Inspection by Attributes – Part 4: Procedures for Assessment of Declared Quality Levels. ISO 2859-4:2002. International Standards Organization, Geneva.

Anon, 2003. Statistical Aspects of Sampling From Bulk Materials – Part 1: General Principles. ISO 11648-1:2003. International Standards Organization, Geneva.

Anon, 2004. Milk and Milk Products – Sampling – Inspection by Attributes. ISO 5538:2004. International Standards Organization, Geneva.

Anon, 2005a. Sampling Procedures for Inspection by Attributes – Part 5: System of Sequential Sampling Plans Indexed by Acceptance Quality Limit (AQL) for Lot-by-Lot Inspection. ISO 2859-5:2005. International Standards Organization, Geneva.

Anon, 2005b. Sampling Procedures for Inspection by Attributes – Part 3: Skip-Lot Sampling Procedures. ISO 2859-3:2005. International Standards Organization, Geneva.

Anon, 2006a. Sampling Procedures for Inspection by Attributes – Part 10: Introduction to the ISO 2859 Series of Standards for Sampling for Inspection by Attributes. ISO 2859-10:2006. International Standards Organization, Geneva.

Anon, 2006b. Sequential Sampling Plans for Inspection by Attributes. ISO 8422:2006. International Standards Organization, Geneva.

Anon, 2009. Random Sampling and Randomisation Procedures. ISO 24153:2009. International Standards Organization, Geneva.

Anon, 2013a. Microbiology of the Food Chain – Preparation of Test Samples, Initial Suspension and Decimal Dilutions for Microbiological Examination – Part 2. Specific Rules for the Preparation of Meat and Meat Products. ISO/DIS 6887-2:2013. International Standards Organization, Geneva.

Anon, 2013b. Sampling Procedures for Inspection by Variables – Part 1: Specification for Single Sampling Plans Indexed by Acceptance Quality Limit (AQL) for Lot-by-Lot Inspection for a Single Quality Characteristic and a Single AQL. ISO 3951-1:2013. International Standards Organization, Geneva.

Anon, 2015. Random Integer Set Generator. <https://www.random.org/integer-sets/> (accessed 13.12.15).

Anon, 2016. FAO/WHO. The Statistical Aspects of Microbiological Criteria Related to Foods: A Risk Managers Guide. Microbiological Risk Assessment Series 24. FAO/WHO, Rome, Geneva, <ftp://ftp.fao.org/codex/meetings/CCFH/CCFH46/FAO%20MC%20draft%20140814a.pdf> (accessed 25.11.15).

ASTM, 2010. Standard Test Method for Attribute Sampling of Metallic and Inorganic Coatings ASTM B602-88. American National Standards Institute, Washington, DC.

Board, R.G., Lovelock, D.W., 1973. Sampling – microbiological monitoring of environments. In: Board, R.G., Lovelock, D.W. (Eds.), SAB Tech. Series 7. Academic Press, London.

Bowker, A.H., 1947. Tolerance limits for Normal distributions. In: Eisenhardt, C., Hastay, M.W., Wallis, W.A. (Eds.), Selected Techniques of Statistical Analysis. McGraw-Hill, New York, pp. 95–109.

Bowker, A.H., Goode, H.P., 1952. Sampling Inspection by Variables. McGraw-Hill, New York.

Bray, D.F., Lyon, D.A., Burr, I.W., 1973. Three class attributes plans in acceptance sampling. Technometrics 15 (3), 575–585.

BSI, 1999. Sampling Procedures for Inspection by Attributes. Sampling Schemes Indexed by Acceptance Quality Limit (AQL) for Lot-by-Lot Inspection. BS 6001-1:1999 + A1:2011 (ISO 2859-1: 1999). British Standards Institute, London.

BSI, 2007. Sampling Procedures for Inspection by Variables. Guide to Single Sampling Plans Indexed by Acceptance Quality Limit (AQL) for Lot-by-Lot Inspection for a Single Quality Characteristic and a Single AQL. BS 6002-1:2007 (ISO 3951-1:2005). British Standards Institute, London.

Clark, D.S., 1978. The International Commission on Microbiological Specifications for Foods. Food Technol. 32 (51–54), 67.

Clopper, C., Pearson, S., 1934. The use of confidence or fiducial limits illustrated in the case of the binomial. Biometrika 26, 404–413.

Gascoigne, J.C., Hill, I.D., 1976. Draft British Standard 6002: "sampling inspection by variables". J. R. Stat. Soc. Ser. A 139, 299–317, (with discussion).

ICMSF, 1986. Microorganisms in Foods. 2. Sampling for Microbiological Analysis: Principles and Specific Applications, second ed. University of Toronto Press, Toronto.

ICMSF, 2002. Microorganisms in Foods. 7. Microbiological Testing in Food Safety Management. Kluwer Academic/Plenum Publishers, New York.

Kilsby, D.C., 1982. Sampling schemes and limits. In: Brown, M.H. (Ed.), Meat Microbiology. Academic Press, London, pp. 387–421.

Kilsby, D.C., Aspinall, L.J., Baird-Parker, A.C., 1979. A system for setting numerical microbiological specifications for foods. J. Appl. Bacteriol. 46, 591–599.

Lowry, R., 2015. The Confidence Interval of a Proportion. <http://vassarstats.net/prop1.html> (accessed 26.06.15).

Malcolm, S., 1984. A note on the use of the non-central *t* distribution in setting numerical microbiological specifications for foods. J. Appl. Bacteriol. 57, 175–177.

Military Standard, 1989. Sampling Procedures and Tables for Inspection by Attributes. MIL-STD-105E. <https://ia700801.us.archive.org/14/items/MIL-STD-105E_1/MIL-STD-105E.pdf-page=5&zoom=auto,-13,7> (accessed 25.06.15).

Newcombe, R.G., 1998. Two-sided confidence intervals for the single proportion: comparison of seven methods. Stat. Med. 17, 857–872.

Pearson, E.S., Hartley, H.O., 1976. Biometrika Tables for Statisticians, third ed., Charles Griffin & Co Ltd for the Biometrika Trustees, Cambridge.

Santos-Fernández, E., Govindaraju, K., Jones, G., 2014. A new variables acceptance sampling plan for food safety. Food Control 44, 249–257.

Steiner, E.H., 1971. Sampling schemes for controlling the average quality of consignments on the proportion lying outside specification. Sci. and Tech. Survey. No. 65. Leatherhead Food Research Association, Leatherhead, UK.

CHAPTER

Errors associated with preparing samples for analysis

6

Estimation of levels of microorganisms in foods involves a series of operations each of which contributes in some measure to the overall 'error' or 'lack of precision' of the analytical result. Before considering the overall errors inherent to methods of enumeration, it is pertinent to consider the errors associated with the stages of laboratory sampling, maceration and dilution that are common to many quantitative microbiological techniques. It is not proposed to discuss the methodology per se since this is dealt with adequately in standard laboratory reference books (ICMSF, 1978; Harrigan and McCance, 1998; Anon, 2005; AOAC, 2012). However, reference to certain aspects cannot be avoided completely when considering the effects of methodology on the precision of laboratory results.

LABORATORY SAMPLING ERRORS

The size of a sample drawn from a 'lot' of food will almost invariably be considerably larger than that required for analytical purposes. Processes such as 'quartering' that are used to obtain a representative laboratory sample for chemical analyses cannot usually be applied in microbiological analysis because of contamination risks, although they may be appropriate for some mycological analyses (Jarvis and Williams, 1987). Standard methods for drawing and subsequent laboratory handling of samples are described in the ISO 6887 series to which reference was made in the previous chapter.

Unless otherwise advised, it is assumed that the sample arriving in a laboratory will have been drawn randomly from a 'lot' or from a retail display of produce. This is the laboratory sample from which test portions will be drawn for examination. The question that now arises relates to the need to decide the size of the test portion and how to ensure that it is representative whilst still remaining a random portion of the target sample. In some cases, it may be appropriate to draw the test portion from different parts of the laboratory sample, but in other cases the laboratory sample may relate only to particular parts, for example, from the neck or breast skin of a chicken. The primary objective is to ensure that test portions are drawn randomly yet are truly representative of the laboratory sample. A detailed discussion of methods for taking and preparing laboratory samples for analysis is given by ICMSF (1978) and in the ISO 6887 series. In general, solid food materials will be macerated during preparation of a suspension for subsequent dilution and examination.

Statistical Aspects of the Microbiological Examination of Foods. http://dx.doi.org/10.1016/B978-0-12-803973-1.00006-1

Table 6.1 Effect of Sample Weight on the Coefficient of Variation (CV) of Weighed Samples

Sample Size (g)	No. of Samples (n)	Mean Weight[a] (g)	SD (g)	SE_M	CV (%)
10	19	9.99	0.071	0.016	0.71
50	20	50.01	0.093	0.021	0.19
100	20	99.99	0.066	0.015	0.07

[a]*Each sample was weighed to ±0.1 g on a top-pan balance and the 'true weight' of the sample was then determined on an analytical balance.*

The size of sample to be taken will depend on the amount of primary sample available and its degree of homogeneity. Ideally, the analytical sample should never be less than 10 g and it should be weighed to the nearest 0.1 g into a blender on a top-pan or other suitable balance. The precision of the weight of material taken increases with increasing sample size as reflected in the percent coefficient of variation (CV; ie, the ratio between the standard deviation and the mean weight; Table 6.1). However, provided that the sample is weighed carefully, the relative contribution of weighing inaccuracies will be small when compared to the contributions due to other sources (see following text) and the errors that might arise through sample contamination.

Differences in analytical results on replicate samples tested in different laboratories, or even between different workers in the same laboratory, may reflect differences in the sample handling procedure. Such differences could arise through inherent differences in the distribution of organisms between test portions but may also reflect the storage and subsequent handling of the primary samples and the way in which test samples are taken.

DILUENT VOLUME ERRORS

It is widely recognised that a major potential source of error in counts is that caused by variations in the volume of diluent used. When dilution blanks are prepared by dispensing before autoclaving, the potential errors are generally larger than those for diluent dispensed after autoclaving, as illustrated in Table 6.2. These data also illustrate the need to ensure accurate setting of automatic dispensing systems (Example 6.1).

Errors in the volumes of diluent used in the primary homogenisation of samples will also affect subsequent dilutions. It is therefore essential to ensure that the volume of diluent added to a primary sample, and that used in subsequent dilutions, is measured as accurately as possible. Diluent volume errors will have a cumulative effect in any serial dilution series and therefore can have a major effect on the derived colony counts. Due attention is required not only to minimise the variation between replicate volumes but also to ensure that the correct volume is accurately dispensed. For instance, whereas 5 × 10-fold serial dilutions give a dilution factor of 10^{-5} (1 in 100,000), 5 × 9.8 mL serial dilutions give a dilution factor of 9.8^{-5} (approximately 1 in 90,400). This is discussed in more detail later.

EXAMPLE 6.1 DILUENT VOLUME ERRORS

What is the effect of autoclaving on the actual dispensed volumes of a diluent to be used in a serial dilution procedure?

A nominal 9 mL of distilled water was dispensed into each of a number of tared universal bottles using an automatic dispensing pipette or a graduated pipette. The weight of water dispensed was determined before and after autoclaving, or after aseptic dispensing of bulk sterilised water. The results are shown in Table 6.2.

The coefficients of variation (CV) for hand-dispensed diluent increased from 0.8% in samples weighed before autoclaving to 2.2% in samples weighed after autoclaving. Similarly, the CV for machine-dispensed volumes increased from 1.5% to 2.5–2.6%, depending on where the bottles were placed during autoclaving. Assessment of volumes dispensed before or after autoclaving were much more similar (CV 1.6% and 1.3%, respectively).

Although calibrated to deliver 9 mL of diluent, the automatic dispensing pump delivered only 8.516 g × 0.997 g/mL (≈8.49 mL); after autoclaving the mean volume was 8.01 mL (8.037 g × 0.997 g/mL). Such errors can lead to significant differences in calculated colony count levels.

In practice, some allowance can be made for evaporation of water during autoclaving, but the volume needs to be checked carefully before use. Accurate aseptic dispensing of diluents after sterilisation is the preferred method of use.

Table 6.2 Errors in Dispensed Nominal 9-mL Volumes of Diluent

Diluent					
Dispensed	**Weighed**	**No. of Replicates**	**Mean Weight (g)**	**SE_M**	**CV (%)**
Automatic dispensing pipette[a]					
A	A	50	8.516	0.124	1.46
A	B_1	25	8.063	0.201	2.49
A	B_2	25	8.010	0.209	2.61
B	B	50	8.502	0.111	1.30
Graduated pipette					
A	A	50	9.277	0.074	0.80
A	B	50	9.008	0.198	2.20

A, before autoclaving; B, after autoclaving; B_1 and B_2, replicates stacked in top or bottom half of basket, respectively.

[a]*Set to deliver a nominal volume of 9 mL.*

PIPETTE VOLUME ERRORS

Over the years, many studies have been made of the variation in volumes of liquid dispensed by bacteriological pipettes of various categories (Examples 6.2 and 6.3). Table 6.4 provides a summary of some typical data.

The variation in the volume of liquid dispensed is dependent on calibration errors, technical (eg, operator) errors in the use of pipettes and the extent to which organisms adhere to the glass (or plastic) of the pipette; in addition, the multiple

EXAMPLE 6.2 CALCULATION OF DILUTION ERRORS USING A SINGLE PIPETTE FOR ALL TRANSFERS

How significant are the errors associated with preparation of a dilution series? Is it better to make six 10-fold dilutions or a 100-fold dilution followed by four 10-fold dilutions?

Assume first that the dilution series is to be prepared from a liquid sample. One millilitre of the sample is transferred by pipette to a 99-mL volume of diluent and after mixing a further pipette is used to prepare the four subsequent 10-fold serial dilutions to 10^{-6}. This gives one 100-fold ($n = 1$) and four 10-fold ($m = 4$) sequential dilutions.

The equation for estimation of dilution error (based on Jennison and Wadsworth, 1940) is as follows: % dilution error $= \pm 100\sqrt{\{[a^2(m+n)^2]/x^2\}+\{[m^2(a^2+b^2)]/u^2\}+\{[n^2(a^2+c^2)]/v^2\}}$

Assume that the SD of the volume of liquid delivered from the 1-mL disposable pipettes $= a = \pm 0.04$ mL, that the SD of the 9-mL diluent volumes $= b = \pm 0.20$ mL and that the SD of the 99-mL diluent volume $= c = \pm 2.0$ mL.

Then for $a = 0.04$, $b = 0.20$, $c = 2.0$, $m = 4$, $n = 1$, $x = 1.0$, $u = 10.0$ and $v = 100.0$.

The error of the 10^{-6} dilution is given by

$$\pm 100\sqrt{\frac{0.04^2(4+1)^2}{1^2}+\frac{4^2(0.04^2+0.2^2)}{(10)^2}+\frac{1^2(0.04^2+2^2)}{(100)^2}}$$
$$=\pm 100\sqrt{\frac{0.0016(25)}{1}+\frac{16(0.0416)}{100}+\frac{1(4.0016)}{10000}}$$
$$=\pm 100\sqrt{0.04+0.00656+0.00040016}$$
$$=\pm 100\sqrt{0.04696}=\pm 100(0.2167)=\pm 21.7\%$$

Assume now that the dilution series is produced using six 10-fold dilutions. For $a = 0.04$, $b = 0.20$, $m = 6$, $n = 0$, $x = 1.0$ and $u = 10.0$. Then the error of the 10^{-6} dilution:

$$=\pm 100\sqrt{\frac{0.04^2 6^2}{1^2}+\frac{6^2(0.04^2+0.2^2)}{10^2}+\frac{0^2(0.04^2+2.0^2)}{100^2}}$$
$$=\pm 100\sqrt{\frac{0.0016(36)}{1}+\frac{36(4.0016)}{10^2}}$$
$$=\pm 100\sqrt{0.0576+0.014976}$$
$$=\pm 100\sqrt{0.072576}=\pm 100(0.2694)=\pm 26.9\%$$

For the range of standard deviations and dilution volumes used in this calculation, the overall percentage dilution error of the 10^{-6} dilution would be higher (26.9%) when prepared using 6 × 10-fold dilutions than if 1 × 100-fold and 4 × 10-fold sequential dilution steps were used (21.7%).

These values are derived using high levels of error for the pipette and diluent volumes in order to illustrate the importance of minimising such errors and do not reflect good laboratory practice!

use of a single pipette to prepare a dilution series introduces covariance errors (Hedges, 1967, 2002, 2003; see following text). It is worthy of note that manufacturers' pipette calibration errors are of two kinds: manufacturing tolerance errors (or inaccuracy) and imprecision (repeatability errors) generally cited as a percent CV. Values for inaccuracy errors are not readily available from suppliers and reflect the extent of the manufacturer's acceptable production tolerance (Hedges, 2002). Ideally all pipettes (including semi-automated pipette systems that use replaceable pipette tips) should be re-calibrated within the laboratory before use (Hedges, 2003).

EXAMPLE 6.3 CALCULATION OF DILUTION ERRORS USING A DIFFERENT PIPETTE FOR EACH TRANSFER

Would the error of the 10^{-6} dilution be less if a separate pipette were used for each stage in the dilution process?

Assume that a dilution series to 10^{-6} is prepared using six 10-fold sequential dilutions. As before (Example 6.2) assume SD of the pipettes = a = 0.04 mL, SD of diluent volume = b = ±0.20 mL, the number of dilution stages = m = 6 and the volume of the inoculated diluent = u = 10; then the error of the 10^{-6} dilution prepared using a different randomly chosen pipette for each stage:

$$= \pm 100\sqrt{\frac{a^2 m}{x^2} + \frac{m(a^2 + c^2)}{u^2}}$$

$$= \pm 100\sqrt{0.04^2(6) + \frac{6(0.04^2 + 0.2^2)}{10^2}}$$

$$= \pm 100\sqrt{0.0096 + 0.002496}$$

$$= \pm 100\sqrt{0.012096} = \pm 11.0\%$$

Hence, the percentage dilution error for the 10^{-6} dilution prepared using 6 × 10 mL serial dilutions with a different randomly drawn pipette for each dilution step is approximately half of that which would be obtained for an equivalent dilution series made using a single pipette for all stages (21.7%; Example 6.2).

Comparative calculated errors for using a single pipette or different pipettes for various serial dilution series are shown in Table 6.3. Note that the standard deviations used for Example 6.2 and this example are different to those used in the table.

Table 6.3 Errors for Various Levels of Dilution Prepared Either With the Same Pipette Throughout or Using a Different, Randomly Selected Pipette for Each Transfer

	% Dilution Error When Dilutions Prepared Using					
	Single Pipette			Different Pipettes		
Dilution Level[a]	A	B	C	A	B	C
10^{-1}	2.2			2.2		
10^{-2}	4.5	2.2		3.2	2.2	
10^{-3}	6.7	4.2		3.9	3.2	
10^{-4}	9.0	6.4	4.5	4.5	3.9	3.2
10^{-5}	11.2	8.6	6.4	5.0	4.5	3.9
10^{-6}	13.5	10.8	8.5	5.5	5.0	4.5

A, only 10-fold dilutions prepared; B, one 100-fold dilution, remainder 10-fold; C, two 100-fold dilutions, remainder 10-fold.

[a]Assumes standard deviations of 1-mL pipettes, 9-mL diluent volumes and 99-mL diluent volumes are 0.02, 0.1 and 1 mL, respectively.

Based on Jennison and Wadsworth (1940) and Hedges (1967).

Table 6.4 Errors in the Volumes of Diluent Delivered Using Different Pipettes

Type of Pipette	Sample	No. of Tests	Mean Weight Delivered (mg)	SE_M	CV (%)	References
Capillary						
L[a]	Water	60	30.0	0.60	2.00	Snyder (1947)
L[a]	Water	10	20.0	0.95	4.75	Spencer (1970)
C[a]	Water	10	33.8	2.87	8.48	
L[b]	Water	20	19.0	0.63	3.33	
C[b]	Water	20	33.3	1.49	4.47	
'Breed' capillary						
C1[b]	Milk	21	10.1	0.30	2.97	Brew (1914)
C2[b]	Milk	12	5.6	0.77	13.75	
C3[a]	Milk	5	9.0	0.70	7.78	
Serological (1 mL disposable)						
C[a]	Water	50	1046.0	37.0	3.49	Jarvis and Whitton (unpublished)
C[a]	Water	50	983.0	28.7	2.92	
C[b]	Water	25	1009.8	45.7	4.53	
Serological (1 mL)						
C[a,c]	Water	59	908.7	9.0	0.99	Snyder (1947)
C[a,d]	Water	60	102.5	2.8	2.73	
Semi-automatic[e]						
C[a]	Water	50	97.04	2.6	2.68	Jarvis and Whitton (unpublished)
C[b]	Water	50	98.2	10.7	10.90	

L, laboratory made and calibrated; C, commercially prepared and calibrated; C1, C2 and C3, different makes of pipette.
[a]Variation within pipette.
[b]Variation between pipettes.
[c]0.9 mL dispensed.
[d]0.1 mL dispensed.
[e]100-µL disposable tips.

OTHER SOURCES OF ERROR

Other interrelated major sources of potential error occur in making sample dilutions. Solid food samples require maceration and homogenisation. Traditionally, laboratory maceration was done using top-drive or bottom-drive homogenisers, although this has now been largely superseded by techniques such as 'stomaching' and 'pulsifying'. A potential source of error with any type of traditional homogeniser is related to the inadequate homogenisation of the sample, such that a heterogeneous suspension results. Extensive maceration with a bottom-drive homogeniser can result in a pronounced temperature rise that may affect the viability of some organisms through sub-lethal damage that may adversely affect colony counts. Studies by Barraud et al. (1967), Kitchell et al. (1973) and others have shown that results varying by about 7% can be obtained with different makes of homogeniser and by 50% when a pestle, mortar and sand were used to grind meat samples. A related source of error is inadequate mixing of the inoculated diluent, such that the organisms are not thoroughly distributed in the suspension, at both the primary and subsequent stages of a serial dilution. The extent of maceration and mixing can affect the size and distribution of cell aggregates within the suspension so that apparent differences are seen in the colony numbers in replicate plates both at a single dilution and between dilution levels.

The introduction of the 'Stomacher™' (Sharpe and Jackson, 1972) provided a totally new approach to the preparation of primary food suspensions that avoids the need to maintain a large number of sterile macerators. In addition, the more gentle 'massaging effect' of the Stomacher separates organisms from the food such that there is less physical breakdown of the food into particles that might interfere in later stages of the test. Comparative studies showed that colony counts on food suspensions prepared by 'stomaching' and by homogenisation were generally comparable, although, in some cases, counts after 'stomaching' may be higher than counts after maceration (Tuttlebee, 1975; Emsviler et al., 1977). A variation in the stomaching procedure uses 'filter bags' into which the food sample is placed before immersion in the diluent so that the suspension contains the organisms without any of the 'stomached' food sample. An alternative procedure uses the 'Pulsifier®', an instrument that combines a high-speed shearing action with intense shock waves to liberate organisms with minimal disruption of the food sample matrix. Described originally by Fung et al. (1998), its use has been described by Sharpe et al. (2000), Kang and Dougherty (2001), Wu et al. (2003) and others. The lower level of both suspended and dissolved solids improves membrane filtration rates and reduces interference in PCR and similar methods.

As pointed out by Mudge and Lawler (1928), amongst others, a further source of 'dilution error' relates to variations in time between preparation of dilutions and plating. Major differences in colony count can result from ignoring this effect; the time between preparation of dilutions and plating should be consistent and as short as possible.

CALCULATION OF THE RELATIVE DILUTION ERROR

It has long been recognised that in seeking to assess the precision of a colony count it is necessary not just to rely on the actual count of colonies but to take account also of the contributions to variance of the preceding steps, especially those of the dilution series. Jennison and Wadsworth (1940) derived a formula to estimate the magnitude of the dilution error knowing the standard deviations of the pipette volume(s), diluent volumes and number of dilutions prepared. For a dilution series consisting of both 1 in 100 and 1 in 10 dilutions, the standard error of the mean of the final dilution is given by

$$SE_M = \pm \frac{x^{m+n}}{u^m v^n} \sqrt{\frac{a^2(m+n)^2}{x^2} + \frac{m^2(a^2+b^2)}{u^2} + \frac{n^2(a^2+c^2)}{v^2}}$$

where SE_M is the standard error of the expected final dilution ($F = x^{m+n}/u^m v^n$), x is the volume measured by pipette (eg, 1 mL) with variance = a^2, u is the volume of inoculum + diluent (eg, 1 + 9 mL = 10 mL) with combined variance = $a^2 + b^2$, b^2 is the variance of the 9 mL diluent volume, v is the volume of inoculum + diluent (eg, 1 + 99 mL = 100 mL) with combined variance = $a^2 + c^2$, c^2 is the variance of 99 mL diluent volume, m is the number of 1 in 10 dilutions and n is the number of 1 in 100 dilutions. Note that this and subsequent equations have been corrected from those given by Jennison and Wadsworth (1940). For simplicity, since $x = 1.0$, $u = 10$ and $v = 100$, F can be expressed as a percentage error:

$$\text{Percentage dilution error} = \pm 100\sqrt{a^2(m+n)^2 + \frac{m^2(a^2+b^2)}{10^2} + \frac{n^2(a^2+c^2)}{100^2}}$$

This generalised equation of the percentage error, as the standard error of the volume, can be used for any series of dilutions with pipette and diluent volumes (a, b and c) for any combination (m and n) of 10- and 100-mL dilution series. Since the absolute value of a^2 should be small, in the order of 0.01 mL, compared to the values of b^2, typically ca. 0.1 mL, and c^2, typically ca. 1 mL, this simplifies to

$$\text{Percentage dilution error} \approx \pm 100\sqrt{(m+n)^2 + m^2 + n^2}$$

But this approximation works only if the assumption of relative volumes is true.

Jennison and Wadsworth (1940) provide a table of dilution errors for various combinations of pipette and diluent blank variances and various combinations of 1 in 10 and 1 in 100 dilutions to 10^{-8}. However, in their original calculation, they assumed that only one pipette would be used in the preparation of a dilution series, that is to say, that the same pipette would be used throughout. Hedges (1967) showed that the basic equation is incorrect if a fresh, randomly selected pipette is used for each stage in preparing a dilution series. Since the latter method is the one adopted by most microbiologists, it is important to recognise the difference in magnitude of the error that results from elimination of the covariance terms contained in the squared values for m, n and $m + n$ in the equations derived earlier. Thus, when a fresh, randomly drawn pipette is used at each stage of preparation of a dilution, the equation given earlier for the percentage dilution error simplifies to

$$\text{Percentage dilution error} = \pm 100\sqrt{(m+n)+n+m} = \pm 100\sqrt{2(m+n)}$$

Percentage dilution errors for dilutions to 10^{-6} are illustrated in Table 6.3 for various combinations of 1 in 100 and 1 in 10 dilutions, prepared with a single pipette or with a fresh pipette for each transfer. The magnitude of the dilution error can be limited to some extent by reducing the number of dilutions prepared, that is, in

Table 6.3 the dilution error of the 10^{-6} dilution (using a separate pipette at each stage) is reduced from 13.4% to 8.5% by preparing two 100-fold and two 10-fold dilutions, rather than six 10-fold dilutions. The calculations for use of different pipettes at each stage of the dilution process show that the errors are lower than in a series for which a single pipette is used throughout. For instance, the dilution error is 5.5% for six 10-fold dilutions made with fresh, randomly drawn pipettes at each stage, cf 13.4% when the same pipette is used throughout.

Hedges (2002) investigated the inherent inaccuracy of pipettes in terms of the pipette manufacturers' quoted calibration and repeatability errors (the latter being a measure of imprecision) and took account also of the impact of the Poisson distribution as the sampling component of variance and the sampling error at each dilution step, which was ignored by Jennison and Wadsworth (1940). As noted previously, Hedges observed that pipette manufacturers' quoted accuracy/inaccuracy represents their acceptable production limits [assumed by Hedges to span ±3 SE_M ie, to be 99.9% confidence limits (CLs)] and cited only an estimate of the imprecision of the pipette, usually as a CV [$CV = 100 \times (s/u)$, where s is the standard deviation and u is the nominal volume of the pipette]; he also assumed that the manufacturers' quoted accuracy/inaccuracy was used to derive its calibration variance. The variance of the delivered volume was then the sum of these two components. In his treatment, Hedges (2002) assumed that a volume (u) of inoculum is pipetted into a volume (v) of diluent to prepare a dilution $u/(u + v)$ and that the process is repeated n times to give a final dilution, with $D = [u/(u+v)]^n$ *as* the nth step. He further supposed two method scenarios: (a) the more traditional approach where separate randomly selected pipettes are used for each of the u volumes and different randomly drawn pipettes are used to dispense each of the v volumes of diluent at each stage; and (b) a more modern approach where the same randomly selected dispenser pipette is used to deliver all u volumes (with a fresh tip used at each step) and another randomly selected pipette is used to deliver all v volumes.

He used Taylor's series to derive equations to determine the dilution components of variance for each serial dilution stage of both methods. Let $D = p/q \equiv u^n/(u+v)^n$, where $p = u^n$ and $q = (u + v)^n$, and $w = u + v$. The volumes u and v are uncorrelated, as also are v and w. For method (b), the cumulative variances associated with the inoculum transfer volumes (u_i) and the dispensed diluent volumes (v_i) are determined for each stage in the dilution process:

$$\mathrm{Var}(p) \equiv \mathrm{Var}[(u)^n] \approx n \times u^{2(n-1)}[\mathrm{Var}(u) + (n-1) \times \mathrm{Covar}(u_i, u_j)]$$

$$\mathrm{Var}(q) \equiv \mathrm{Var}[(w)^n] \approx n \times w^{2(n-1)}[\mathrm{Var}(w) + (n-1) \times \mathrm{Covar}(w_i, w_j)]$$

whence the variance of the dilution process is given by

$$\mathrm{Var}(D) \equiv \mathrm{Var}(p/q) \approx D^2\{[\mathrm{Var}(p)/p^2] + [\mathrm{Var}(q)/q^2] - [2\mathrm{Covar}(p,q)/pq]\}.$$

However, for method (a), the variances of p and q are not affected by covariances, so the equations are simplified by the exclusion of the covariance factors. Hedges also introduced an estimate of the sampling component due to the random distribution of organisms at the sequential stages of the dilution series. Table 6.5 summarises

Table 6.5 Dilution and Sampling Components of Variance for Serial Dilutions Made Using Separate Pipettes at Each Step [Method (a)] and the Same Pipette at Each Step [Method (b)]

	Dilution Component[a]		
Step (n)	Method (a)	Method (b)	Sampling Component[b]
1	2.0479×10^{-7}	2.0479×10^{-7}	1.0×10^{-2}
2	4.0958×10^{-9}	5.3644×10^{-9}	1.1×10^{-3}
3	6.1436×10^{-11}	9.9532×10^{-11}	1.11×10^{-4}
4	8.1915×10^{-13}	1.5815×10^{-12}	1.111×10^{-5}
5	1.0239×10^{-14}	2.2951×10^{-14}	1.1111×10^{-6}
6	1.2287×10^{-16}	3.1361×10^{-16}	1.11111×10^{-7}

[a]*Values to be multiplied by $(N^*)^2$, where N^* is the estimate of the colony count.*
[b]*Values to be multiplied by N^* and u'^2, where u' is the volume delivered to the plate.*
Modified from Hedges (2002) and reproduced by permission of the author and Elsevier.

the calculated dilution and sampling components and the correlation coefficients for derivation of covariances are shown in Table 6.6. The derivations of the %CVs of the colony count for dilution methods (a) and (b) are shown in Table 6.7. The overall percent error of a count of 223 colonies at the 10^{-6} dilution (using the same pipette throughout) estimated by the method of Hedges (2002) was 7.3%. The procedures to be followed to derive the dilution components are illustrated in Example 6.4.

In a subsequent paper, Hedges (2003) assessed the impact on the precision of serial dilutions and colony counts following re-calibration of pipettes in the laboratory, a require-

Table 6.6 Correlation Coefficients for Covariates in a Serial Dilution Scheme

	Correlation Coefficients (r) for	
Covariates	Method (a)[a]	Method (b)[a]
u_i, u_j	$= 0.00$	$\sqrt{\frac{\text{cal}(u)}{\text{Var}(u_i)} \times \frac{\text{cal}(u)}{\text{Var}(u_j)}} = 0.313$
v_i, v_j	$= 0.00$	$\sqrt{\frac{\text{cal}(v)}{\text{Var}(v_i)} \times \frac{\text{cal}(v)}{\text{Var}(v_j)}} = 0.500$
u_i, v_i	$= 0.00$	$= 0.00$
w_i, w_j	$= 0.00$	$\sqrt{\frac{\text{cal}(u)+\text{cal}(v)}{\text{Var}(w_i)} \times \frac{\text{cal}(u)+\text{cal}(v)}{\text{Var}(w_j)}} = 0.476$
$u_i, w_i\ (\equiv p_n, q_n)$	$\sqrt{\text{Var}(u)/\text{Var}(w)} = 0.355$	$\sqrt{\frac{\text{Var}(u)}{\text{Var}(w)}} = 0.355$

[a]*Methods (a) and (b) are described in Example 6.4.*
Modified from Hedges (2002) and reproduced by permission of the author and Elsevier.

Table 6.7 Examples of the Calculated Variance for Colony Counts Done by the Standard Plate Count Method

	Method	
	Separate Pipettes	**Single Pipette**
Dilution error[a]	1.2287×10^{-16}	3.1361×10^{-16}
Final pipetting error[b]	7.2684	16.7537
Total sampling error[c]	247.7778	247.7778
Var(X) = sum of pipetting and sampling errors	255.046	264.532
Coefficient of variation of *X* (CV, %)[d]	7.162	7.293
Pipetting error as a % Var(X)	2.85	6.33
Distribution error as a % Var(X)	97.15	93.67

Colonies counted = X = 223; dilution level = D = 10^{-6}. Volume of diluent plated = u′ = 1.0 mL; number of dilutions = n = 6; colony count (N) = 2.23×10^8 cfu/mL.*
[a]From Table 6.5.
[b]Final pipetting error = $(N^*)^2 \times Var(P)$, where $Var(P) = [u'^2 \times Var(D) + (D^2 \times Var(u'))]^2$.
[c]Total sampling error = $[u'^2 \times N^* \times$ sampling component value (Table 6.5) $+ X]$.
[d]CV (%) = $100\sqrt{Var(X)}/X$ = (eg) $100\sqrt{255.046}/223$.
Modified from Hedges (2002) and reproduced by permission of the author and Elsevier.

EXAMPLE 6.4 DETERMINATION OF THE PRECISION OF SERIAL DILUTIONS AND THE COEFFICIENT OF VARIATION OF A COLONY COUNT USING THE PROCEDURE OF HEDGES (2002)

Suppose that you need to compare the overall precision of a 10-fold serial dilution series to 10^{-6}, assuming (1) use of different randomly selected 1- and 10-mL pipettes for each dilution stage and (2) use of the same 1- and 10-mL pipettes throughout, but with random selection of pipette tips.

Although this procedure seems complex, it is relatively straightforward providing that the calculations are done in the following sequence:

1. Calculate the basic pipette variances and covariances.
2. Calculate the variance of the dilution series due to pipette 'errors' (by separating the numerator and the denominator of the final dilution fraction).
3. Add in the pipette error due to the final delivery to the culture plate.
4. Calculate the Poisson sampling error due (a) to 'sampling' during dilution and (b) to the final delivery to the plate.
5. Add the two sources of error to obtain final variance.

The following terms are used in this example:

cal(·) and rep(·) mean the pipette calibration and pipette repeatability variances for u and v;

u is the volume of inoculum transferred by pipette from dilution i to the next dilution j;

u' is the volume of inoculum transferred by pipette to the culture plate;

v is the pipetted volume of diluent to be inoculated at each step;

$w = u + v$ is the total volume of inoculated diluent at each step;

$p = u^n$ and $q = w^n = (u+v)^n$, where n is the number of dilutions;

$D = p/q \equiv u^n/(u+v)^n$ is the dilution at step n;

X is the number of colonies counted on plate at dilution D;

Var(·) is the variance, where (·) means u, v, w, D, X;

Covar(·) is the covariance between two factors, for example, Covar(u, v) = covariance between u and v.

Pipette variances and covariances

Hedges (2002) shows that the variance of the inoculum volumes at step n of a dilution series is given by

$$\text{Var}(p) \equiv \text{Var}[(u)^n] \approx n \times u^{2(n-1)}[\text{Var}(u) + (n-1) \times \text{Covar}(u_i, u_j)]$$

Similarly, the variance of the diluent volumes at step n is given by

$$\text{Var}(q) \equiv \text{Var}[(w)^n] \approx n \times w^{2(n-1)}[\text{Var}(w) + (n-1) \times \text{Covar}(w_i, w_j)]$$

whence the variance of the dilution (D) = the variance of (p/q) is given by

$$\text{Var}(D) \equiv \text{Var}\left(\frac{p}{q}\right) \approx D^2\left[\frac{\text{Var}(p)}{p^2} + \frac{\text{Var}(q)}{q^2} - \frac{2\text{Covar}(p,q)}{pq}\right]$$

Now $\text{Var}(p/p^2)$ is the square of the coefficient of variation of p, that is, $(\text{CV}_p)^2$; similarly $\text{Var}(q/q^2)$ is the square of the coefficient of variation of q, that is, $(\text{CV}_q)^2$.

So, we have

$$\text{Var}(D) = D^2\left[(\text{CV}_p)^2 + (\text{CV}_q)^2 - \frac{2\text{Covar}(p,q)}{pq}\right]$$

For both methods of dilution, the volumes u and v are uncorrelated, and the variance of the volume of inoculum delivered u is given by Var(u) = cal(u) + rep(u); similarly the variance of the volume of diluent (v) is Var(v) = cal(v) + rep(v), so the total pipetting variance Var(w) is given by Var(w) = Var(u) + Var(v).

For our 10-fold dilution series we use 1-mL (u = 1) pipettes that have a calibration accuracy of 0.81% and an imprecision of 0.4% and that diluent is dispensed in 9-mL (v = 9) volumes with pipettes having a calibration inaccuracy of 0.3% and an imprecision of 0.1%, as provided by the manufacturer and defined in the text.

Then, using the equations from Hedges (2002), the calibration variance of the 1-mL pipette = cal(u) = $[(u \times \text{accuracy})/(3 \times 100)]^2 = [(1 \times 0.81)/300]^2 = 7.29 \times 10^{-6}$; similarly, the calibration variance of the 9-mL pipette = cal(v) = $[(9 \times 0.3)/300]^2 = 8.1 \times 10^{-5}$.

The repeatability variance of the 1-mL pipette = rep(u) = $[(u \times \text{imprecision})/100]^2 = [(1 \times 0.4)/100]^2 = 1.60 \times 10^{-5}$; similarly, the repeatability variance of the 9-mL pipette = rep(v) = $[(9 \times 0.1)/100]^2 = 8.1 \times 10^{-5}$.

The variance of the volume transferred at each step is given by

$$\text{Var}(u) = (7.29 \times 10^{-6}) + (1.60 \times 10^{-5}) = 2.329 \times 10^{-5}$$

and the variance of the diluent volumes is given by

$$\text{Var}(v) = (8.1 \times 10^{-5}) + (8.1 \times 10^{-5}) = 1.62 \times 10^{-4}$$

so the total pipetting error is given by

$$\text{Var}(w) = \text{Var}(u) + \text{Var}(v) = (2.329 \times 10^{-5}) + (1.62 \times 10^{-4}) = 1.853 \times 10^{-4}$$

However, for any dilution step a covariance factor ($\text{Covar}_{u,w}$) may be necessary to allow for correlated errors in both the inoculum volumes (u) and the diluent volume (v).

To determine covariance, the relation $\text{Covar}(s,t) = \sqrt{r_{s,t}^2 \cdot [\text{Var}(s) + \text{Var}(t)]}$ is used, where $r_{s,t}$ is the Pearson correlation coefficient, for which r^2 gives the proportion of the total variance shared linearly by the two covariates. The correlation coefficients for the various possible covariates are given in Table 6.6.

Hedges (2002) shows that for method (b) the value of the correlation coefficient for u,w was 0.355 (a value supported by simulation studies), so, for the pipette volumes referred to earlier, the covariance of u and w for method (a) is as follows:

$$\text{Covar}(u,w) = \text{Covar}(p,q) = 0.355\sqrt{\text{Var}(p)\times\text{Var}(q)}$$
$$= 0.355\sqrt{(1.397\times10^{-4})\times(1.118\times10^{-7})} = 1.4030\times10^{-5}$$

Pipette variances

For method (a), where a different randomly drawn pipette is used at each step, the only covariance is $\text{Covar}_{u,w}$; then the total variance contribution to the 10^{-6} dilution for the 1-mL pipette volume is given by

$$\text{Var}(p) \equiv \text{Var}[(u)^n] \approx n\times u^{2(n-1)}[\text{Var}(u)] = 6\times1^{2(5)}(2.329\times10^{-5})$$
$$= 6\times(2.329\times10^{-5}) = 1.397\times10^{-4}$$

Similarly the total variance contribution to the 10^{-6} dilution for the 9-mL pipette volume is given by

$$\text{Var}(q) \equiv \text{Var}[(w)^n] \approx n\times w^{2(n-1)}[\text{Var}(w)] = 6\times(10)^{2\times5}(1.853\times10^{-4})$$
$$= 6\times10^{10}\times(1.853\times10^{-4}) = 1.1118\times10^{7}$$

The value of the Pearson correlation coefficient for $u,w = 0.355$ (Hedges, 2002), so, for the pipette volumes referred to earlier, the covariance of u and w for method (a) is as follows:

$$\text{Covar}(u,w) = \text{Covar}(p,q) = 0.355\sqrt{\text{Var}(p)\times\text{Var}(q)}$$
$$= 0.355\sqrt{(1.397\times10^{-4})\times(1.118\times10^{-7})} = 14.030\times10^{-6}$$

and the total variance of the 10^{-6} dilution (D) is given by

$$\text{Var}(D) \equiv \text{Var}\left(\frac{p}{q}\right) \approx D^2\left[\frac{\text{Var}(p)}{p^2}+\frac{\text{Var}(q)}{q^2}-\frac{2\text{Covar}(p,q)}{pq}\right]$$
$$= 10^{-12}\left[\frac{1.397\times10^{-4}}{1}+\frac{1.118\times10^{7}}{10^{12}}-\frac{2(14.030)}{10^{6}}\right]$$
$$= 10^{-12}[(1.397\times10^{-4})+(1.118\times10^{-5})-(28.060\times10^{-6})] = 1.228\times10^{-16}$$

For method (b), covariance factors are required at all stages since the same pipettes are used to deliver each of the u and v volumes. Hence, the variance of p is given by

$$\text{Var}(p) \equiv \text{Var}[(u)^n] \approx n\times u^{2(n-1)}[\text{Var}(u)+(n-1)\times\text{Covar}(u_i,u_j)]$$
$$= 6\times1^{2(6-1)}[2.329\times10^{-5}+(6-1)(7.290\times10^{-6})]$$
$$= 6\times[(2.329\times10^{-5})+(3.645\times10^{-5})]$$
$$= 6\times5.974\times10^{-5} = 3.584\times10^{-4}$$

Similarly, the covariance of q is given by

$$\text{Var}(q) \equiv \text{Var}[(w)^n] \approx n\times w^{2(n-1)}[\text{Var}(w)+(n-1)\times\text{Covar}(w_i,w_j)]$$
$$= 6\times10^{2\times5}[(1.853\times10^{-4})+5(8.820\times10^{-5})]$$
$$= 6\times10^{10}\times6.263\times10^{-4} = 3.758\times10^{7}$$

Now the covariances are given by

$$\text{Covar}(u,w) = \text{Covar}(p,q) = 0.355\sqrt{\text{Var}(p)\times\text{Var}(q)}$$
$$= 0.355\sqrt{(3.584\times10^{-4})\times(3.758\times10^{-7})} = 41.199$$

$$\text{Covar}(u_i,u_j) = 0.313\sqrt{(2.329\times10^{-5})^2} = 7.290\times10^{-6}$$

and

$$\text{Covar}(w_i,w_j) = 0.476\sqrt{(1.853\times10^{-4})^2} = 8.820\times10^{-5}$$

Hence, the total variance of the 10^{-6} dilution is given by

$$\mathrm{Var}(D) \equiv \mathrm{Var}\left(\frac{p}{q}\right) \approx D^2\left[\frac{\mathrm{Var}(p)}{p^2}+\frac{\mathrm{Var}(q)}{q^2}-\frac{2\mathrm{Covar}(p,q)}{pq}\right]$$
$$=10^{-12}\left[\frac{3.584\times10^{-4}}{1}+\frac{3.758\times10^{7}}{10^{2}}-\frac{2(41.199)}{10^{6}}\right]$$
$$=10^{-12}[(3.584\times10^{-4})+(3.758\times10^{-5})+(8.2398\times10^{-5})]=3.136\times10^{-16}$$

Poisson sampling error

Next, we need to take account of the Poisson sampling variance [Var(*P*)] at each step throughout the series up to and including the volume delivered to the culture plate. Again, Hedges (2002) provides an approximation for this variance, where u' represents the volume delivered to the plate:

$$\mathrm{Var}(P)=[u'^2\times\mathrm{Var}(D)]+[D^2\times\mathrm{Var}(u')]$$

For method (a):

$$\mathrm{Var}(P)=(1\times1.228\times10^{-16})+(10^{-12}\times2.329\times10^{-5})$$
$$=[(1.228\times10^{-16})+(2.329\times10^{-17})]=1.461\times10^{-16}$$

For method (b):

$$\mathrm{Var}(P)=(1\times3.136\times10^{-16})+(2.329\times10^{-17})=3.369\times10^{-16}$$

The final pipetting error

The pipetting error of the observed count of *X* colonies per volume (u') at step *n* is estimated by $(N^*)^2\times\mathrm{Var}(P)$, where N^* is the estimate of the original count (*N*): $N^*=X\times(1/D)\times(1/u')$.

If $X=223$, then at step $n=6$, $N^*=223\times10^6$ and the final pipetting error for:

method (a) = $(N^*)^2\times\mathrm{Var}(P)=(223\times10^6)^2\times(1.4609\times10^{-16})=7.265$;

method (b) = $(223\times10^6)\times(3.3689\times10^{-16})=16.7532$.

The total sampling error

There now remains the estimate of the total (Poisson) sampling error, which is the same for both methods. There are two error components: the first is the delivery of the final diluted inoculum to the plate and the second is the cumulative sampling error during the preparation of the dilution series. The first is estimated by the number of colonies on the plate (*X*); the second is derived as follows:

$$u'^2\times N^*\times\sum_{i=n+1}^{i=2n}\frac{1}{z^i}=u'^2\times N^*\times[\text{the sampling component (Table 6.5)}]$$

where $(1/z^i)$ is size of the dilution step (eg, 1 in 10).

For this example, where $n=6$, the sampling component is 1.11111×10^{-7} and the total sampling error is $(1^2\times223\times10^6\times1.11111\times10^{-7})+223=24.77+223=247.778$.

Combined variance of the colony count

The variance of (*X*) = sum of the pipetting and sampling errors; for:

method (a), the combined variance of (*X*) is 7.265 + 247.778 = 255.043;

method (b), the combined variance of (*X*) is 16.753 + 247.778 = 264.531.

The standard error and the coefficient of variation of the colony count at the 10^{-6} dilution using the two different methods of preparing the dilution series are:

method (a), $SE_M=\sqrt{255.043}=15.97$ and %CV = 100(15.97/223) = 7.161;

method (b), $SE_M=\sqrt{264.531}=16.264$ and %CV = 100(16.26/223) = 7.293.

It is worthy of note that for method (a), the pipetting error is only 2.85% of the combined error of the colony count whilst for method (b) it is 6.33%. These data are summarised in Table 6.7.

Some of the text is based on Hedges (2002) and is reproduced by permission of the author and Elsevier.

ment for accredited laboratories (Anon, 2005). He concluded that although re-calibration improves the precision of the methods, owing to the dominant effect of the final sampling variance, the overall improvement in the precision of colony counts is small.

Augustin and Carlier (2006) applied the methods of Hedges (2002, 2003) in their evaluation of repeatability variance for data from laboratory proficiency testing as part of a 'bottom-up' approach to estimating 'uncertainty' (see Chapter 11). For ease of calculation, they converted some of the equations into CVs. For instance, the CV for the volume (u) of inoculum transferred between dilutions is given by $CV^2_{(p)} = \mathrm{Var}(p)/p^2 = \mathrm{Var}(u^n)/u^{2n}$, where Var is the variance and $p = u^n$ is the volume of inoculum transferred at dilution n. They then applied the various equations to determine the overall CV of the derived colony count, which they showed to be related more closely to the number of colonies counted than to any other parameter examined. This again indicates the overwhelming importance of the statistical distribution of organisms both in the original foodstuff and in the final countable test plates.

EFFECTS OF GROSS DILUTION SERIES ERRORS ON THE DERIVED COLONY COUNT

Gross errors in the volumes of inoculum and diluent affect the overall dilution level and will therefore influence the apparent colony count. For instance, the sequential inoculation of a 1-mL volume into, for example, 8.5-mL volumes will give a dilution level of 1 in 9.5, not 1 in 10 as intended. Hence, the error associated with the 10^{-6} dilution, assuming use of different pipettes with variance 0.02 mL and diluent blanks with variance 0.1 mL (Table 6.3), would be ±5.5% for six 10-fold dilutions, but would be ±5.7% for six 9.5-fold dilutions. Possibly more important is the impact on the calculated number of organisms.

For instance, an average colony count of 100 colonies from the 10^{-6} dilution would be recorded as 100×10^6 cfu/unit of sample. But if we take due note of the volume errors, the derived count should be 100×9.5^6, that is, $100 \times 735{,}100 = 73.5 \times 10^6$ cfu/unit of sample. The difference in these counts is highly significant since a 'true' count of 74 million would have been recorded as 100 million. Such differences could affect the likelihood that a colony count result on a sample might, or might not, comply with a microbiological criterion and justifies the necessity for calibration of dispensing equipment and checking of dispensed volumes as part of laboratory quality monitoring.

REFERENCES

Anon, 2005. General Requirements for the Competence of Testing and Calibration Laboratories. ISO/IEC 17025:2005. International Standards Organisation, Geneva.

AOAC, 2012. Official Methods of Analysis, 19th ed. Association of Official Analytical Chemists Inc., Rockville, MA, USA.

Augustin, J.-C., Carlier, V., 2006. Lessons from the organization of a proficiency testing program in food microbiology by interlaboratory comparison: analytical methods in use, impact of methods on bacterial counts and measurement uncertainty of bacterial counts. Food Microbiol. 23, 1–38.

Barraud, C., Kitchell, A.G., Labots, H., Reuter, G., Simonsen, B., 1967. Standardisation of the total aerobic count of bacteria in meat and meat products. Fleischwirtschaft 12, 1313–1318.

Brew, J.D., 1914. A comparison of the microscopical method and the plate method of counting bacteria in milk. N. Y. Agric. Exp. Station, Geneva, NY, Bull. 373, 1–38.

Emsviler, B.S., Pierson, L.J., Kotula, A.W., 1977. Stomaching versus blending. Food Technol. 31(10), 40–42.

Fung, D.Y.C., Sharpe, A.N., Hart, B.C., Liu, Y., 1998. The Pulsifier® a new instrument for preparing food suspensions for microbiological analysis. J. Rapid Methods Autom. Microbiol. 6, 43–49.

Harrigan, W.F., McCance, M.E., 1998. Laboratory Methods in Microbiology, third ed. Academic Press, London.

Hedges, A.J., 1967. On the dilution errors involved in estimating bacterial numbers by the plating method. Biometrics 23, 158–159.

Hedges, A.J., 2002. Estimating the precision of serial dilutions and viable bacterial counts. Int. J. Food Microbiol. 76, 207–214.

Hedges, A.J., 2003. Estimating the precision of serial dilutions and colony counts: contribution of laboratory re-calibration of pipettes. Int. J Food Microbiol. 87, 181–185.

ICMSF, 1978. Microorganisms in Foods. 1. Their Significance and Methods of Enumeration, second ed. University of Toronto Press, Toronto.

Jarvis, B., Williams, A.P., 1987. Methods for detecting fungi in foods and beverages. In: Beuchat, L.R. (Ed.), Food and Beverage Mycology. second ed. AVI Publishers, New York, pp. 599–636.

Jennison, M.W., Wadsworth, G.P., 1940. Evaluation of the errors involved in estimating bacterial numbers by the plating method. J. Bacteriol. 39, 389–397.

Kang, D.H., Dougherty, R.H., 2001. Comparisons of Pulsifier® and Stomacher® to detach microorganisms from lean meat tissues. J. Rapid Methods Autom. Microbiol. 9, 27–32.

Kitchell, A.G., Ingram, G.C., Hudson, W.R., 1973. Microbiological sampling in abattoirs. In: Board, R.G., Lovelock, D.W. (Eds.), Sampling – Microbiological Monitoring of Environments. SAB Technical Series No. 7. Academic Press, London, pp. 43–61.

Mudge, C.S., Lawler, B.M., 1928. Is the statistical method applicable to the bacterial plate count? J. Bacteriol. 15, 207–221.

Sharpe, A.N., Jackson, A.K., 1972. Stomaching: a new concept in bacteriological sample preparation. Appl. Microbiol. 24, 175–178.

Sharpe, A.N., Hearn, E.M., Kovacs-Nolan, J., 2000. Comparison of membrane filtration rates and hydrophobic grid membrane filter coliform and *Escherichia coli* counts in food suspensions using paddle-type and Pulsifier sample preparation procedures. J Food Prot. 62, 126–130.

Snyder, T.L., 1947. The relative errors of bacteriological plate counting methods. J. Bacteriol. 54, 641–654.

Spencer, R., 1970. Variability in colony counts of food poisoning clostridia. Research Report No. 151. Leatherhead Food Research Association, Leatherhead, UK.

Tuttlebee, J.W., 1975. The Stomacher – its use for homogenisation in food microbiology. J. Food Technol. 10, 113–123.

Wu, V.C.H., Jitareerat, P., Fung, D.Y.C., 2003. Comparison of the Pulsifier® and the Stomacher® for recovering microorganisms in vegetables. J. Rapid Methods Autom. Microbiol. 11, 145–152.

CHAPTER

Errors associated with colony count procedures

7

In all colony count techniques, replicate volumes of each of several serial dilutions are dispersed in, or on, a nutrient medium. A count of colonies is made after incubation at an appropriate temperature and the level of organisms per unit of sample is derived from the mean number of colonies counted and the appropriate dilution factor.

The assumption is frequently made that each colony arises from a single viable organism, but since a colony can arise also from a clump, or aggregate, of organisms, the colony count procedure gives an estimate of numbers of colony-forming units (cfu), not of total viable organisms *per se*. Furthermore, colony count methods will provide only an estimate of those organisms able to grow in, or on, the specific culture medium in the conditions of incubation used in the test. No colony count procedure should be expected, therefore, to provide a true estimate of the total viable population of microorganisms.

All colony count methods are subject to errors, some of which are common, whilst others are specific to a particular method or group of methods. Errors common to all procedures include: (1) the sampling and dilution errors, discussed previously (Chapter 6); (2) errors in pipetting volumes of diluted sample; (3) microbial distribution errors; (4) errors of counting and recording colony numbers; and (5) errors of calculation. Details of alternative forms of methodology are given in standard texts such as Harrigan and McCance (1998) and ICMSF (1978).

SPECIFIC TECHNICAL ERRORS

POUR PLATE AND SIMILAR METHODS

Any method in which the inoculum is mixed with molten agar may result in a heat shock to the organisms, the extent of which is dependent on the thermal sensitivity of the organisms and the temperature of the molten agar. Careful tempering of the agar is obviously of paramount importance. In all methods of this type, thorough mixing of the inoculum with the medium is vital in order to obtain an even distribution of organisms. Errors can arise through, for example, splashing agar onto the lid of a Petri dish, or partial setting of agar before adequate mixing has been undertaken. Colonies that grow in the depth of the medium may be obscured by other colonies on the surface (this is particularly the case when 'spreaders' occur).

Statistical Aspects of the Microbiological Examination of Foods. http://dx.doi.org/10.1016/B978-0-12-803973-1.00007-3

In general, colonies in the depths of the agar will be exposed to a slightly lower oxygen tension than will colonies on the surface and marked differences in colony size may result after a finite incubation period. Consequently organisms that are particularly sensitive to thermal shock, lowered oxygen tension or both should not be counted by these methods.

The Colworth droplette method (Sharpe and Kilsby, 1971), in which dilutions are prepared in molten agar and replicate drops of the molten agar are dispensed into a Petri dish, is reported not to be affected by limitations on oxygen diffusion, but will be subject to problems of thermal shock and to the factors affecting drop counts (see later).

Deep agar counts (ie, tube counts, black-rod counts, agar in plastic bags, etc.) will provide an estimate of the facultative and obligate anaerobic organisms in a sample. Since the oxygen and redox potential (E_h) tolerance of such organisms will vary, different organisms may grow at different depths of medium. Clearly, this may not give an accurate estimate of the numbers of potential and facultative anaerobic organisms. Where pre-reduced media are used in deep tubes, one sometimes sees stratified growth where organisms in the bottom of a tube are inhibited by a very low E_h, and organisms in the upper layers of agar are inhibited by oxygen diffusion leading to elevated E_h levels. The latter problem can be overcome by pouring a plug of non-inoculated agar onto the top of the solidified, inoculated agar.

A further problem sometimes experienced with deep agar counts is splitting of the agar due to gas production. Such effects often occur in deep agar tube counts of proteolytic clostridia and make accurate counting of colonies very difficult.

SURFACE PLATING METHODS

In surface plating, a volume of inoculum is pipetted onto the 'dried' surface of an agar medium. A problem common to all methods of this type relates to the extent of surface drying of the agar; inadequate drying will lead to delay in absorption of the inoculum whilst excessive drying will cause 'case-hardening' and, possibly, a reduction in the water activity (a_w) of the surface layer. Case-hardening will permit drops of inoculum to 'run' across the plate and will generally result in a smaller area of inoculum spread – this is of importance in the drop count (Miles et al., 1938) methods. Restriction of a_w may also affect the rate of growth of hydrophilic organisms.

When the inoculum is spread across the surface of the agar (as in the whole or one-quarter-plate spread technique), a proportion of the inoculum may adhere to the surface of the spreader, thereby giving a falsely low colony count (Thomas et al., 2012). However, the extent of this error is likely to be low when compared with the other intrinsic errors of colony count procedures. Many studies (eg, Jarvis et al., 1977) have shown good correlation between this and other procedures, provided that the technique is appropriately standardised. The spiral plate (SP; Gilchrist et al., 1973) and similar semi-automated methods reduce the errors of dilution since an increasingly reducing inoculum is distributed across the surface of the agar plate in the form of an Archimedes' spiral to give the equivalent of a 3 log dilution of organisms on a single plate (see Plate 7.1). Possible technical errors include malfunction of the continuous pipette. A more frequent problem is that the agar surface is not completely level; problems of uneven or sloping surface cause an abnormal distribution pattern of organisms and computation errors will result.

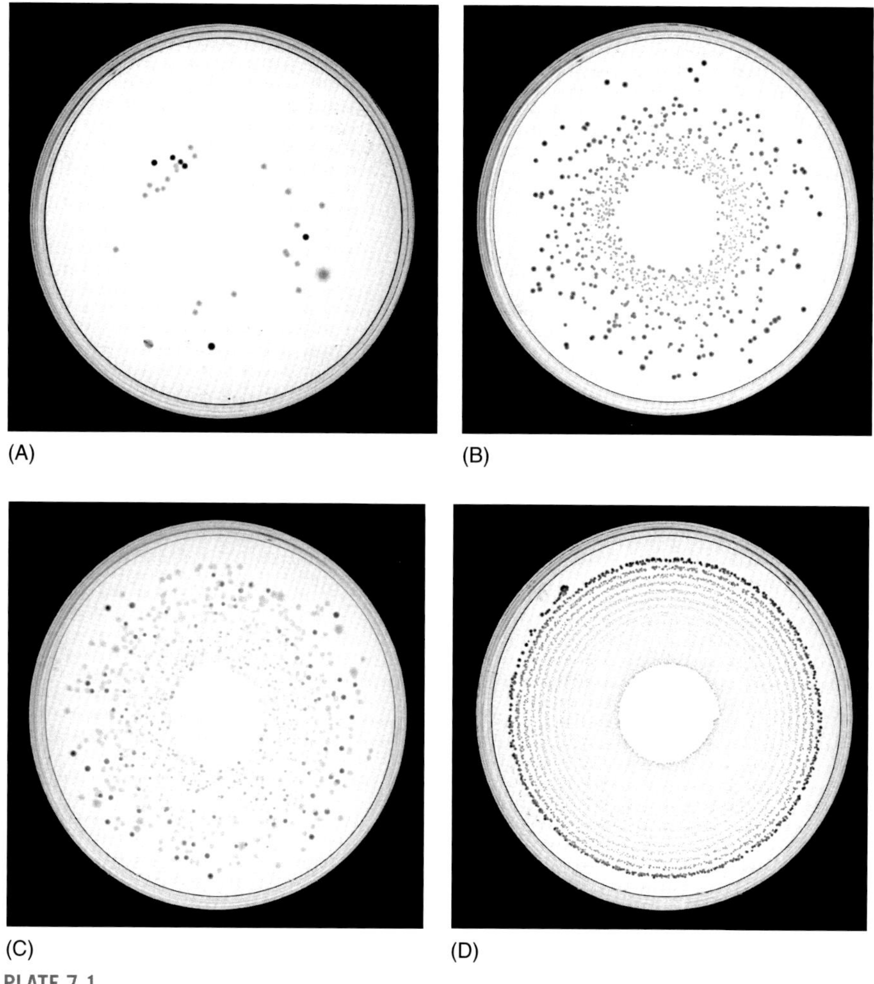

(A) (B) (C) (D)

PLATE 7.1

Distribution of bacterial colonies from raw milk on agar plates using the spiral plating system, to show (A and B) low and medium numbers of colonies for counting either manually or using an automated counting system, (C) medium–high numbers of colonies suitable for counting using an automated counting system and (D) unacceptably high numbers of colonies. Counts on plates (B) and (C) are typically done by scoring the number of colonies in each of several sectors of the plate

Images supplied by and reproduced with the kind permission of Don Whitley Scientific Ltd.

PIPETTING AND DISTRIBUTION ERRORS

PIPETTE ERRORS

The accuracy of pipettes has been discussed in Chapter 6. Since volumes of each dilution used for preparing plates (or tube) counts are also pipetted, an error factor for the variance of pipettes used must be included in the overall assessment of colony count accuracy.

DISTRIBUTION ERRORS

It is usually assumed (Fisher et al., 1922; Wilson, 1922; Snyder, 1947; Reed and Reed, 1949; Badger and Pankhurst, 1960) that the distribution of cfu in a Petri dish, or in a 'drop' of inoculum, follows a Poisson series (see also Chapter 4). However, other workers (eg, Eisenhart and Wilson, 1943) have demonstrated that although pure cultures of bacteria generally follow a Poisson series, mixed cultures may deviate from Poisson and demonstrate either regular or contagious distribution. This effect was illustrated in Example 4.4 for colony counts of organisms surviving disinfection. In such a situation, the organisms detected are those that are not sub-lethally damaged to an extent precluding growth in the recovery conditions used. Hence, one would expect deviation of colony counts from Poisson whenever sub-lethal cell damage has occurred (eg, after heating, freezing, treatment with chemicals) since the susceptibility of individual organisms and strains will vary, especially if they form part of a cell clump or aggregate. Contagious distribution can be seen also in many situations where the inoculum consists of a mixture of viable cells and cell aggregates.

The Index of Dispersion test of Fisher et al. (1922) has been suggested as a means of testing observed plate count results for agreement with a Poisson distribution (see also Chapter 4) using χ^2 as the test criterion:

$$\chi^2 = \sum_{i=1}^{n} \frac{(x_i - \bar{x})^2}{\bar{x}}$$

where $\bar{x}$ is the mean colony count from n plates, x_i is the colony count on the ith plate and χ^2 has $\nu = n - 1$ degrees of freedom.

Fisher et al. (1922) demonstrated that abnormally large variations in colony numbers are associated with factors such as antibiosis between colonies growing on a plate and that subnormal variations (ie, regular distributions) were associated with culture media defects. Eisenhart and Wilson (1943) proposed the use of a control chart to test whether counts are in statistical control. Since the critical value of the Fisher χ^2 test is dependent on the number of replicate counts, this procedure cannot be used unless at least duplicate counts are available at the same dilution level. An example of a χ^2 control chart is shown in Fig. 7.1 for colony counts of clostridia (Spencer, 1970). It can be seen that 1 of the χ^2 values on the first 22 tests (drop counts) slightly exceeded the $P = 0.025$ level and none fell below the value for $P = 0.975$. Similarly, only 1 of the 10 tube counts gave a χ^2 value below $P = 0.975$. If the counts are 'in control',

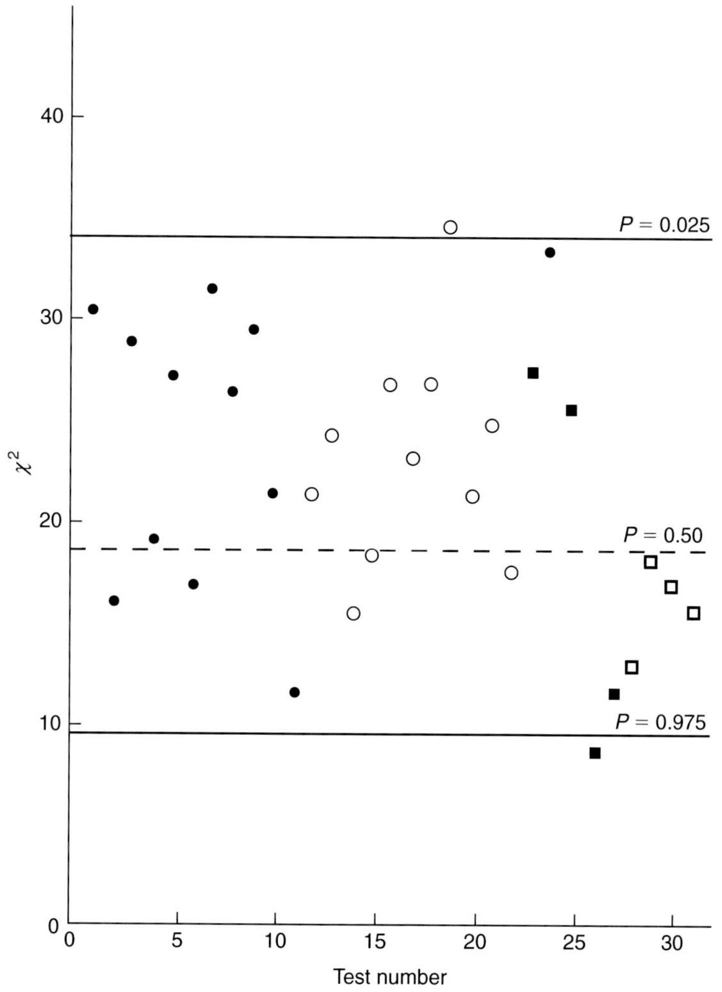

FIGURE 7.1

Control chart for colony counts of *Clostridium botulinum* (○, □) and *Clostridium perfringens* (●, ■) by 'drop' count (○, ●) and tube count (□, ■) methods, with 20 replicate counts/test

(data from Spencer, 1970). The P-values show the 0.025, 0.50 and 0.975 probability limits for the χ^2 distribution used to assess the Index of Dispersion test; values of χ^2 above P = 0.025 or below P = 0.975 are considered to be 'out of control'

on average not more than 1 in 20 tests should lie outside the critical values for χ^2; hence, Spencer's (1970) tube counts were in control and the distribution of colonies could not be shown to differ significantly from Poisson. For the tube counts, there are insufficient data to accept or reject the hypothesis that counts were 'in control', but there is a reasonable presumption that they were. However, these counts were on suspensions of pure cultures.

Colony counts determined in parallel by the drop count and spiral plate maker (SPM) methods on a range of foods (Jarvis et al., 1977) were examined similarly (Figs 7.2 and 7.3). The control chart for 20 sets of drop counts (Fig. 7.2) shows that the upper limit ($\chi^2 = 5.02$ for $P = 0.025$) was exceeded in two cases whilst χ^2 values for 10 data sets were below the lower limit ($\chi^2 = 0.45$ for $P = 0.975$). For these data only 8 of the 20 sets appeared to conform to Poisson. However, for the

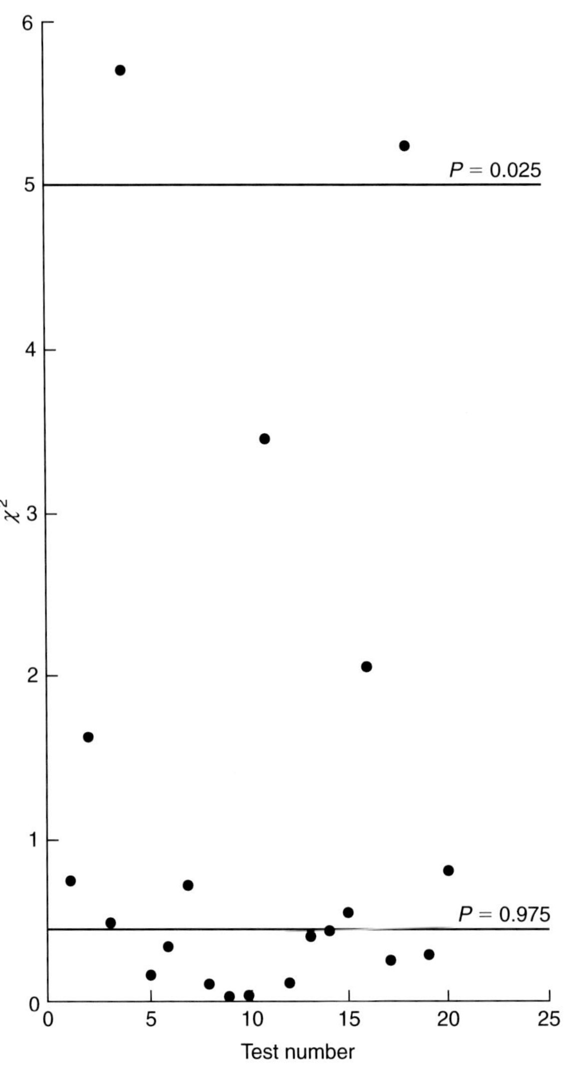

FIGURE 7.2

Control Chart for Colony Counts on Food Samples Using Drop Count Method; Interpretation as in Fig. 7.1

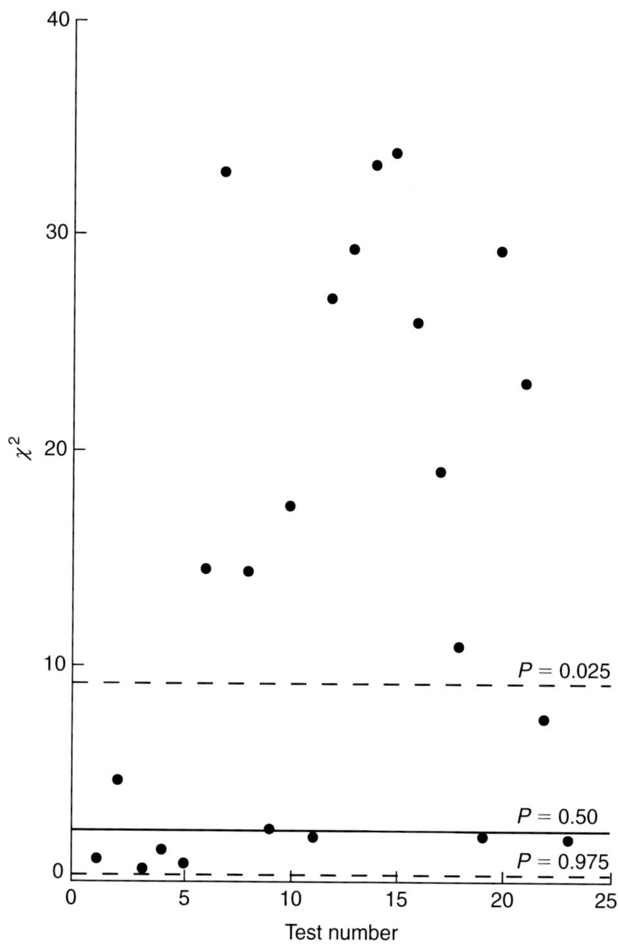

FIGURE 7.3

Control Chart for Colony Counts on Food Samples Using Spiral Plate Method (3 Degrees of Freedom); Interpretation as in Fig. 7.1

20 counts taken together, the cumulative $\chi^2 = 23.53$, for which $0.50 > P > 0.25$; hence, overall, the hypothesis could not be rejected that counts conform to a Poisson distribution. Similarly, the control chart for SPM counts (Fig. 7.3) indicates 13 tests with $\chi^2 < 0.025$ and 1 test with $\chi^2 > 0.975$; only 9 of the 20 counts conformed to Poisson. The cumulative χ^2 value was 334.1 with 69 degrees of freedom for which $P < 0.001$; hence, it may be concluded that these counts made by SPM are subject to contagion and probably conform to a negative binomial distribution rather than Poisson. Both methods were used in parallel to test the same dilutions of the test samples, although the SPM counts were done on lower dilutions than were the drop counts.

The latter observation was unexpected, since SPM colony counts on whole plates and segments from SPM counts of pure cultures conform to Poisson (Jarvis, unpublished data). The contagious distribution detected may be related to cell aggregates associated with the presence of microcolonies in the food samples. Further tests on other SPM data for food samples indicated that counts on higher dilutions which gave colonies distributed across the whole plate conformed well with Poisson $\left(\sum \chi^2 = 19.983 \text{ for } \nu = 33, \text{ giving } P \simeq 0.95\right)$. However, colony counts for lower dilutions of the samples, derived from the numbers of colonies developing in specific sectors of a plate, demonstrated a contagious distribution. Of 37 individual sets of counts analysed, 18 demonstrated contagion, 1 a regular and 18 a Poisson distribution. The cumulative χ^2 of 633.8 for $\nu = 36$ gave $P \ll 0.001$. Since many of the counts were replicates, at lower dilutions, of counts at higher dilution that had not been rejected as conforming to the Poisson distribution, one must suppose either that the effect of close development of colonies on the SPM plates leads to interactive effects or, more likely, that at low dilution organisms are not randomly distributed in the diluent. Comparison of mean colony count data for different dilutions, however, indicated little practical difference in the results obtained.

LIMITING PRECISION AND CONFIDENCE LIMITS OF THE COLONY COUNT

For data distributed according to a Poisson series, the population variance (σ^2) should equal the mean value ($\lambda = \mu$) and therefore the mean count itself provides an approximate measure of the precision of the test. When a large number of sample units is tested ($n \geq 30$) and the samples are drawn at random, the derived mean is one of many possible means distributed around the true population mean. Similarly, when a mean colony count is used as an estimate of numbers of cfu in parallel dilution sets from a sample or from a set of replicate samples, the derived mean cfu will be Normally distributed around the true population mean of that dilution. In a population that conforms to a Normal distribution, 95% of the values will lie within the range ±1.96 standard deviations of the mean. Therefore, 95% of the sample means ($\bar{x}$) will be expected to lie within the range of ±1.96 standard errors of the population mean (μ).

The lower bound of the 95% confidence interval of the mean is given approximately by $\bar{x} - 1.96\text{SE}_\text{M}$ and the upper bound by $\bar{x} + 1.96\text{SE}_\text{M}$. More precisely, the 95% confidence interval is given by $\bar{x} \pm t\sqrt{s^2/n}$, where $\sqrt{s^2/n}$ is the standard error of the mean and t, the value for Student's t distribution, is dependent on the number of degrees of freedom ($\nu = n - 1$). In all cases where σ^2 is not known, Student's t should be used since the sample variance (s^2) provides only an estimate of σ^2. Values of t decrease as n increases, but for the 95% confidence limits (CLs), since for $\nu > 10$ values of t are close to 2, the latter value may be used as an approximation. In a like manner, CLs for $P = 0.99$, or any other probability value, may be derived (Table 7.1).

For small samples ($n < 30$) from a Poisson series, the Normal approximation cannot be applied unless the product nm is greater than 30 (m is the estimate of the

Table 7.1 Percentage Points for the Two-Sided *t* Distribution Used to Determine 95% (1 − $\alpha = 0.95$) and 99% (1 − $\alpha = 0.99$) Confidence Limits at Different Degrees of Freedom (v)

$v = n - 1$	Probability	
	$\alpha = 0.05$	$\alpha = 0.01$
1	12.706	65.657
2	4.303	9.925
3	3.082	5.841
4	2.776	4.604
5	2.571	4.032
6	2.447	3.707
7	2.365	3.499
8	2.306	3.355
9	2.262	3.250
10	2.228	3.169
15	2.131	2.947
20	2.086	2.845
25	2.060	2.787
30	2.042	2.750
40	2.021	2.704
60	2.000	2.660
120	1.980	2.617
∞	1.960	2.576

Poisson parameter λ). As $\lambda = \sigma^2$ in the Poisson series (see Chapter 3), the analytical mean ($\bar{x}$) provides an estimate of both m and s^2; hence, the estimate of m is given by $\bar{x} \pm \sqrt{\bar{x}/n}$, where $\sqrt{\bar{x}/n}$ is the standard error of the mean.

CLs for the estimate ($s = \bar{x}$) of the population mean ($\mu = \bar{x}$) are given by

$$\bar{x} - t\sqrt{\frac{\bar{x}}{n}} \text{ to } \bar{x} + t\sqrt{\frac{\bar{x}}{n}}$$

where t is approximately 2 (for $\upsilon \geq 10$) or may be obtained from Pearson and Hartley (1976), their Table 12 for $2Q = 0.05$ (for the 95% limits) or $2Q = 0.01$ (for the 99% limits). The calculation is illustrated in Example 7.1.

When $nm < 30$, CLs for a Poisson variable can be obtained from standard tables [Pearson and Hartley, 1976; their Table 40 for $1 - 2\alpha = 0.95$ provides the 95% CLs for a value c, which can be either a single count or an estimate ($\bar{x}$) of the statistic m].

For small samples ($n < 30$) from a binomial distribution, the Normal approximation can be used, when the observed proportion, $P = 0.4$–0.6 and $n = 10$–30, or if the proportion is $P = 0.1$–0.9 and $n > 30$, where n is the maximum possible level of occurrence of individuals. When $n < 10$ or $n = 10$–30, approximate CL levels can be obtained from standard tables or charts of P (eg, Pearson and Hartley, 1976; their

EXAMPLE 7.1 CALCULATION OF THE 95% CONFIDENCE INTERVAL FOR A SMALL SAMPLE ($N < 30$) FROM A POISSON SERIES

The colony counts used in Example 4.1 (section 'Apparent Under-Dispersion') had a mean (m) = $\bar{x}$ = 76.7 for which $s^2 = 49.8$; $n = 10$. What is the CI for this count?

Agreement with a Poisson series was not rejected ($P > 0.05$); therefore dispersion of the population was accepted to be random. Since the product $nm > 30$, the estimate of the standard error of the mean (m) (and hence of $\bar{x}$) is given by

$$\text{SE}_{\text{M}} = \pm\sqrt{m} = \pm\sqrt{76.7} = \pm 8.76$$

The value of Student's t (Table 7.1) for $v = n - 1 = 9$ degrees of freedom and $\alpha = 0.05$ (for 95% limits) is: $t = 2.262$.

So the estimate of the approximate bounds for the 95% confidence interval around the population mean (λ) is given by

$$\left(m - t\sqrt{m}\right) \text{ and } \left(m + t\sqrt{m}\right) = 76.7 - (2.262 \times 8.76) \text{ and } 76.7 + (2.262 \times 8.76)$$
$$= 56.77 \text{ and } 95.77$$

Hence, for these data, the estimate of the population mean (76.7 cfu/g) would be expected to lie between 57 and 96 cfu/g with 95% probability. The method of Garwood (1936) gives an 'exact' estimate of the 95% CI bounds as 60.8 and 95.1 cfu/g.

Table 41). In the latter, values of n are printed along the curves and values of P (as c/n) on the abscissa. CLs are read from the ordinate. As $P = \bar{x}/n$, the limits for $\bar{x}$ are derived from limits for nP. Such limits are rarely needed in microbiology except in relation to sampling schemes (Chapter 5). Approximate limits for the Poisson variable will always be wider and more skewed than for the corresponding binomial variable, for example, 95% limits for $\bar{x} = 14$ and $n = 20$ range from 8 to 24 for a Poisson distribution and from 9.5 to 17.8 for a binomial distribution. The Wilson method described in Example 5.1 can provide more precise CLs for proportions.

For small samples ($n < 30$) from a contagious (eg, a negative binomial) distribution with $s^2 > \bar{x}$, it is not possible to use the Normal distribution to derive CLs until the data have been transformed. The choice of transformation depends on the values of $\bar{x}$ and $\hat{k}$ [Table 3.6 and method (3) for estimating $\hat{k}$]. Each value of x is replaced by a transformed value y, where $y = \log[x + (\hat{k}/2)]$ or $y = \sinh^{-1}\sqrt{(x + 0.375)/(k - 0.75)}$. Taking the simple case, the mean transformed count ($\bar{y}$) is given by $\bar{y} = \left\{\sum \log_{10}[x + (\hat{k}/2)]\right\}/n$, where n is the number of counts.

As the distribution of the transformed counts is approximately Normal and the expected variance is 0.1886 trigamma $\hat{k}$ (Table 4.3), the 95% CLs are given by $\bar{y} \pm t\sqrt{0.1886 \text{ trigamma}(\hat{k}/n)}$, where t is given by Student's distribution (Table 7.1; Pearson and Hartley, 1976, Table 12). These limits are transformed back to the original scale to give CLs for the population mean:

$$\text{antilog}\left(\bar{y} \pm t\sqrt{0.1886 \text{ trigamma} \frac{\hat{k}}{n}}\right) - \frac{\hat{k}}{2}$$

However, for small values of n ($n < 10$) the estimate of CLs will be only very approximate. If the more complex transformation is used, the 95% CLs are given by $\bar{y} \pm t\sqrt{0.25\,\text{trigamma}(\hat{k}/n)}$ but it is doubtful whether the greater accuracy of this complex transformation is necessary. In many instances, the Poisson approximation can be used if $\bar{x}$ is small (<5) and $\hat{k}$ is large (>5). The logarithmic transformations [$y = \log x$, or $y = \log(x + 1)$] can be used if $\hat{k}$ is close to 2 – the addition of 1 is required if some of the results are 0 since it is not possible to determine the logarithmic value of 0. An example of a calculation of 95% CLs for a negative binomial is given in Example 7.2 and for a logarithmic transformation in Example 7.3. Table 7.2 summarises the 95% CLs for various colony levels, assuming Poisson distribution. From the data on limiting precision, it can be seen that the precision of the count increases as the number of colonies counted increases.

EXAMPLE 7.2 CALCULATION OF A 95% CONFIDENCE INTERVAL FOR A SMALL SAMPLE FROM A NEGATIVE BINOMIAL DISTRIBUTION

The mean colony count (20.8 cfu/mL) from Example 4.4 is transformed using $y = \log[x + (\hat{k}/2)]$, *so the mean transformed count* $= \bar{y} = 1.3683$, *with* $\hat{k} = 5.1$ *and* $n = 10$. *The geometric mean count* $(\bar{x}) = \text{antilog}\,(\bar{y}) = 23.35$.

From Table 4.3, the expected variance $(0.1886\,\text{trigamma}\,\hat{k}) = 0.0408$.

The value of $t = 2.262$ for $v = n - 1 = 9$ and $\alpha = 0.05$ (Table 7.1).

Therefore the bounds of the 95% confidence interval of the derived geometric mean (m) of 23.35 are given by

$$\text{antilog}\left[\bar{y} \pm \left(t\sqrt{0.1886\,\text{trigamma}\frac{\hat{k}}{n}}\right)\right] - \frac{\hat{k}}{2} = \text{antilog}\left[1.3683 \pm \left(2.262\sqrt{\frac{0.0408}{10}}\right)\right] - 2.55$$

$$= \text{antilog}(1.3683 \pm 0.1445) - 2.55 = 14.19 \text{ to } 30.02$$

It should be noted that the confidence interval is distributed asymmetrically around the geometric mean value, that is, the width of the lower bound of the confidence interval below the mean is $23.35 - 14.19 = 9.16$, and that of the upper bound above the mean is $30.02 - 23.35 = 6.67$. Asymmetry is always found when a reverse transformation is done on values calculated using transformed data.

EXAMPLE 7.3 CALCULATION OF 95% CONFIDENCE INTERVALS FOLLOWING LOGARITHMIC TRANSFORMATION

Suppose that we have transformed colony counts ($y = \log x$) *of 4.2695, 4.9258, 4.8293, 4.8887 and 4.9165 log cfu/g. The mean transformed count* $(\bar{y}) = 4.7660$ *and the variance of the transformed counts* $(s_y^2) = 0.07844$.

For $v = n - 1 = 4$ degrees of freedom, and $\alpha = 0.05$, the value of $t = 2.776$ from Table 7.1.

The 95% bounds of the confidence interval for $\bar{y}$ are given by

$$\bar{y} \pm t\sqrt{\frac{s_y^2}{n}} = 4.766 \pm 2.776\sqrt{\frac{0.07844}{5}} = 4.766 \pm 0.3477 = 4.4183\text{–}5.1137 \approx 4.42 \text{ to } 5.11 \log \text{cfu/g}$$

The derived (geometric) mean = antilog 4.766 = 58,345 cfu/g and the derived 95% CI is 26,200 to 129,927 cfu/g. Again, note the asymmetry of the confidence interval after back transformation.

Table 7.2 Approximate and "Exact" 95% Confidence Intervals for Number of Colonies Counted Assuming Agreement With a Poisson Series

No. of Colonies Counted (m)	Limiting Precision[a] (to Nearest Percentage)	95% Poisson CI	
		Approximate[b]	'Exact'[c]
500	±9	455–545	457.1–545.8
400	±10	360–440	361.8–441.2
320	±11	284–356	285.9–357.1
200	±14	172–228	173.2–229.7
100	±20	80–120	81.4–121.6
80	±22	62–98	63.4–99.6
50	±28	36–64	37.1–65.9
30	±37	19–41	20.2–42.8
20	±45	11–29	12.2–30.9
16	±50	8–24	9.1–26.0
10	±63	4–16	4.8–18.4
6	±82	1–11	2.2–13.1

CI, confidence interval.

[a]*Limiting precision defined as* $\pm\left(2\sqrt{m}/m\right) \times 100$, *where* m *is the number of colonies counted.*

[b]*The approximate CIs are derived from* $m \pm \left(2\sqrt{m}\right)$ *rounded to the nearest integer.*

[c]*Based on the method of Garwood (1936) and shown to one decimal place. Note that the 'exact' CIs are all slightly wider than the approximate CIs.*

For drop counts it is possible to count colonies until a pre-determined minimum count is reached, by plating a greater number of drops of each dilution than might normally be plated (Badger and Pankhurst, 1960). This has the advantage that the precision of the colony count is kept constant in relation to the distribution of organisms in the inoculum. For most purposes a count of about 100 colonies provides a precision of ±20%, although Gaudy et al. (1963) recommended lower and upper limits for acceptable colony counts by the drop count method of 100 and 300 colonies. For pour plate, spread plate and similar methods of enumeration, this procedure cannot readily be applied. It is frequently recommended (Harrigan and McCance, 1998; ICMSF, 1978; Anon, 2005a,b) that colony numbers are counted only on plates having between 30 and 300 colonies (or between 25 and 250), thereby giving counts with a precision of about ±37 to ±11%. It was noted by Cowell and Morisetti (1969) that use of other than 10-fold serial dilution series could lead to improved precision in the plate count. For instance, they suggested that if a threefold dilution series were used and counts were made only on plates with 80 to 320 colonies, the precision of the count would be increased to ±22% or 11%, respectively. Furthermore, such a procedure would avoid the situation where anomalies occur, for instance, no plates in a 10-fold series having more than 30 and less than 300 colonies, or plates for sequential dilutions having between 30 and 300 colonies.

When plates at more than one dilution level are counted, it is possible merely to derive an arithmetic mean count at, for example, two dilution levels (ICMSF, 1978,

p. 116). However, it has been recommended for many years that a weighted mean count be derived (Farmiloe et al., 1954); in this method, all colonies are counted at each dilution level providing countable plates and the mean count is weighted to take account of the differing relative levels of precision of the counts, using the following formula:

$$\bar{x} = \frac{1}{d_1} \cdot \left[\frac{\sum C_1}{n_1} + \frac{\sum C_2}{(n_2/a)} + \cdots \frac{\sum C_z}{(n_z/a^{(z-1)})} \right]$$

where $\sum C_1$ is the total colony count on all (n_1) plates at the dilution with the largest countable number (d_1); $\sum C_2$ is the total colony count on all (n_2) plates at the next countable dilution (d_2) with dilution factor = a, etc. The use of these methods is illustrated in Example 7.4. It is perhaps simpler to consider that, in a 10-fold series,

EXAMPLE 7.4 CALCULATION OF WEIGHTED AND SIMPLE MEAN COUNTS

Assume colony counts have been determined as follows on a series of dilution plates:

	Number of Colonies (C)			
Dilution (d)	***Per Plate***	***Total ($\sum C$)***	***Mean***	***CV (%)***[a]
10^{-4}	*320, 286, 291*	*897*	*299*	*5.8*
10^{-5}	*45, 38, 43*	*126*	*42*	*15.4*
10^{-6}	*3, 5, 4*	*12*	*4*	*50.0*

[a]*Coefficient of variation as a measure of limiting precision of mean count assuming Poisson distribution.*

The arithmetic mean count of cfu at 10^{-4} dilution is given by

$$\left(\frac{1}{d}\right)\left(\frac{\sum C}{n}\right) = 10^4 \times \left(\frac{897}{3}\right) = 2.99 \times 10^6 \text{ cfu/mL}$$

Similarly, the arithmetic mean counts of cfu at the 10^{-5} and 10^{-6} dilutions are 4.2×10^6 and 4×10^6, respectively. Hence, depending on the dilution level used, the simple mean colony count will vary from 3.0 to 4.2×10^6. ICMSF (1978) recommended use of the arithmetic mean value, that is, $[(4.2 \times 10^6) + (2.99 \times 10^6)]/2 = 3.6 \times 10^6$ cfu per unit of sample. Note that the counts at the 10^{-6} dilution are ignored as being below the lower acceptable limit.

This value is, however, based on counts ranging in precision from ±5.8% to ±15.4%. If counts on the highest dilution were also included, the mean value would rise to 3.73×10^6, but the precision range would widen to ±5.8% to ±50%.

Using the formula of Farmiloe et al. (1954), the weighted mean count at all dilutions tested would be as follows:

$$\bar{x} = \left(\frac{1}{d_1}\right)\left[\frac{\sum C_1 + \sum C_2 + \cdots + \sum C_z}{n_1 + (n_2/a) + \cdots + (n_z/a^{z-1})}\right] = \frac{1}{10^{-4}} \times \frac{897 + 126 + 12}{3.33} = 10^4 \times \frac{1035}{3.33} = 3.1 \times 10^6$$

This calculation gives one-tenth of the weight to the colony count from the 10^{-5} dilution (precision ±34%) and one-hundredth of the weight to the 10^{-6} dilution count (precision ±73%), whilst acknowledging that the overall precision of the mean Poisson-distributed count is increased by counting all countable colonies. Omission of the 10^{-6} counts would not affect the weighted mean value for these data.

the colonies (cfu) on a plate prepared at the first countable dilution level (d_1) are equivalent to 1 mL of the diluted sample; the cfu at the next dilution represents 0.1 mL and at the next dilution 0.01 mL of the diluted sample. Hence, the total number of cfu counted on a single plate at each of three sequential dilutions is divided by 1.11 to give the derived number of cfu at the first of those dilution levels. If two or three parallel plates are counted at each dilution, then the total count of colonies is divided by 2.22 or 3.33, respectively.

Since a larger number of colonies is used to derive the weighted mean, the limiting precision of the mean is often better than for either the mean count of colonies on the first countable plates or the arithmetic mean of colonies from two countable dilutions. However, Hedges (2002) disputed this and showed that increased dilution and sampling outweighed the other benefits. Assuming a Poisson distribution, the weighted mean is also the maximum likelihood estimate of the colony count. For the colony counts, used in Example 7.4, the 95% CLs of the mean counts of colonies, and their limiting precisions, are shown in Table 7.2. In the example, the weighted mean and CLs of colony counts at two or three dilution levels do not differ from the arithmetic mean but the precision of the weighted mean count is slightly better (±13.5%) than that using the arithmetic mean (the best is ±14.4%). Where arithmetic means are derived from counts at two or more dilution levels, the CLs can be no better than the limits for the least accurate count; hence, arithmetic averaging is not to be recommended. However, experience suggests that the improved precision is rarely necessary or justified.

GENERAL TECHNICAL ERRORS

INCUBATION ERRORS

The temperature profile within a stack of Petri dishes in an incubator will be dependent on the temperature differential between the incubator air and the Petri dish contents, the number of Petri dishes in a stack, the degree to which stacks of dishes are crowded together, the extent to which the incubator is filled and whether or not the incubator has an air circulator. Differences in count may occur in replicate dishes as a result of overcrowding of plates, poor air circulation, etc., since different dishes may effectively be incubated at different temperatures. Where a temperature gradient is suspected within an incubator, duplicate plates should be incubated on different shelves in order to try to cancel out any temperature effects.

These effects may be exacerbated by problems of inadequate oxygen transfer, resulting in oxygen starvation of strict aerobes, unless vented Petri dishes are used. Similarly, anaerobic cultures may be adversely affected by failure of anaerobe cultivation systems, etc.

COUNTING ERRORS

The ability to count accurately the number of colonies in a Petri dish (or other system) is dependent on many factors, not least of which is the ability of the worker to discriminate colonies of different sizes both from one another and from food debris or imperfections in the agar. This ability depends not only on the individual worker, *per se*, but also on the worker's state of mental and physical health, and whether distractions occur which may affect the precise counting and recording of colony numbers. The efficiency of colony counting will generally be lower with increasing number of plates counted and will also be lower when very large numbers of colonies occur on a plate, such that only an estimated colony count can be derived (ICMSF, 1978).

However, low precision may also occur where few colonies appear on a plate. In a study of the counting efficiency of 6 workers, Fruin et al. (1977) demonstrated that only 51% of colony counts (over the range <5 to 400 colonies/plate) lay within 5% of the real count (determined from photographs of the plates) and only 82% lay within 10% of the 'photo count'. Their data, summarised in Table 7.3, show that the efficiency of counting increased slightly when more than 20 colonies/plate were counted. Analysts tended to count fewer colonies than were actually on the plate and the mean deviation by all analysts from the photo count was −2.5% with a range of −0.1% to −4.8%. However, good laboratory practice requires that it is better generally not to use low counts except if these occur on the first countable plates.

The number of colonies counted may also be affected by factors such as coalescence. Hedges et al. (1978), using the Miles and Misra (MM; Miles et al., 1938) technique, demonstrated that for *Escherichia coli*, the average number of colonies when counted after 9-h incubation was 55 per drop (range 45–63), whereas the average number counted after 18 h was 45 (range 38–52; $n = 16$). Similar effects were not reported for counts of *Staphylococcus aureus*, which produces more compact colonies, or for *E. coli* when counted by the Colworth 'droplette' method (Sharpe and Kilsby, 1971) or the SP method (Gilchrist et al., 1973).

The use of instrumental methods to enumerate colonies on plates can reduce the errors due to operator bias and fatigue. Yet counts made by such methods may not always be comparable with 'manual' counts because of differences in the discriminatory power of the eye and the electronic counting systems (Jarvis and Lach, 1975; Jarvis et al., 1978). For instance, the Fisher Bacterial Colony Counter can detect the

Table 7.3 Percentage Efficiency of Colony Counting by Six Workers

Colony Count Range Per Plate	No. of Plates Counted Per Worker	Percentage of Workers' Counts	
		Within 5% of Photo Count	Within 10% of Photo Count
10–100	321	52[a]	78[a]
20–200	313	60[b]	85[b]
30–300	314	60[b]	88[b,c]
40–400	315	61[b]	89[c]

Mean values within a column followed by the same letter (a–c) are not significantly different (P > 0.05).
Modified from Tables 1 and 2 of Fruin et al. (1977).

presence of colonies that cannot readily be counted by eye, yet it cannot discriminate between food particles and bacterial colonies, or imperfections in the agar or the Petri dish. Large colonies may be counted more than once, whilst colonies at different depths of the agar may be ignored because colonies on the surface have already been 'scored'. These faults are common to many electronic counting systems. For counting SPs, a Laser Bacterial Colony Counter is available which has been shown to give good agreement with 'manual' counts in both laboratory and factory evaluations (Jarvis et al., 1978; Kramer et al., 1979).

Whereas all colonies are normally recorded only for 'countable' plates, tubes and 'drops', the counting system for the SPM assumes that only part of each plate will be counted since the dilution is made on the plate, except when the colony count is low (Plate 7.1). In manual counting of SPM plates, colonies in two diametrically opposed sectors are counted and the number is related to the area counted and hence the amount of inoculum dispensed in those areas. The Laser Bacterial Colony Counter displays the area occupied by a pre-determined number of colonies or, if that number is not reached, the total number of colonies on the plate is shown. By counting to a pre-determined colony level, the distribution error is kept constant, thereby ensuring comparability in precision between replicate series of plates. Hedges et al. (1978) showed that the precision of SPM plate counts is similar whether counts are made on the whole plate or only on a 'sector'.

WORKER'S ERROR

Courtney (1956), APHA (2001) and ICMSF (1978) recommend that all workers should be able to repeat their own counts of colonies within 5% and the counts of other workers within 10%. Fowler et al. (1978) determined a coefficient of variation (CV) of 7.7% for individuals to reproduce their own results and 18.2% for workers to reproduce counts made by other persons; these errors include preparation of dilutions and plates, as well as counting errors. Donnelly et al. (1960) had previously shown that 7 of 21 analysts produced colony count data for milk with a markedly consistent bias.

COMPARABILITY OF COLONY COUNT METHODS

Because of the increasing interest being paid to microbiological quality monitoring and the present-day interest in microbiological criteria for foods (Chapter 14), assessment of the comparability of various methods of analysis is necessary from time to time. In certain circumstances, specific techniques may be used since these will favour the growth of particular groups of organisms (eg, spread plates for psychrotrophs on meat and poultry; Barnes and Thornley, 1966).

In a comparison of the pour plate and spread plate methods for mesophilic aerobes on meat, Nottingham et al. (1975) reported that higher counts were generally obtained by the spread plate method. This confirmed the earlier observations by Barraud et al. (1967).

In contrast, close comparability of counts was found by Donnelly et al. (1976), Jarvis et al. (1977) and Kramer and Gilbert (1978) for various colony count methods when applied to a range of food products. Jarvis et al. (1977) were unable to demonstrate any statistically significant differences ($P > 0.05$) between counts determined by four workers using the pour plate, spread plate, drop count and SP methods on sausages, minced beef, cream and coleslaw. Other studies on milk have shown a high degree of correlation between use of standard methods and the SP method (Donnelly et al., 1976). In a further investigation, Kramer and Gilbert (1978) compared the pour plate, spread plate, drop count (Miles et al., 1938), the Colworth 'droplette' (Sharpe and Kilsby, 1971) and micro-dilution (Kramer, 1977) methods on a wide range of food products and were also unable to determine any significant differences. Correlation between all the methods was high ($r = 0.979–0.994$; $P < 0.001$).

It has rightly been stated by Hedges et al. (1978) that such studies have been concerned more with comparability than with precision. In their work, Hedges et al. demonstrated that the SP method provided colony counts of bacteria in pure suspension that were at least as precise as those obtained under comparable conditions by the Colworth 'droplette' and the MM methods. The CVs were 1.50% (SP), 0.96% (MM) and 4.41% (droplette) for 25 sets of replicate counts; where 5 replicates were tested, the CV for each method was about 2.1%. Earlier studies using the pour plate method (Ziegler and Halvorson, 1935) had shown a CV for counts on pure cultures ranging from about 3.5% (100 replicates) up to 24%, with a majority at the lower end.

OVERALL ERROR OF COLONY COUNT METHODS

Table 7.4 summarises the types of error that can occur in colony count procedures. Jennison and Wadsworth (1940) confined consideration of the overall error to (1) the distribution, or sampling, error and (2) the dilution error. They defined it mathematically as follows:

$$\%\ \text{total error} = \pm\sqrt{(\%\ \text{distribution error})^2 + (\%\ \text{dilution error})^2}$$

Such measurement errors do not take into account the errors associated with examination of multiple samples or multiple test portions, which were shown by Jarvis et al. (2012) to comprise between 50% and 90% of the overall variance of colony counts; the extent of such variance is dependent on the composition and nature of the sample materials. Procedures such as weighing, diluting and plating (Hedges, 2002) contribute to the distribution error that also includes the worker's error (in counting, recording, calculating, etc.). Hence, the overall error of colony count procedures should be derived as follows:

$$\%\ \text{total error} = \pm\sqrt{(A)^2 + (B)^2 + (C)^2}$$

where A is the % sampling error, B is the % distribution error and C is the % dilution error.

Table 7.4 Sources of Error in Colony Count Procedures

Source of Error	Includes Errors Due to
Sampling error	Heterogeneous distribution of organisms
	Weighing
	Maceration
Dilution error	Pipette imprecision
	Pipette calibration errors
	Dispensed diluent volumes
Plating error	Pipetting errors
	Culture medium faults
	Incubation faults
Distribution error	Non-randomness of propagules
	Counting errors
Calculation error	'Recording' errors
	Mathematical errors

It is frequently stated that the overall results obtainable by colony count procedures should be repeatable within ±10% of the value obtained by another worker, using the same techniques and the same samples. It has already been shown (Chapter 6) that dilution errors alone can exceed 10% in some circumstances, depending on the relative accuracy of pipette and dilution volumes, and the number of dilutions prepared. Distribution errors can also exceed ±10% when only small numbers of colonies (<100) are counted.

We can derive estimates for the typical overall error of the colony count method, assuming a sampling error of ±5% (*A*), a distribution error of ±10% (100 colonies counted) (*B*) and a dilution error of ±5.5% (*C*) (from Table 6.4, dilution to 10^{-6} with different pipettes); then the overall error (%) is given by

$$\text{Overall error (\%)} = \pm\sqrt{5^2 + 10^2 + 5.5^2} = \pm\sqrt{155.25} \approx \pm 12.5\%$$

So if it were assumed that the cultural and other conditions are satisfactory, and that the distribution of organisms follows a Poisson series (ie, a hypothesis of randomness is not rejected), then the approximate 95% CLs on a count of 100×10^6 would be ±25%. Table 7.5 illustrates the CLs for a range of colony count levels (and of their logarithmic transformations).

Donnelly et al. (1960) recommended that the variance amongst analysts in state laboratories testing milk samples containing 5000–150,000 cfu/mL should not exceed 0.012 (in terms of log count); hence, on 95% of occasions, the difference between two counts should not exceed $2\sqrt{2(0.012)}$, that is, 0.31 log units. This value is not dissimilar to that derived in Table 7.5 for a mean count of 200 colonies/plate. However, none of these data takes account of abnormal variation in distribution of organisms within a food sample, since milk is easily mixed and the theoretical calculation assumes random distribution.

Table 7.5 Approximate Confidence Intervals (CI) for Counts of Colonies Based on Poisson Distribution, Assuming Sampling Error of 5% and a Dilution Error (to 10^{-6}) of 5.5%

Total Number of Colonies Counted	Mean[a] Colony Count ($\times 10^{-6}$)	Overall Error[b] (%, ±)	Bounds of the 95% CI on the Colony Count ($\times 10^{-6}$)	Mean log Colony Count	Bounds of the 95% CI on log Mean Count[c]
600	300	8.48	274–326	8.48	8.44–8.51
400	200	8.96	164–236	8.30	8.21–8.37
200	100	10.26	89–111	8.00	7.95–8.05
100	50	14.89	42–58	7.70	7.62–7.76
60	30	12.91	22–38	7.48	7.34–7.58
40	20	17.47	16–24	7.30	7.20–7.38
30	15	19.71	9–21	7.18	6.95–7.32

[a]*Assumes duplicate plates.*
[b]*Assumes six 10-fold dilutions, using different pipettes for each stage; error increases disproportionately with reducing numbers of colonies counted.*
[c]*N.B. The 95% CI bounds around the log mean count are asymmetrical.*

Hall (1977) showed that the variance of log-transformed colony counts done by different workers in different laboratories varied from 0.015 to 0.39, with an overall variance of 0.033 (almost treble that recommended by Donnelly et al., 1960) giving overall confidence ranges from duplicate analyses of 0.51 log units. Variance between laboratories was 0.057 log units, giving reproducibility limits of 0.68 log units for duplicate analyses.

Data obtained in analyses of various types of frozen vegetables (Hall, 1977) showed 95% CLs for aerobic plate counts ranging from ±0.16 to ±0.56 log units (Table 7.6). These limits take account of the inter-sample variation, in addition to

Table 7.6 95% Confidence Intervals for Aerobic Colony Counts on Various Frozen Vegetables

Type of Vegetable	Mean Colony Count ($\log_{10}$ cfu/g)	95% CI[a] ($\log_{10}$ cfu/g)
Sliced green beans	4.52	±0.21
Peas	4.27	±0.22
Diced swede	4.33	±0.11
Diced carrots	5.02	±0.13
Diced celery	4.95	±0.22
Spinach leaf I	5.71	±1.03
Spinach leaf II	5.16	±0.49
Spinach leaf III	5.28	±1.05

CI, confidence interval.
[a]*The recalculated CIs differ in some cases from those shown in Tables 1–8 of Hall (1977).*
Recalculated from Hall (1977).

the dilution and distribution errors. In similar tests using differential colony counts for coliforms, *E. coli* and *S. aureus*, Hall (1977) reported 95% CLs ranging up to ±1.5 log units. Clearly, in a situation where sub-lethal damage causes variable recovery of organisms, especially on selective–diagnostic media, the degree of confidence that can be placed in the results may be limited.

It has been stated elsewhere (Jarvis et al., 1977; Kramer and Gilbert, 1978) that the expected 95% CLs about the mean for aerobic colony counts are of the order of ±0.5 log cycle. This is not dissimilar to some of the limits shown in Table 7.6 or to values derived from other experimental data. The data of Hall (1977) indicate that much lower precision can occur frequently. Some contribution towards these wide CLs is related to worker error and to non-uniform experimental technique, including variable resuscitation. However, Jarvis et al. (2012) showed that estimates of reproducibility of colony counts are influenced significantly by the nature of the food matrix as well as intra- and inter-laboratory factors (see Chapter 12).

REFERENCES

Anon, 2005a. Milk and Milk Products–Quality Control in Microbiological Laboratories–Part 1 Analyst Performance Assessment for Colony Counts. ISO 14461-1:2005. International Standards Organisation, Geneva.

Anon, 2005b. Milk and Milk Products–Quality Control in Microbiological Laboratories–Part 2 Determination of the Reliability of Colony Counts of Parallel Plates and Subsequent Dilution Steps. ISO 14461-2:2005. International Standards Organisation, Geneva.

APHA, 2001. In: Downes, F.P., Ito, K. (Eds.), Compendium of Methods for the Microbiological Examination of Foods. fourth ed. American Public Health Association, Washington, DC.

Badger, E.M., Pankhurst, E.S., 1960. Experiments on the accuracy of surface drop bacterial counts. J. Appl. Bacteriol. 23, 28–36.

Barnes, E.M., Thornley, M.J., 1966. The spoilage flora of eviscerated chickens stored at different temperatures. J. Food Technol. 1, 113–119.

Barraud, C., Kitchell, A.G., Labots, H., Reuter, G., Simonsen, B., 1967. Standardisation of the total aerobic count of bacteria in meat and meat products. Fleischwirtschaft 12, 1313–1318.

Courtney, J.L., 1956. The relationship of average standard plate count and ratios to employee proficiency in plating dairy products. J. Milk Food Technol. 10, 336–344.

Cowell, N.D., Morisetti, M.D., 1969. Microbiological techniques – some statistical methods. J. Sci. Food Agric. 20, 573–579.

Donnelly, C.B., Harris, E.K., Black, L.A., Lenis, K.H., 1960. Statistical analysis of standard plate counts of milk samples split with state laboratories. J. Milk Food Technol. 23, 315–319.

Donnelly, C.B., Gilchrist, J.E., Peeler, J.T., Campbell, J.E., 1976. Spiral plate count method for examination of raw and pasteurised milk. Appl. Environ. Microbiol. 32, 21–27.

Eisenhart, C., Wilson, P.W., 1943. Statistical methods and control in bacteriology. Bacteriol. Rev. 7, 57–137.

Farmiloe, F.J., Cornford, S.J., Coppock, J.B.M., Ingram, M., 1954. The survival of *Bacillus subtilis* spores in the baking of bread. J. Sci. Food Agric. 5, 292–304.

Fisher, R.A., Thornton, H.G., MacKenzie, W.A., 1922. The accuracy of the plating method of estimating the density of bacterial populations with particular reference to the use of Thornton's agar medium with soil samples. Ann. Appl. Biol. 9, 325–359.

Fowler, J.L., Clark, W.S., Foster, J.F., Hopkins, A., 1978. Analyst variation in doing the standard plate count as described in 'standard methods for the examination of dairy products'. J. Food Prot. 41, 4–7.

Fruin, J.T., Hill, T.M., Clarke, J.B., Fowler, J.L., Guthertz, L.S., 1977. Accuracy and speed in counting agar plates. J. Food Prot. 40, 596–599.

Garwood, F., 1936. Fiducial limits for the Poisson distribution. Biometrika 28, 437–442.

Gaudy, Jr., A.F., Abu-Niaaj, F., Gaudy, E.T., 1963. Statistical study of the spot-plate technique for viable-cell counts. Appl. Microbiol. 11, 305–309.

Gilchrist, J.E., Campbell, J.E., Donnelly, C.B., Peeler, J.T., Delaney, J.M., 1973. Spiral plate method for bacterial determination. Appl. Microbiol. 25, 244–252.

Hall, L.P., 1977. A study of the degree of variation occurring in results of microbiological analyses of frozen vegetables. Report No. 182. Campden & Chorleywood Food Research Association, Chipping Campden, UK.

Harrigan, W.F., McCance, M.E., 1998. Laboratory Methods in Food Microbiology, third ed. Academic Press, London.

Hedges, A.J., 2002. Estimating the precision of serial dilutions and viable bacterial counts. Int. J Food Microbiol. 76, 207–214.

Hedges, A.J., Shannon, R., Hobbs, R.P., 1978. Comparison of the precision obtained in counting viable bacteria by the "spiral plate maker", the "droplette", and the "Miles & Misra" methods. J. Appl. Bacteriol. 45, 57–65.

ICMSF, 1978. Microorganisms in Foods 1. Their Significance and Methods of Enumeration, second ed. University of Toronto Press, Toronto.

Jarvis, B., Lach, V., 1975. Evaluation of the Fisher bacterial colony counter. Technical Circular No. 600. Leatherhead Food Research Association, Leatherhead.

Jarvis, B., Lach, V.H., Wood, J.M., 1977. Evaluation of the spiral plate maker for the enumeration of microorganisms in foods. J. Appl. Bacteriol. 43, 149–157.

Jarvis, B., Lach, V.H., Wood, J.M., 1978. Evaluation of the laser bacterial colony counter. Research Report No. 290. Leatherhead Food Research Association, Leatherhead.

Jarvis, B., Hedges, A.J., Corry, J.E.L., 2012. The contribution of sampling uncertainty to total measurement uncertainty in the enumeration of microorganisms in foods. Food Microbiol. 30, 362–371.

Jennison, M.W., Wadsworth, G.P., 1940. Evaluation of the errors involved in estimating bacterial numbers by the plating method. J. Bacteriol. 39, 389–397.

Kramer, J., 1977. A rapid microdilution technique for counting viable bacteria in food. Lab. Pract. 26, 676.

Kramer, J.M., Gilbert, R.J., 1978. Enumeration of microorganisms in food: a comparative study of five methods. J. Hyg. (Camb.) 61, 151–159.

Kramer, J.M., Kendall, M., Gilbert, R.J., 1979. Evaluation of the spiral plate and laser colony counting techniques for the enumeration of bacteria in foods. Eur. J. Appl. Microbiol. Biotechnol. 6, 289–299.

Miles, A.A., Misra, S.S., Irwin, J.O., 1938. The estimation of the bacteriological power of blood. J. Hyg. (Lond.) 38, 732–749.

Nottingham, P.M., Rushbrock, A.J., Jury, K.E., 1975. The effect of plating technique and incubation temperature on bacterial counts. J. Food Technol. 10, 273–279.

Pearson, E.S., Hartley, H.O., 1976. Biometrika Tables for Statisticians, third ed., Griffin, High Wycombe for The Biometrika Trustees.

Reed, R.W., Reed, G.B., 1949. "Drop plate" method for counting viable bacteria. Can. J. Res. E 26, 317–326.

Sharpe, A.N., Kilsby, D.C., 1971. A rapid inexpensive bacterial count technique using agar droplets. J. Appl. Bacteriol. 34, 435–440.

Snyder, T.L., 1947. The relative errors of bacteriological plate counting methods. J. Bacteriol. 54, 641–654.

Spencer, R., 1970. Variability in colony counts of food poisoning clostridia. Research Report No. 151. Leatherhead Food Research Association, Leatherhead.

Thomas, P., Sekhar, A.C., Mujawar, M.M., 2012. Non-recovery of varying proportions of viable bacteria during spread plating governed by the extent of spreader usage and proposal for an alternate spotting-spreading approach to maximize the CFU. J. Appl. Microbiol. 113, 339–350.

Wilson, G.S., 1922. The proportion of viable bacteria in young cultures with especial reference to the technique employed in counting. J. Bacteriol. 7, 405–446.

Ziegler, N.R., Halvorson, H.O., 1935. Application of statistics to problems in bacteriology. IV. Experimental comparison of the dilution method, the plate count and the direct count for the determination of bacterial populations. J. Bacteriol. 29, 609–634.

CHAPTER

Errors associated with quantal response methods

A quantal test is one that gives an 'all-or-nothing' response, for example, growth (+) or non-growth (−) in a suitable culture medium. Microbiological quantal procedures are those used to detect the 'presence or absence' of a specific target organism, for example, a pathogen such as salmonella, *Listeria* or *Escherichia coli* in foods, or specific spoilage organisms, for example, *Alicyclobacillus* spp. in fruit concentrates. Procedures include the simple detection of microbial growth, following inoculation into a suitable culture medium, by the development of turbidity or by biochemical/biophysical change in the medium, for example, production of acid and gas in MacConkey broth or a change in an electrical signal in an impedance culture system. In other cases, inoculation of pre-enrichment media followed by selective enrichment, diagnostic plating and identification of organisms may be required. Modern developments in quantal methodology include the use of immuno-magnetic beads (for reviews see Safarik et al., 1995; Stevens and Jaykus, 2004) to separate target organisms from a mixed flora following the pre-enrichment stage, and the use of genetic techniques such as the PCR tests for detection of DNA or RNA from selected target species (see, eg, Anon, 2005a,b, 2006a,b). No matter the approach used, all such procedures are based on the concept of quantal response.

DILUTION SERIES AND MOST PROBABLE NUMBER COUNTS

In its simplest form, the dilution count method consists of serial dilution of an initial food sample or a homogenate followed by inoculation of an appropriate volume of each dilution into one or more tubes of culture medium (agar or broth). The number of tests that show evidence of growth is recorded after incubation. Cultures showing no evidence of growth are assumed not to have received a viable inoculum, that is, to have received no viable cells or cell aggregates, an assumption that may not always be correct. Similarly, any culture showing growth is assumed to have received at least one viable cell or cell aggregate.

When a culture of microorganisms is inoculated at each of several dilution levels, a graded response would be expected with more tests positive at the lower dilution levels and none, or only a few, positive tests at the higher dilutions tested. This is illustrated in Table 8.1 for dilutions from 10^{-3} to 10^{-7} of a culture initially containing 2×10^5 viable cells/mL and 1, 5 or 10 replicate tests at each dilution level.

Statistically, such quantal responses are described by the binomial distribution where the proportion of positive reactions is given by p and negative reactions by

Statistical Aspects of the Microbiological Examination of Foods. http://dx.doi.org/10.1016/B978-0-12-803973-1.00008-5

Table 8.1 Expected Proportional Responses in a Multiple-Tube Dilution Series

		Expected Proportion of Tests[b]		Expected Number of Positives for Number of Tests Inoculated		
Dilution Tested	**Mean Inoculum Cells/Test**[a]	**Positive** $P_{(x>0)}$	**Negative** $P_{(x=0)}$	**1**	**5**	**10**
10^{-3}	200	0.999	0.001	1	5	10
10^{-4}	20	0.999	0.001	1	5	10
10^{-5}	2	0.865	0.135	1	4	9
10^{-6}	0.2	0.181	0.819	0	1	2
10^{-7}	0.02	0.020	0.980	0	0	0

Initial inoculum 2×10^5 viable cells/mL; dilution factor = 10.
[a]Assumes 1 mL of sample dilution per test.
[b]Assumes Poisson distribution.

q, and $q = 1 - p$ (see Chapter 3). In practice, the binomial distribution proves very unwieldy if many tests are used and the Poisson distribution provides a simpler approach. If it is assumed that each viable cell is capable of producing a growth response, then the probability that a culture will receive no inoculum ($P_{x=0}$), after inoculation with a suitable volume of a dilution containing a mean inoculum level of m cells, is given by the first term of the Poisson expansion (see Table 3.2), that is: $P_{x=0} = e^{-m} = q$. Similarly, the probability for a positive (ie, growth) response is given by $P_{x\geq 1} = 1 - P_{x=0} = 1 - e^{-m} = p$.

A plot of the expected dose–response curve, based on Poisson distribution, is shown in Fig. 8.1. Eisenhart and Wilson (1943) showed that, in practice, the actual dose–response curve may be less steep and Meynell and Meynell (1970) suggest that this could arise both through systematic technical error and as a result of contagious distribution, which is a consequence of cell clumping, sub-lethal damage, etc.

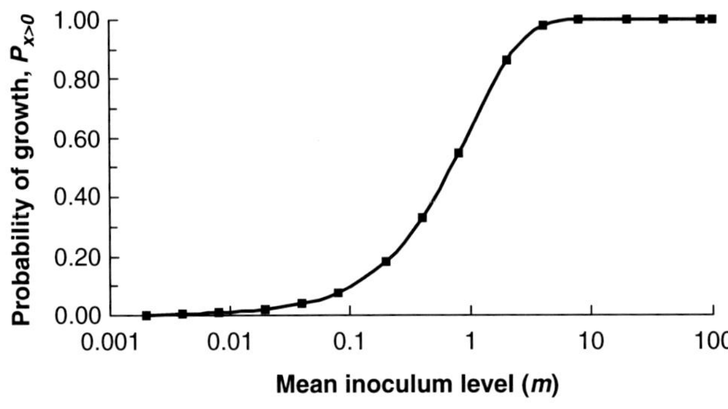

FIGURE 8.1 Dose–Response Curve

Probability of growth (ie, a positive response) at various inoculum levels (m)

SINGLE-TUBE DILUTION TESTS

If only one test is done at a given dilution level, only two responses are possible, that is, growth (positive) or no growth (negative). When single tests are done at several dilution levels, it is not uncommon for 'skips' to occur, for example:

Dilution	10^{-1}	10^{-2}	10^{-3}	10^{-4}	10^{-5}
Response	+	+	+	−	+

The probability for a negative response (ie, no viable organisms/test) is dependent on the level of organisms in the original sample. For instance, if the series shown above were derived from a sample containing, say, 5×10^3 viable cells/mL, then, assuming 1 mL inoculum/test, the probability of occurrence of the observed negative result at the 10^{-4} dilution would be 0.61 and at the 10^{-5} dilution the probability would be 0.95, respectively, so the probability of a positive result at the 10^{-5} dilution would be only 0.05. Hence, the result obtained on the 10^{-4} dilution might be expected to occur on about 6 out of 10 occasions but the result on the 10^{-5} dilution would be expected to occur only on about 1 out of 20 occasions. Neither result would be impossible, but the positive result on the 10^{-5} dilution would be most unlikely.

However, if the original sample contained only 1×10^3 viable cells/mL, the probabilities for the observed results in the 10^{-4} and 10^{-5} dilutions would be 0.90 and <0.01, respectively. Hence, the result at the 10^{-4} dilution would be highly probable since it might be expected to occur in 9 out of 10 tests, but the result at the 10^{-5} dilution would be highly improbable since it might reasonably be expected to occur in fewer than 1 in 100 tests.

The probabilities for such an observation are listed for various inocula levels (m) in Table 8.2, from which it can be seen that a 'skip' could occur by chance at least once in every 20 tests for mean inoculum levels between 0.5 and 3 viable cells/test at the 10^{-z} dilution. When such skips occur, the analyst needs to decide whether the observed result could have occurred merely by chance (ie, $P > 0.05$), in which case it should be ignored, or whether its occurrence should be accepted. Whichever the decision, the occurrence of a 'skip' should always be recorded.

Only an approximate indication of the initial contamination level can be derived from a single-test assay, as illustrated in Example 8.1, and the occurrence of a 'skip' can affect the confidence that can be placed in the result. It is doubtful whether it is worthwhile to derive confidence intervals (CIs) for probable contamination levels determined by single-tube dilution tests.

It should be noted that the calculation of the mean and CIs assumes a Poisson distribution (which may not be correct) and ignores any effects of dilution and other technical errors. In the single-tube dilution tests, as in multiple-tube series, the CIs for the results can be improved by reducing the dilution ratio (ie, a 1 in 2 dilution is better than 1 in 4, which is better than 1 in 10), since the response curve obtained would be smoother with the lower dilution factor.

Table 8.2 95% Probability of Occurrence of 'Skips' in a Single-Tube Dilution Series for Various Inocula Levels

	Probability[b] for			
	Negative Test $P_{(x=0)}$	Positive Test $P_{(x>0)}$	Negative Test $P_{(x=0)}$	Positive Test $P_{(x>0)}$
Mean Inoculum[a] Level in Dilution 10^{-z} (m)	At Dilution 10^{-z}		At Dilution $10^{-(z+1)}$	
0.05	0.95	0.05	>0.99	<0.01
0.1	0.90	0.10	0.99	0.01
0.2	0.82	0.18	0.98	0.02
0.3	0.74	0.26	0.97	0.03
0.4	0.67	0.33	0.96	0.04
0.5	0.61	0.39	0.95	0.05
0.6	0.55	0.45	0.94	0.06
0.7	0.50	0.50	0.93	0.07
0.8	0.45	0.55	0.92	0.08
0.9	0.41	0.59	0.91	0.09
1.0	0.37	0.63	0.90	0.10
2.0	0.14	0.86	0.82	0.18
3.0	0.05	0.95	0.74	0.26
4.0	0.02	0.98	0.67	0.33
5.0	<0.01	>0.99	061	0.39
6.0	<0.01	>0.99	0.55	0.45
7.0	<0.01	>0.99	0.50	0.50
8.0	<0.01	>0.99	0.45	0.55
9.0	≪0.01	≫0.99	0.41	0.59

The shaded area shows the cell concentrations where a positive result at dilution 10^{-z} might be followed by a positive result at the next higher dilution $10^{-(z+1)}$. The unshaded area indicates that it is less likely that a negative result at dilution 10^{-z} would be followed by a positive result at dilution $10^{-(z+1)}$.
[a] 10^{-z}, dilution level, for example, 10^{-1}; $10^{-(z+1)}$, next dilution level, for example, 10^{-2}.
[b] Based on Poisson distribution and shown to two significant places.

EXAMPLE 8.1 CALCULATION OF INITIAL CONTAMINATION LEVEL FROM A SINGLE-TEST DILUTION SERIES

Suppose that we have a suspension of bacteria that we expect to contain between 10^3 and 10^4 viable cells/mL. To check the concentration we prepare a 10-fold serial dilution to 10^{-6} and inoculate 1 mL of each dilution into a suitable broth medium (sample A). A second sample (B) is also tested in parallel. The following results are obtained:

	Result[a] at Dilution 10^x, Where x =				
	−2	−3	−4	−5	−6
Sample A	+	+	−	−	−
Sample B	+	+	−	+	−

[a] +, growth; −, no growth.

What is the probability that the result on sample B is spurious?

Sample A: To get growth each 1 mL of diluted culture must contain at least one viable cell. The positive result at the −3 dilution shows that the initial viable cell density was at least 1×10^3 viable organisms/mL of sample and the negative result at the −4 dilution that the cell density was $<1 \times 10^4$/mL. The statistical probability of no viable cells in an inoculum is given by $P_{(x=0)} = e^{-m}$, where m is the average cell density. So for $m = 1$ cell in the 10^{-3} dilution, the probability for 0 cell in 1 mL of inoculum is $P_{(x=0)} = e^{-1} = 0.37$. Hence, the chance for no viable cell is about a one in three and so, conversely, there are about two in three chances of at least one viable cell in 1 mL of the −3 dilution, since $P_{(x\geq 1)} = [1 - P_{(x=0)}] = 0.63$. Similarly, the chance of obtaining a negative result at the −4 dilution is given by $P_{(x=0)} = e^{-0.1} = 0.905$, so the probability of a positive result ($P_{x>0} = 0.095 \approx 0.1$) is only about 1 in 10.

However, if the original cell density had been, say, 9×10^3 cells/mL, the probability of growth at the −3 dilution would be $P > 0.99$ and at the −4 dilution $P = 0.59$. So it would be possible to get growth at both the −3 and −4 dilutions in these circumstances. But the negative result obtained at the −4 dilution suggests that this scenario is not likely.

The results obtained on sample B show a probability of $P < 0.01$ for growth at the −5 dilution with an expected cell density of 0.01 cell/mL; even if the original suspension contained 9×10^3 cells/mL the cell density in the −5 dilution would be only 0.09 cells/mL, so $P_{(x=0)} = 0.91$ and $P_{(x\geq 1)} = 0.09$. So we can conclude that the chances of a positive result at the −5 dilution level are extremely unlikely, given the negative result at the −4 dilution.

Table 8.2 shows that the 95% probability for a negative result $[P_{(x=0)}]$ at dilution 10^{-z} would be associated with a mean inoculum level of 0.05 (or fewer) organisms in 1 mL of that dilution and the probability for a positive result to occur at the $10^{-(z+1)}$ dilution is <0.01 for that level of viable cells per mL inoculum. The occurrence of 'skips' of this type is not infrequent in a simple dilution assay – the problem is to know whether to ignore the result! A more extensive explanation and method of statistical analysis is presented by Myers et al. (1994), who used a binomial approach to come to the same conclusion.

MULTIPLE TESTS AT A SINGLE DILUTION LEVEL

If more than one test is inoculated at a given dilution level, the growth response will range from all positive (at a high inoculum level) to all negative (with a very low inoculum level), with a graded response between in which some tubes remain sterile whilst others exhibit growth (see last columns, Table 8.1). For a single dilution level, the probability that no viable organisms will occur in a sample of volume V when the true density = m cells/unit volume is given by $P_{(x=0)} = e^{-Vm}$. If n portions each of volume V are tested and, of these, S samples are sterile, then the proportion of sterile tests is S/n and the estimated probability ($\hat{P}$) will be given by $\hat{P} = S/n = e^{-Vm}$. This can be rearranged to give an estimate ($\hat{m}$) of the mean number of organisms (m) in volume V:

$$\hat{m} = -\left(\frac{1}{V}\right)\ln\left(\frac{S}{n}\right) = -\left(\frac{2.303}{V}\right)\log\left(\frac{S}{n}\right)$$

where ln and log are logarithms to bases e and 10, respectively. The value $\hat{m}$ is the 'probable' estimated number of organisms per millilitre.

At a probability of P_0 that a sample is 'sterile', the probability that S of n samples are sterile is given by the binomial expansion of $\left(\frac{n!}{S!(n-S)!}\right)P_0^S(1-P_0)^{n-S}$, but since

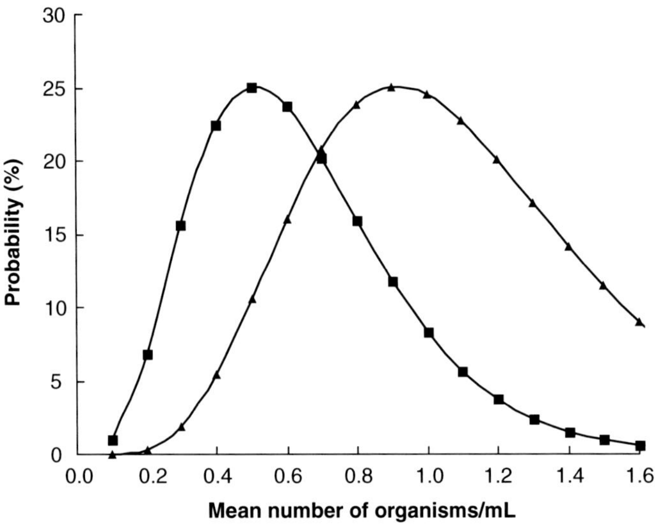

FIGURE 8.2

Probability distribution curves for two multiple-tube dilution tests at a single dilution level, with $n = 10$ replicates, and 6 (■) or 4 (▲) sterile tests, derived using the following equation: $\{n!/[S!(n-S)!]\}e^{-VmS}(1-e^{-Vm})^{n-S}$. The maximum likelihood values are ca. 0.5 organisms/mL for 6 sterile tests, and ca. 0.9 organisms/mL for 4 sterile tests

$P_0 = e^{-Vm}$, the expression may be rewritten as follows: $\left(\frac{n!}{S!(n-S)!}\right)e^{-VmS}\left(1-e^{-VmS}\right)^{n-S}$ (Cochran, 1950).

The terms derived from the expansion of this expression permit the construction of a curve showing the probability (P) of occurrence against the true density of organisms (m). The value of $\hat{m}$ is the Poisson estimate of the mean that corresponds to the highest probability value and provides an estimate of the probable number. Fig. 8.2 presents curves for two values of S/n, assuming $V = 1$ mL. If $S/n = 0.6$, then the probable number of organisms inoculated ($\hat{m}$) = 0.51 cells/mL but if $S/n = 0.4$, $\hat{m} = 0.92$ cells/mL. These values are the same as those that would be derived using the equation $\hat{m} = -(1/V)\cdot\ln(S/n)$. However, from the shapes of the curves it is clear that many other densities of organisms could also give the observed ratios of S/n (ie, 0.4 and 0.6), though at a lower probability.

MULTIPLE TEST DILUTION SERIES

MULTIPLE TESTS AT SEVERAL DILUTION LEVELS

The precision of multiple dilution tests is very poor when the number of organisms inoculated is likely to give either all positive or all negative results. If all replicate tests are positive, the maximum probability for the estimated density is infinity and if all results are negative, the estimated density is zero. Thus, a test using replicates at a single dilution is of value only if the number of organisms inoculated (mV) is chosen to give a graduated response with some tests positive and others negative.

When more than one dilution level is used, an estimate of population density ($\hat{m}$) can be derived for each dilution level. If a series of multiple tests (n_i) is set up at several serial dilutions (i), each with an inoculum volume V_i, and S_i sterile tests occur, then the estimated density of organisms for the ith dilution is given by

$$\hat{m}_i = -\left(\frac{2.303}{V_i}\right)\log\left(\frac{S_i}{n_i}\right)$$

However, the best estimate at each dilution will vary in precision, so it is wrong to take the arithmetic mean of the estimates for each dilution level in order to derive an overall estimate. Two primary methods exist for deriving the best estimate.

STEVENS' METHOD FOR MULTIPLE DILUTION LEVELS

Fisher and Yates (1963) published tables based on the mean fertile level (x) and the mean sterile level (y), where x is the number of positive cultures at all dilution levels divided by the number of cultures (n) at each level and y is the number of levels (i) minus x. Using the same notation as before (ie, S_i is the number of negative results out of n_i tests at the ith dilution level), we have

$$x = \frac{\sum (n_1 - S_1) + (n_2 - S_2) + \cdots + (n_i - S_i)}{n}$$

and

$$y = i - x$$

Using Table 8.3 (from Fisher and Yates, 1963) we determine the value of a factor K for values of x or y according to the number of dilution levels (i) and the dilution ratio α. The estimated number of organisms is then given by $\log_{10}\lambda = x(\log_{10}\alpha) - K$, where λ is the estimated number of organisms ($\hat{m}$). For 10-fold dilutions this simplifies to $\log_{10}\lambda = x - K$. The variance of the estimate of log λ is given by $s^2 = (1/n)(\log_{10} 2)(\log_{10}\alpha)$. For 10-fold dilutions: $s^2_{\log\lambda} = \frac{1}{n}(\log_{10} 2)(\log_{10} 10) = 0.3010/n$. This method is applicable to 2-, 4- and 10-fold dilution ratios with not less than 4, 4 or 3 dilution levels, respectively. Unfortunately, the method does not discriminate in any way between valid and invalid series of results (see following text).

THE MOST PROBABLE NUMBER METHOD FOR MULTIPLE DILUTION LEVELS

This is an extension of the concept used earlier for multiple tubes at a single level. Suppose that for i dilution levels the proportions of sterile (S_i) cultures are given by

$$\frac{S_1}{n_1}, \frac{S_2}{n_2}, \frac{S_3}{n_3}, \ldots, \frac{S_i}{n_i}$$

Table 8.3 Determination of *K* Values for the Estimation of the Density of Organisms by the Dilution Method

	Value of *K* for				
	4-Fold Dilutions at Levels			10-Fold Dilutions[b] at 3 or More Levels	
Value, *x*[a]	**4**	**5**	**$\geq$6**	**$x \leq 1$**	**$x > 1$**
0.0					0.763
0.1					0.768
0.2					0.768
0.3					0.760
0.4	0.704	0.706	0.707	0.761	0.747
0.5				0.740	0.736
0.6	0.615	0.617	0.618	0.733	0.733
0.7				0.736	0.736
0.8	0.573	0.576	0.577	0.744	0.749
0.9				0.753	0.753
1.0	0.555	0.558	0.559	0.763	0.763
1.5	0.545	0.551	0.553		
2.0	0.537	0.548	0.551		
2.5		0.545	0.552		

[a]*x is the number of positive tests.*
[b]*When x > 1, enter table with decimal part of x only.*
Reproduced with permission of Pearson Education Ltd from Fisher and Yates (1963, Table VIII.2), Statistical Tables for Biological Agricultural and Medical Research, *published by Oliver & Boyd, Edinburgh.*

The probability that these events should all happen at once can be derived from the product of the probabilities that each individual event would occur. As before, a graph of probability against m shows a single maximum likelihood value that is referred to as the most probable number (MPN, $\hat{m}$). As pointed out by Cochran (1950), the value of $\hat{m}$ cannot be written down explicitly, but can be derived from the following equation:

$$S_1V_1 + S_2V_2 + \cdots + S_iV_i = \frac{(n_1 - S_1)V_1\,e^{-V_1m}}{1 - e^{-V_1m}} + \frac{(n_2 - S_2)V_2\,e^{-V_2m}}{1 - e^{-V_2m}} + \cdots + \frac{(n_i - S_i)V_i\,e^{-V_im}}{1 - e^{-V_im}}$$

where V_i is the volume of culture inoculated, S_i is the number of sterile tubes out of n_i tests prepared at dilution i and m is the MPN.

Methods for solving this equation by iteration have been given by Halvorson and Ziegler (1933), Finney (1947) and Hurley and Roscoe (1983). De Man (1975, 1983) derived composite probability distribution curves for various combinations of results, dilution levels and culture numbers, using computer techniques. Hurley and Roscoe (1983) provided a simple 'BASIC' language computer programme to calculate MPN values and the standard error of the $\log_{10}$ MPN. More recently, Jarvis et al. (2010) used a 'maximum likelihood' approach to determine MPNs for any combination of tube numbers and dilutions (see below).

For routine purposes, standard tables can be used to derive from test responses the MPN of organisms in a sample provided that the numbers of dilutions and the number of replicate test at each dilution conform to a standard format. Some MPN tables do not provide data on the CIs that can be applied to the MPN values and, in some instances, include many unacceptable and improbable results (ie, nonsense combinations; De Man, 1975). However, reliable MPN tables based on De Man (1975) are given, for instance, by ICMSF (1978), Blodgett (2002), Anon (2005c) and AOAC (2012). The approach of Jarvis et al. (2010) has been recommended for all ISO standards; the spreadsheet can be downloaded from http://standards.iso.org/iso/7218/.

Woodward (1957), De Man (1975) and Hurley and Roscoe (1983) all pointed out that the occurrence of improbable results indicates a malfunction in the performance of the MPN test. As an extreme example, De Man (1975) cites the case of a result 0, 0 and 10 positive results from a 10-tube, 3-dilution series (10^0 to 10^{-2}). The MPN for such a result can be calculated as 0.9 organisms/mL with 95% CIs of 0.5–1.7 but the probability of obtaining such a result is $<10^{-24}$ for a mean inoculum level of 0.9 organisms/mL. Furthermore, it should be noted that the method of Stevens would give the same estimate of the number of organisms [$\lambda = 1.73$; 95% confidence limit (CL) = 0.78–3.85] for this combination (0–0–10) as for the combinations 0–10–0 and 10–0–0, even though only the last example would be acceptable, giving an MPN of 2.3 organisms/mL with a 95% CI of 1.2–5.8 organisms/mL (De Man, 1975).

In order to assess whether an improbable result has been obtained, the χ^2 test can be used to assess the goodness of fit of observed and expected results. Moran (1954a,b, 1958) proposed a more severe test but it is applicable only under defined circumstances Example 8.2).

Cochran (1950) pointed out that in planning an MPN test it is necessary to decide on: (1) the range of volumes to be tested; (2) the dilution factor to be used; and (3) the number of tests to be inoculated at each dilution level. The aim of the test is to obtain equal relative precision across a number of possible levels of cell density, that is, the ratio of the standard error to the true density should be at least 1. When the density of organisms is expected to be reasonably constant (eg, in analysis of a water supply of known quality), it is possible to select inoculum levels such that the lowest dilution volume (ie, largest volume of inoculum) should contain at least one viable organism and that the highest dilution (ie, smallest volume) should contain not more than two viable organisms. However, when the upper density of organisms nears the limit of the test for the chosen dilutions, the probability of obtaining all positive cultures is greater for a small number of replicates at each dilution level than for a large number of replicates. Hence, it is safer to use the rule that the maximum number of organisms should

EXAMPLE 8.2 USE OF MORAN'S TEST FOR THE VALIDITY OF MULTIPLE-TUBE DILUTION COUNTS

Assume the following results have been obtained in multiple-tube dilution assays on 2 samples, using 10 tubes at each dilution level and a 10-fold dilution ratio:

	No Tests Positive at Dilution 10^z, Where $z =$				
	−1	−2	−3	−4	−5
Sample A	10	10	8	4	0
Sample B	10	9	7	5	0

Is each of these sets of results a valid series for calculation of a most probable number?

Moran's (1954a,b, 1958) test for the validity of results from a series of tests requires the calculation of a value $T_{obs} = \sum f_m(n_m - f_m)$, where f_m is the number of tests showing growth at each dilution tested and N is the number of tests at each dilution. The equation can be expanded as follows:

$$T_{obs} = \sum[f_{m1}(n_1 - f_{m1}) + f_{m2}(n_2 - f_{m2}) + \cdots + f_{mn}(n_{mn} - f_{mn})]$$

For sample A:

$$T_{obs} = 10(10-10) + 10(10-10) + 8(10-8) + 4(10-4) + 0(10-0)$$
$$= 0 + 0 + 16 + 24 + 0 = 40$$

Entering Table 8.4 for $n = 10$ and with a 10-fold dilution factor, we get

$$T_{exp} = 27.09 \quad \text{and} \quad \text{SE}_T = 9.34$$

A value M is calculated as the difference between the observed value for T_{obs} and the expected value T_{exp} divided by the SE_T. For sample A:

$$M_A = \frac{T_{obs} - T_{exp}}{\text{SE}_T} = \frac{40 - 27.09}{9.34} = 1.38$$

Since $M_A < 1.645$ (the critical value for the test), the observed results for sample A do not differ significantly ($P > 0.05$) from those expected and the series is valid for an MPN calculation.

For sample B:

$$T_{obs} = 10(10-10) + 9(10-9) + 7(10-7) + 5(10-5) + 0(10-0)$$
$$= 0 + 9 + 21 + 25 + 0 = 54$$

Again, from Table 8.4 for 10 tests at 10-fold dilutions, $T_{exp} = 27.09$ and $\text{SE}_T = 9.34$, so

$$M_B = \frac{54 - 27.09}{9.34} = 2.88$$

Since $M_B > 2.326$, the observed number of positive tubes differs significantly ($P < 0.01$) from the expected number and therefore the results on sample B are improbable and should not be used to calculate a MPN.

Note that these conclusions were checked using the spreadsheet procedure of Jarvis et al. (2010). The results for sample A have a 'rarity index' of 0.16 and are therefore category 1 whilst those for sample B have a rarity index of <0.0001 and are category 3.

Table 8.4 Expected Values of T (T_{exp}) and Standard Error of T (SE_T) for Moran's Test

	T_{exp} at Dilution Ratio			SE_T at Dilution Ratio		
n	**2**	**4**	**10**	**2**	**4**	**10**
5	20	10	6.02	5.69	4.02	3.12
6	30	15	9.03	7.63	5.40	4.19
7	42	21	12.64	9.75	6.89	5.35
8	56	28	16.86	12.02	8.50	6.60
9	72	36	21.67	14.46	10.22	7.93
10	90	45	27.09	17.03	12.04	9.34
11	110	55	33.11	19.74	13.96	10.83
12	132	66	39.74	22.58	15.97	12.39
13	156	78	46.96	25.55	18.07	14.02
14	182	91	54.79	28.63	20.24	15.71
15	210	105	63.22	31.83	22.51	17.74
16	240	120	72.25	35.14	24.85	19.28
17	272	136	81.88	38.56	27.27	21.16
18	306	153	92.12	42.07	29.75	23.08
19	342	171	102.95	45.70	32.31	25.08
20	380	190	114.39	49.41	34.94	27.11
30	870	435	261.90	91.53	64.72	50.22
40	1560	780	469.61	141.49	100.05	77.63

n, *number of tests inoculated at each dilution.*
Modified from Meynell and Meynell (1970).

not exceed one viable cell per volume of the highest dilution, by estimating the upper and lower limits ($\hat{m}_H$ and $\hat{m}_L$) between which the true density would be expected to lie. Then the largest volume (V_H; ie, the volume at lowest dilution level) and the smallest volume (V_L; the volume at highest dilution level) can be derived from: $V_H \geq 1/\hat{m}_L$ and $V_L \geq 1/\hat{m}_H$.

For example, if the probable true density (m) is expected to lie between 10 and 800 organisms/mL, the largest volume to be inoculated per test should not exceed 1/10 mL (1 mL of a 10^{-1} dilution) and the smallest volume should not exceed 1/800 mL (1.25 mL of a 10^{-3} dilution). In practice, this range could be covered by testing 1 mL of each of three 10-fold dilutions (ie, 1/10, 1/100 and 1/1000) or by testing 1 mL of, say, one 10-fold and four 4-fold dilutions (ie, 1/10, 1/40, 1/160, 1/640, 1/2560).

The choice of dilution ratio affects the test precision significantly over the range of dilution ratios between 2 and 10 only if the total number of tests is not kept constant. However, the precision is more constant throughout the range of

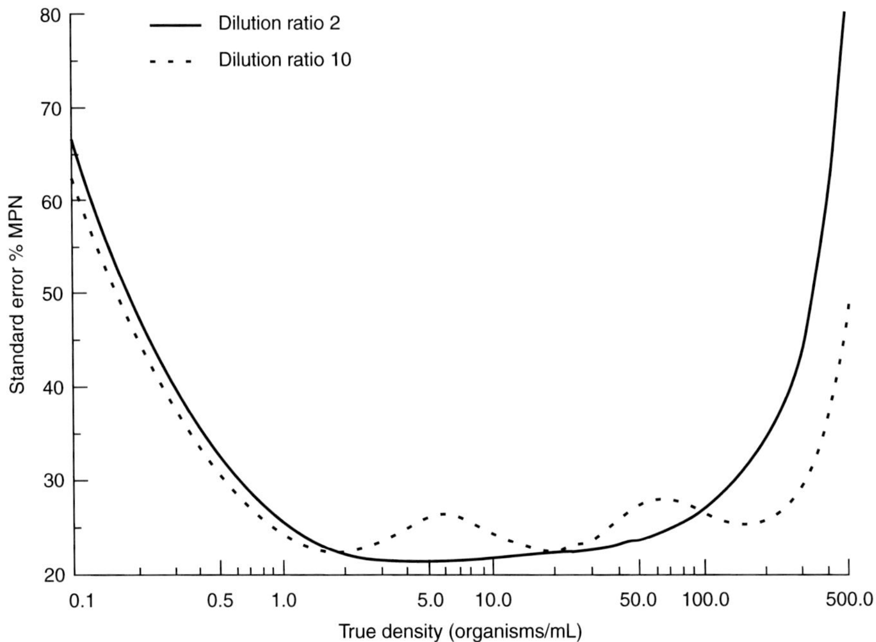

FIGURE 8.3 Changes in the Standard Error of MPN Values for Dilution Ratios 2 and 10

Reproduced from Cochran (1950), by permission of the International Biometric Society

densities tested (ie, $1/V_H$ to $1/V_L$) if a lower dilution ratio is used. This is illustrated in Fig. 8.3 for dilution ratios of 2 and 10. The standard error of the MPN can be derived (Cochran, 1950) from the following equation (assuming a cell density between $1/V_H$ and $1/V_L$, a defined number of tests at each dilution and a dilution ratio of 5 or less):

$$\mathrm{SE}_{\log \hat{m}} = 0.55\sqrt{\frac{\log \alpha}{n}}$$

where α is the dilution ratio and n is the number of tests at each dilution. For 10-fold dilutions, where the standard error may peak at the value for $\hat{m}$, a more conservative estimate is preferred:

$$\mathrm{SE}_{\log \hat{m}} = 0.58\sqrt{\frac{\log \alpha}{n}} = 0.58\sqrt{\frac{1}{n}}$$

Values for various levels of α and n, together with factors for deriving standard errors and 95% CIs, are given in Table 8.5 and in the paper by Hurley and Roscoe (1983).

Table 8.5 Standard Error of log$\hat{m}$ and Factor for Confidence Intervals

No. of Samples Per Dilution, *n*	$SE_{(\log_{10}\hat{m})}$ for Dilution Ratio (α)				Factor[a] for 95% Confidence Intervals With Dilution Ratio (α)			
	2	4	5	10	2	4	5	10
1	0.301	0.427	0.460	0.580	4.00	7.14	8.32	14.45
2	0.213	0.302	0.325	0.410	2.67	4.00	4.47	6.61
3	0.174	0.246	0.265	0.335	2.23	3.10	3.39	4.68
4	0.150	0.214	0.230	0.290	2.00	2.68	2.88	3.80
5	0.135	0.191	0.206	0.259	1.86	2.41	2.58	3.30
6	0.123	0.174	0.188	0.237	1.76	2.23	2.38	2.98
7	0.114	0.161	0.174	0.219	1.69	2.10	2.23	2.74
8	0.107	0.151	0.163	0.205	1.64	2.00	2.12	2.57
9	0.100	0.142	0.153	0.193	1.58	1.92	2.02	2.43
10	0.095	0.135	0.145	0.183	1.55	1.86	1.95	2.32

[a]*In deriving the 95% CIs the MPN should be multiplied or divided by the appropriate factor. For example, if the MPN for a 5-test dilution series with a 10-fold dilution ratio (α) is 22 organisms/mL, the 95% CIs are given by 22/3.30 and 22 × 3.30, that is, 6.7 and 72.5.*

Reproduced from Cochran (1950), by permission of the International Biometric Society

The standard error increases (precision decreases) when the likelihood of all positive or all negative results increases. The precision is greatest when approximately equal numbers of positive and negative results are most likely to occur. The optimum mean inoculum level (about 1.6 organisms/culture) is more likely to be missed by chance in a 10-fold than in a 2-fold dilution series; hence, the graph for 10-fold dilutions fluctuates more than that for 2-fold dilutions (Fig. 8.3).

These CLs assume that the logarithm of the MPN is Normally distributed and differ from those cited by De Man (1975), who derived CLs from the computer-printed histograms of the distribution functions for different combinations of test numbers and positive results. However, De Man's tables (Tables 8.6 and 8.7) cover only 10-fold dilution ratios; the data in Table 8.5 provide a way of assigning CIs for MPNs derived from other dilution series, as does the computer programme given by Hurley and Roscoe (1983).

More recently, Jarvis et al. (2010) revisited the concept of determination of MPN values and used a 'best likelihood approach' to determine the estimate of the MPN. The primary objective was to ensure the availability of both printed tables (see, eg, Tables 8.6 and 8.7) and a simple computer spreadsheet method to determine MPN values. The latter is particularly appropriate for those who wish to use unusual dilution levels and/or replicate test numbers since standard MPN tables for such combinations are not available. A secondary objective was to provide an estimate of the standard deviation and the 95% CIs for calculated MPN values and to introduce the concept of 'rarity values', based on original work by Blodgett (2002). This provides a simpler and more credible approach to understanding that some data combinations are highly improbable [cf De Man's (1975) nonsense combinations]. The 'rarity value' (r) lies within the range 0–1: a category 1 result (highly probable) is obtained if $0.05 < r < 1.0$; a category 2 result (low probability) is given by $0.01 < r < 0.05$; and a category 3 result (unacceptably low probability) by $0 < r < 0.01$.

Table 8.6 MPN Table for 3 × 1, 3 × 0.1 and 3 × 0.01 g (or mL) Inoculum

Number of Positive Results for Inoculum Volume (mL or g)			MPN (/mL or /g)	$\log_{10}$ MPN	SD of $\log_{10}$ MPN	95% Confidence Intervals		Rarity Index	Category
1.00	0.10	0.01				Lower	Upper		
0	0	0	0	NA*	NA	0	1.1	1.000	1
0	1	0	0.30	−0.52	0.43	0.041	2.3	0.087	1
1	0	0	0.36	−0.45	0.44	0.048	2.7	1.000	1
1	0	1	0.72	−0.14	0.31	0.17	3.0	0.021	2
1	1	0	0.74	−0.13	0.31	0.18	3.1	0.211	1
1	2	0	1.1	0.056	0.26	0.35	3.7	0.021	2
2	0	0	0.92	−0.037	0.32	0.21	4.0	1.000	1
2	0	1	1.4	0.16	0.26	0.42	4.8	0.041	2
2	1	0	1.5	0.17	0.27	0.43	5.0	0.426	1
2	1	1	2.0	0.31	0.23	0.69	6.0	0.019	2
2	2	0	2.1	0.32	0.24	0.71	6.2	0.069	1
3	0	0	2.3	0.36	0.31	0.55	9.7	1.000	1
3	0	1	3.8	0.59	0.31	0.93	16	0.084	1
3	1	0	4.3	0.63	0.33	0.95	19	1.000	1
3	1	1	7.5	0.87	0.30	1.9	30	0.209	1
3	1	2	12	1.1	0.26	3.6	37	0.021	2
3	2	0	9.3	0.97	0.32	2.2	40	1.000	1
3	2	1	15	1.2	0.27	4.4	51	0.420	1
3	2	2	21	1.3	0.24	7.2	64	0.068	1
3	3	0	24	1.4	0.32	5.6	100	1.000	1
3	3	1	46	1.7	0.34	9.6	220	1.000	1
3	3	2	110	2.0	0.32	25	480	1.000	1
3	3	3	∞	NA	NA	36	∞	1.000	1

For interpretation see footnotes to Table 8.7.
NA is not applicable.
Reproduced from Jarvis et al (2010) by permission of the authors and John Wiley & Sons, publishers

Table 8.7 MPN Table for 5 × 1, 5 × 0.1 and 5 × 0.01 g (or mL) Inoculum

Number of Positive Results for			MPN (/mL or /g)	log_{10} MPN	SD of log_{10} MPN	95% Confidence Intervals		Rarity Index	Rarity Category
1.00	0.10	0.01				Lower	Upper		
0	0	0	0			0	0.66	1.000	1
0	1	0	0.18	−0.74	0.43	0.03	1.3	0.092	1
1	0	0	0.20	−0.70	0.44	0.03	1.5	1.000	1
1	0	1	0.40	−0.40	0.31	0.10	1.6	0.020	2
1	1	0	0.40	−0.39	0.31	0.10	1.7	0.205	1
1	2	0	0.61	−0.21	0.25	0.19	2.0	0.024	2
2	0	0	0.45	−0.35	0.31	0.11	1.9	1.000	1
2	0	1	0.68	−0.17	0.25	0.21	2.2	0.034	2
2	1	0	0.68	−0.16	0.25	0.21	2.2	0.354	1
2	1	1	092	−0.036	0.22	0.33	2.5	0.015	2
2	2	0	0.93	−0.031	0.22	0.34	2.6	0.062	1
3	0	0	0.78	−0.11	0.26	0.24	2.5	1.000	1
3	0	1	1.1	0.024	0.22	0.38	3.0	0.053	1
3	1	0	1.1	0.030	0.22	0.38	3.0	0.565	1
3	1	1	1.4	0.14	0.20	0.54	3.5	0.034	2
3	2	0	1.4	0.14	0.20	0.54	3.5	0.147	1
3	2	1	1.7	0.23	0.19	0.72	4.0	0.013	2
3	3	0	1.7	0.24	0.19	0.73	4.1	0.029	2
4	0	0	1.3	0.11	0.23	0.44	3.7	1.000	1
4	0	1	1.7	0.22	0.21	0.62	4.4	0.084	1
4	1	0	1.7	0.23	0.21	0.63	4.5	0.920	1
4	1	1	2.1	0.33	0.20	0.85	5.3	0.073	1
4	2	0	2.2	0.33	0.20	0.86	5.4	0.314	1
4	2	1	2.6	0.42	0.19	1.1	6.3	0.031	2
4	3	0	2.7	0.43	0.19	1.1	6.5	0.069	1
4	4	0	3.4	0.53	0.18	1.4	7.8	0.011	2
5	0	0	2.3	0.36	0.24	0.76	7.0	0.769	1
5	0	1	3.1	0.50	0.24	1.0	9.4	0.086	1
5	1	0	3.3	0.52	0.24	1.1	10	1.000	1
5	1	1	4.6	0.66	0.25	1.5	14	0.202	1
5	1	2	6.3	0.80	0.24	2.1	19	0.024	2
5	2	0	4.9	0.69	0.26	1.5	16	1.000	1
5	2	1	7.0	0.84	0.24	2.3	22	0.358	1
5	2	2	9.4	0.97	0.22	3.4	26	0.062	1

(Continued)

Table 8.7 MPN Table for 5 × 1, 5 × 0.1 and 5 × 0.01 g (or mL) Inoculum (*cont.*)

Number of Positive Results for			MPN (/mL or /g)	$\log_{10}$ MPN	SD of $\log_{10}$ MPN	95% Confidence Intervals		Rarity Index	Rarity Category
1.00	0.10	0.01				Lower	Upper		
5	3	0	7.9	0.90	0.25	2.5	25	1.000	1
5	3	1	11	1.0	0.22	3.9	31	0.574	1
5	3	2	14	1.1	0.20	5.5	36	0.148	1
5	3	3	17	1.2	0.19	7.4	42	0.029	2
5	4	0	13	1.1	0.23	4.4	38	1.000	1
5	4	1	17	1.2	0.21	6.4	46	0.941	1
5	4	2	22	1.3	0.20	8.8	56	0.304	1
5	4	3	28	1.4	0.19	12	67	0.068	1
5	4	4	35	1.5	0.18	15	81	0.011	2
5	5	0	24	1.4	0.24	7.8	74	0.738	1
5	5	1	35	1.5	0.25	11	110	1.000	1
5	5	2	54	1.7	0.27	16	190	1.000	1
5	5	3	92	2.0	0.26	28	300	1.000	1
5	5	4	160	2.2	0.24	53	490	1.000	1
5	5	5	∞			65	∞	1.000	1

Columns 1–3, the number of positive tests from three replicates at each of the dilutions tested.
Column 4, the most probable number (MPN) per millilitre or gram of the first dilution tested.[a]
Column 5, the $\log_{10}$ of the MPN.
Column 6, the standard deviation of the derived MPN.
Columns 7 and 8, the lower and upper bounds of the 95% CI for the MPN – these relate to the MPN derived for the first dilution level tested.
Columns 9 and 10, the rarity index indicates the statistical probability for a result. A category value of 1 indicates that the result is highly probable ($P > 0.05$); a category 2 result is less likely ($0.05 > P > 0.01$) and should be treated with caution; a category 3 result is highly unlikely ($P < 0.01$) and is not acceptable. In these tables all category 3 results have been deleted but will be found in the output of the software programme (http://standards.iso.org/iso/7218/; accessed 07.07.15).
[a]*Suppose for a 5–5–5-tube series that each of the first 5 tubes was inoculated with 1 mL of a 10^{-2} dilution, each of the second five tubes with 1 mL of 10^{-3} dilution and each of the third set with 1 mL of 10^{-4} dilution, and a test result of 2–1–0 was obtained. The MPN (0.68) derived from the table relates to the inoculum of 1 mL of the 10^{-2} dilution and must be multiplied by 100 to get the MPN per millilitre or gram of original sample. Hence, the MPN for the original sample would be $0.68 \times 100 = 68\ mL^{-1}$ or g^{-1}. The recorded $\log_{10}$ MPN of −0.16 relates to the derived value of 0.68; the $\log_{10}$ MPN for the original sample is as follows: 2.0–0.16 = 1.84. The lower and upper bounds of the 95% CI for the derived MPN of 0.68 are 0.21 and 2.2, respectively, so for an original sample the CIs around the MPN range from 21 to 220.*
Based on Jarvis et al. (2010).

DIFFERENCES BETWEEN MPN VALUES

From time to time it may be necessary to establish the significance of a difference between two MPN estimations. This can be determined using the combined standard errors of the two MPNs and standard statistical tables of Student's t distribution:

$$t = \frac{\log\hat{m}_1 - \log\hat{m}_2}{0.58\sqrt{[(\log\alpha_1)/n_1)] + [(\log\alpha_2)/n_2]}}$$

where $\hat{m}_1$ and $\hat{m}_2$ are the MPN values for two series of tests (1) and (2), with dilution ratios α_1 and α_2, respectively, and n_1 and n_2 are the numbers of replicate tests for each series. If the dilution ratio (α) is 10-fold, the equation simplifies to

$$t = \frac{\log\hat{m}_1 - \log\hat{m}_2}{0.58\sqrt{(1/n_1) + (1/n_2)}}$$

The use of the method is illustrated in Example 8.3. However, the availability of standard deviations, CI limits and estimates of validity simplify the procedures since these are calculated automatically on the spreadsheet of Jarvis et al. (2010).

EXAMPLE 8.3 CALCULATION OF THE MOST PROBABLE NUMBERS FOR TWO DILUTION SERIES AND THE SIGNIFICANCE OF THE DIFFERENCES BETWEEN THE MPN ESTIMATES

In the first test (A), 5 tubes were each inoculated with 1 mL of a series of 10-fold dilutions and, in the second test (B), 3 tubes were each inoculated with 1 mL of a series of 10-fold dilutions. The results were as follows:

		Number of Positive Tests at Dilution 10^z, Where z =			
Test	**No. of Tests at Each Dilution**	**−1**	**−2**	**−3**	**−4**
A	5	5	2	1	0
B	3	3	3	0	0

What are the MPN values for these samples and are the estimates statistically different?

From Table 8.7, using the sequence 5–2–1 for sample A, the MPN for the 10^{-1} dilution = 7.0, so the MPN for the undiluted sample is 7.0 × 10 = 70 organisms/mL, with 95% confidence limits of 23 and 220 organisms/mL. Using Table 8.6 for sample B with the sequence 3–3–0, the MPN = 24 × 10 = 240 organisms/mL of undiluted sample, with 95% CL of 56 and 1000 organisms/mL. Both are category 1 results. The two test series provide MPNs of 70 and 240 organisms/mL but the 95% CLs of the MPNs overlap. Is the difference in MPN estimates significant?

We define two hypotheses: H_0 is the null hypothesis is that the results do not differ and H_1 is that the results do differ. We use Cochran's (1950) test to assess the significance of the difference between the results estimating Student's t value, with 6 degrees of freedom ($v = 5 + 3 - 2 = 6$):

$$t = \frac{\log\hat{m}_A - \log\hat{m}_B}{0.58\sqrt{[(\log\alpha_A)/n_A] + [(\log\alpha_B)/n_B]}}$$

where $\hat{m}_A$ and $\hat{m}_B$ are the MPN values given by tests A (70) and B (240), respectively; the dilution ratio = $\alpha_A = \alpha_B = 10$; n_A and n_B are the number of tubes at each dilution level, respectively (for test A, $n_A = 5$ and for test B, $n_B = 3$). Then:

$$t = \frac{\log 240 - \log 70}{0.58\sqrt{[(\log 10)/5] + [(\log 10)/3]}} = \frac{2.3802 - 1.8451}{0.58\sqrt{0.2 + 0.33}} = \frac{0.5351}{0.4223} = 1.2665$$

We now refer to the percentage points of the t distribution (eg, Table 12 in Pearson and Hartley, 1976), and note that for 6 degrees of freedom, the observed value for t (1.2665) is much lower than the tabulated two-tailed value (2.447) for $P = 0.05$. So the observed difference in MPN levels would be expected to occur by chance ($P > 0.05$) and the null hypothesis (H_0) that the two results do not differ significantly is not rejected. This example demonstrates the low level of precision associated with MPN tests.

The two-tailed value of t is used because the alternative hypothesis stated that the two results differ; if the alternative hypothesis had been that the result on sample A was lower than that on sample B, then we would have used a one-tailed t-test, but the outcome would have been the same.

SPECIAL APPLICATIONS OF MULTIPLE-TUBE DILUTION TESTS

For quality control purposes it may be necessary to make a decision to accept or reject a consignment of food based on the result of an MPN test. For such purposes it is possible to construct an operating characteristics (OC) curve for acceptance or rejection at specific confidence levels of tube dilution counts. The best control system would be based on the use of a limit permitting 'acceptance of the "lot" if less than c out of n tests at dilution i show evidence of growth' (ie, if the number of sterile tests at dilution i is $>c_i$).

The shape of the OC curve will be dependent on the values c and n for the dilution level giving the critical density of organisms. On the assumption that log m is Normally distributed when n is large, the steepest part of the OC curve will occur at the 50% acceptance point (approximate value of $c = 0.2n$). Lower percentage acceptance would be associated with $c < 0.2n$ and vice versa. The appropriate choice of c will therefore lie in the range 0–0.5n, assuming that the cumulative binomial distribution is a valid model.

Aspinall and Kilsby (1979) proposed a quality control procedure based on this concept. In their scheme a critical value (c) is chosen such that when: (1) the cell density exceeds c, the probability of rejection is at least 0.95; and (2) if the cell density is less than 0.2c, the probability of acceptance is 0.95.

For a critical value (c) of 10^{z+1} cfu/g, the test sample is diluted 10^{z-1}-fold and 0.5 mL of this suspension is inoculated into each of 2 tubes of culture medium. A further 10-fold dilution of the suspension (ie, a 10^{-z}-fold dilution) is prepared and 0.5 mL is added to each of 7 tubes of media. After incubation, the MPN and CI can be derived from the data in Table 8.8; it should be noted that the CIs cited are wider than those cited in the original paper of Aspinall and Kilsby (1979). If all tubes give a positive response, the material is rejected; if a result is obtained which is not shown in Table 8.8, the test must be repeated. The scheme is illustrated in Example 8.4.

Table 8.8 Estimates of MPN for a Defined Quality Control Multitube Test Assuming an Inoculum of 0.5 mL of Each Relevant Dilution/Tube

Number of Positive Tests at Dilution			95% Confidence Intervals	
10^{z-1}	10^{z}	MPN /g, at $10^{-(z-1)}$	Lower, at $10^{-(z-1)}$	Upper, at $10^{-(z-1)}$
0	0	<0.75 (0)	—	2.7
0	1	0.75	0.01	5.6
0	2	1.50	0.37	6.3
1	0	0.93	0.12	7.0
1	1	1.9	0.46	8.2
1	2	3.0	0.92	10.0
1	3	4.3	1.50	12.0
2	0	2.7	0.59	12.0
2	1	4.6	1.20	17.0
2	2	7.4	2.20	24.0
2	3	11.0	3.70	35.0
2	4	17.0	6.1	47.0
2	5	25.0	9.7	65.0
2	6	39.0	15.0	100.0
2	7	>39 (∞)	18.0	∞

NB: *Results for 0–3 to 0–7, 1–4 to 1–7 are deliberately omitted from this table since such results are improbable (rarity categories 2 or 3).*
Modified from Aspinall and Kilsby (1979), using the spreadsheet calculator of Jarvis et al. (2010) available for download from http://standards.iso.org/iso/7218/.

EXAMPLE 8.4 USE OF MULTIPLE-TUBE MPN FOR QUALITY ACCEPTANCE/REJECTION PURPOSES

An MPN scheme using different levels of test at two dilutions of the sample has been devised for acceptance or rejection of sample lots (Aspinall and Kilsby, 1979)*. Suppose a manufacturer's quality specification for a product requires rejection if the product contains more than a critical limit (c) of a defined organism (eg, c = 10^3 coliforms/g) but a lower level will permit acceptance for quality assurance purposes.*

Serial dilutions of product samples from each of three lots (A, B and C) are prepared at 10^{-1} and 10^{-2} levels. Each of two tubes of culture medium is inoculated with 0.5 mL of the 10^{-1} dilution and each of seven tubes is inoculated with 0.5 mL of the 10^{-2} dilution. Following incubation, the results are recorded and MPN values determined.

The following results were recorded for the three samples of product:

Product Lot	Result at Dilution of 10^{-1}	10^{-2}	MPN (/g)[a]	Decision
A	1+	5+	Not valid	Repeat test
B	1+	1+	$1.9 \times 10^1 = 19$	Accept lot
C	2+	7+	$>39 \times 10^1 = >390$	Reject lot

[a]From Table 8.8.

An invalid result (see footnote to Table 8.8) requires that the test be repeated (as in case A) whilst a low MPN result (case B) allows acceptance of the product. For product C, all nine tests results are positive (ie, 2+, 7+); the calculated MPN is >390 /g, exceeds the criterion (*c*) based on the upper bound of the 95% CI which is ∞, that is, $>10^3$ /g (Table 8.8). Note, however, that if only eight positive tubes had occurred (eg, 2+, 6+), the product would be accepted with an MPN of 390 /g and an upper CI bound of 10^3 /g (Table 8.8).

This scheme is less likely to reject satisfactory material than is a three-level three-test scheme (3, 3, 3) such as is frequently recommended for control purposes (eg, ICMSF, 1978). However, it should be noted that several potential results (eg, for 0–3 to 0–7 and 1–4 to 1–7) are excluded since they are highly improbable (based on the MPN calculator of Jarvis et al., 2010).

QUANTIFICATION BASED ON RELATIVE PREVALENCE OF DEFECTIVES

The use of the binomial distribution to describe the OC curve for sample plans was described in Chapter 5. This relationship, describing the probability of occurrence of defective samples, can be used also to quantify the prevalence of defective units and the probable level of occurrence of target organisms in a food lot, as illustrated in Example 5.1.

From the data in Table 8.9, it can be seen that in testing for a specific target organism (eg, salmonella) using an appropriate method, a negative test result on a single sample (ie, $n = 1$) of 25 g will merely indicate that at the upper 95% CL the lot will contain no more than 38 salmonellae/kg. However, there is a 5% chance that the true prevalence will exceed this value by an unknown amount. Increasing the number of samples tested increases the level of confidence that the prevalence of the salmonellae

Table 8.9 One-Sided Upper Bound of the Confidence Interval (CI) for Maximum Percentage Defective 25 g Sample Units, Assuming Random Distribution of Target Organism and No Positives Detected

No. of Sample Units Tested	% Defectives[a] at		Estimated Mean Number of Organisms/kg[b] at	
(*n*)	*P* = 0.95	*P* = 0.99	*P* = 0.95	*P* = 0.99
1	95.0	99.0	38	40
2	77.6	90.0	31	36
3	63.2	78.5	25	31
4	52.7	68.4	21	27
5	45.1	60.2	18	24
10	25.9	36.9	10	15
20	13.9	20.6	6	8
30	9.5	14.2	4	6
50	5.8	8.8	2	4
100	3.0	4.5	1	2

[a]*Rounded to the first decimal place.*
[b]*Rounded to the nearest whole number.*

is low since if all test results were negative, the upper 95% CL on, for example, 10 or 20 tests would be not more than 11, nor more than 6, salmonellae/kg, respectively. The data also indicate that the 99% upper CLs for prevalence would be 40, 15 or 8 salmonellae/kg, respectively, for 1, 10 or 20 samples tested with 'negative' results. Note that no matter how many sample units are tested one can never give a guarantee that the 'lot' from which the samples were drawn is totally free from salmonellae, or other potential pathogens, even when the test results are all negative.

For multiple tests at a single dilution, a mean contamination level can be estimated but this value is only a 'probable number', *not* a 'Most probable number'. Using the following equation: $\hat{m} = -(1/V) \cdot \ln(S/n)$, a single positive result from 10 tests, each on 25 g, estimates a mean value of 4.2 salmonellae/kg, with 95% CIs ranging from 0.1 to 17.8 organisms/kg. As would be expected, the upper bound of the 95% CI is higher than that obtained (10 salmonellae/kg) if no positive results had been obtained.

OTHER STATISTICAL ASPECTS OF MULTISTAGE TESTS

THE TEST PROCEDURE

When samples are tested for the presence or absence of specific organisms, the analytical procedures frequently involve four stages: pre-enrichment; selective enrichment; plating onto diagnostic agar; and identification of 'typical' colonies. Different methodological procedures are sometimes recommended in relation to the media to be used and the incubation times and temperatures during the pre-enrichment stage.

The ISO standard method (Anon, 2007) for detection of salmonellae in foods and feeds requires that the initial 25-g sample should be inoculated into 9 volumes of a liquid pre-enrichment medium (buffered peptone water) that is incubated for 18 ± 2 h at 37°C before transferring 1 mL into 10 mL of each of two selective enrichment media [Rappaport Vassiliadis medium with Soya (RVS) and Muller-Kauffmann tetrathionate/novobiocin broth (MK)]. The RVS is incubated for 24 ± 3 h at 41.5°C and the MK for 24 ± 3 h at 37°C. Thereafter, one 5-mm loopful (about 0.02 mL) from each enrichment culture is streaked on to one plate of each of two selective diagnostic media that are incubated for 24 ± 3 h at 37°C, prior to selection of typical colonies for typing.

Hence, three dilution stages occur. The first stage results in a 10-fold dilution of the microorganisms in the test portion. The numbers of both target and competitor organisms that grow during incubation will depend on (1) the initial contamination levels and dispersion of both the target organism and competitive organisms; (2) the condition of the organisms (ie, whether cells have been 'stressed' by the environmental conditions); (3) the duration of the lag phase for the organisms; and (4) the population doubling time for the target organism in the conditions of the test (ie, culture medium composition and the incubation temperature and time). Growth of target organisms during the second and third stages (selective enrichment and diagnostic culture) will also be dependent on the specific cultural conditions. The fourth stage (identification of discrete colonies) is considered later.

The possibilities for isolation of specific organisms will depend not only on the initial prevalence of contamination but also on the manner in which the test is done. A period of pre-enrichment that is too short may not result in recovery of the stressed organisms; this reduces the probability of transferring resuscitated organisms to the enrichment broth and subsequently to the diagnostic culture. Too long a period in a non-selective pre-enrichment may result in overgrowth by competitive organisms, leading to suppression of the target organism. An illustration of the effects of incubation time is shown for a hypothetical example in Examples 8.5 and 8.6, and in Fig. 8.4.

EXAMPLE 8.5 EFFECT OF INCUBATION TIME IN PRE-ENRICHMENT ON NUMBERS OF SALMONELLAE FOR SUBCULTURE

Assume that:

1. *25 g dry food sample is inoculated into 225 mL of peptone broth for pre-enrichment;*
2. *the initial contamination level = 1 salmonella/10 g sample;*
3. *the (hypothetical) length of lag phase (L) for resuscitation = 6 h;*
4. *the population doubling time (T_d) in the pre-enrichment broth is 30 min (0.5 h), at the appropriate incubation temperature; and, for simplicity,*
5. *synchronous growth will occur.*

The relative numbers of organisms can be determined using the growth kinetics equation: $N_T = N_0 2^{(T-L)/T_d}$, where N_T is the number of salmonellae per millilitre after time T (h), N_0 is the initial number of salmonellae per mL, L is the length of lag phase (h) and T_d is the doubling time (h).

Since the initial food sample contained 2.5 salmonellae in 25 g, the initial cell number in the pre-enrichment culture $N_0 = 2.5/250 = 0.01$ organism/mL. If $L = 6$ h, then the theoretical number of viable salmonellae after incubation for 6 h will not have changed (assuming that no cells die). We derive values for other incubation periods:

$$
\begin{aligned}
&9\,\text{h}, \quad N_T = 0.01 \times 2^{(9-6)/0.5} = 0.01 \times 2^6 = 0.64 \\
&12\,\text{h}, \quad N_T = 0.01 \times 2^{(12-6)/0.5} = 0.01 \times 2^{12} = 41.0 \\
&18\,h, \quad N_T = 0.01 \times 2^{(18-6)/0.5} = 0.01 \times 2^{24} = 1.68 \times 10^5 \\
&24\,\text{h}, \quad N_T = 0.01 \times 2^{(24-6)/0.5} = 0.01 \times 2^{36} = 6.87 \times 10^8
\end{aligned}
$$

Then, assuming random distribution and synchronous growth, the expected number of organisms likely to be transferred in an inoculum of pre-enrichment medium and the probability of transferring at least one viable organism can be calculated for any of these, or other, time intervals:

Incubation Time (h)[a]	Estimated Mean Number of Organisms in 1-mL Medium[b]	Probability (P) of ≥ 1 Organism in	
		1 mL	10 mL
6	0.01	0.01	0.095
9	0.64	0.47	0.99
12	41	0.99	0.99
18	1.7×10^3	0.99	0.99
24	6.9×10^8	0.99	0.99

[a]In pre-enrichment, assuming conditions as given previously.
[b]Assuming 0.1 salmonella/g food sample and random distribution.

Once sufficient growth has occurred during pre-enrichment, the 95% probability of transferring at least 1 viable organism in 1 mL of culture requires at least 3 viable salmonellae/mL, assuming Poisson distribution [$\hat{m} = 3 : P_{x=0} = 0.05$; hence, $P_{(x\geq 1)} = 0.95$]. The table (above) shows that, for the conditions and growth kinetics used, the pre-enrichment culture would need to be incubated for at least 12 h to provide a 99% probability of transferring a viable inoculum in 1 mL of the culture.

At lower viable cell numbers, the chances of transferring at least one viable organism will be improved by inoculation of a larger volume of pre-enrichment culture (eg, 10 mL) into the enrichment culture, or by using procedures such as immuno-magnetic beads to recover the target organism from a large volume of the pre-enrichment culture. In media where resuscitation, but no growth, occurs during pre-enrichment it is essential to use the entire pre-enrichment culture to inoculate double-strength enrichment medium, although the selectivity may then be affected by food constituents.

EXAMPLE 8.6 PROBABILITY OF TRANSFERRING SALMONELLAE DURING ENRICHMENT AND ISOLATION

A question often asked about presence or absence tests for pathogens is, 'What is the probability of detecting salmonellae, if present in a food sample, using pre-enrichment and enrichment cultures prior to streaking onto diagnostic agars to isolate any organisms that may be present?'

Assume that:

1. *the initial mean contamination level = 1 salmonella/10 g sample;*
2. *pre-enrichment for 12 h, yielding about 41 salmonellae/mL (as Example 8.5);*
3. *one-millilitre pre-enrichment medium inoculated into 9-mL selective enrichment medium;*
4. *lag time (L) in enrichment medium = 4 h with doubling time = 60 min = 1 h;*
5. *culture period 18 or 24 h;*
6. *0.02 mL (one loopful) inoculated onto diagnostic medium.*

From Example 8.5, the 12-h pre-enrichment culture would contain 41 salmonellae/mL. Since the pre-enrichment is diluted 1 in 10 into a selective enrichment medium, the initial level of salmonellae would be $41/10 = 4.1/\text{mL} = N'_0$. Assuming synchronous growth and other factors summarised earlier, then after incubation for 18 h:

$$N'_T = 4.1\times 2^{(18-4)/1} = 4.1\times 2^{14} \approx 6.7\times 10^4 \text{ salmonellae/mL of enrichment culture}$$

Similarly, after 24 h:

$$N'_T = 4.1\times 2^{(24-4)/1} = 4.1\times 2^{20} \cong 4.3\times 10^6 \text{ salmonellae/mL of enrichment culture}$$

Hence, a standard 0.02-mL loop would be expected to transfer ca. 1340 salmonellae after 18-h incubation or 86,000 salmonellae after 24-h incubation. Hence, the statistical probability of transferring at least one viable salmonella to the isolation medium would be $\gg 0.99$.

However, if the enrichment medium had been inoculated after 6-h pre-enrichment (see Example 8.6), during which time resuscitation but no cell growth had occurred, then the level of inoculum transferred to the enrichment medium would have been only 0.1 cell/10 mL of inoculum. It would therefore have been extremely unlikely ($P \ll 0.01$) that salmonellae would have been detected in the enrichment culture even after 24 h. These effects are illustrated diagrammatically in Fig. 8.4.

COMPOSITING OF SAMPLES

When multiple samples are to be tested, it is sometimes recommended that the samples be composited in order to reduce the workload on the laboratory. Suppose that a test procedure requires the testing of 10 × 25 g samples for the presence of salmonellae. What are the options for compositing samples? Several alternative approaches

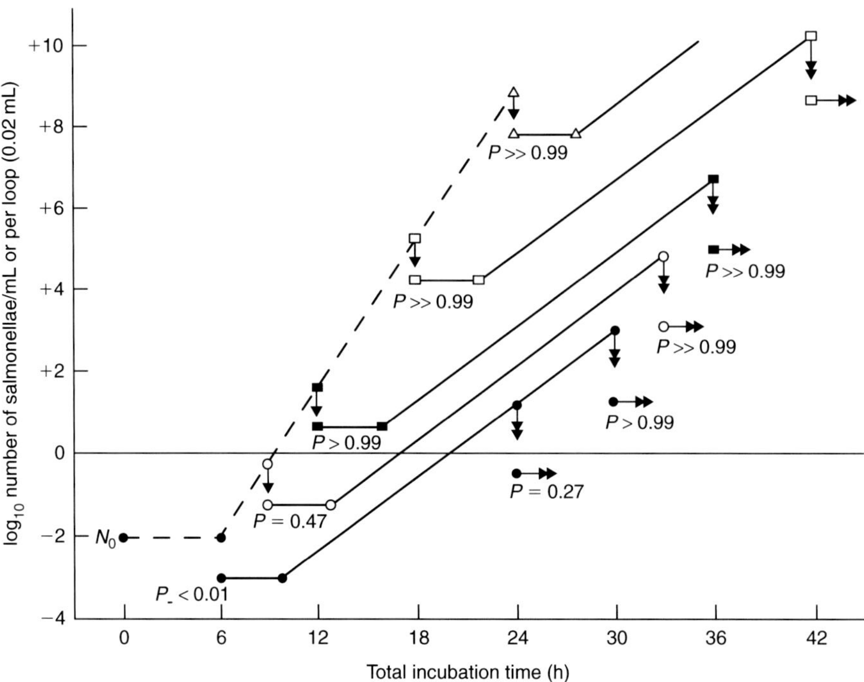

FIGURE 8.4 Hypothetical Illustration of Changes in Numbers of Viable Salmonellae During Incubation in Pre-Enrichment (----) and Enrichment (——) Cultures, and the Probability of Transfer of At Least One Salmonella Cell After Specific Periods of Time

have been recommended, but various critical issues must be resolved before choosing a procedure for use (Jarvis, 2007):

1. *An unacceptable approach*: The 10 × 25 g samples are composited into a single 250-g quantity from which a single 25-g subsample is taken for testing. This method, which is used widely in chemical analyses to ensure greater homogeneity of the material sampled, is potentially prone to cross-contamination. More importantly, in statistical terms, only a single 25-g portion of the sample has been tested (not 10 samples), so that the results will provide no more information about the lot from which the original 10 samples were taken than would testing of any single 25-g sample, albeit the combined test sample will be more representative of the lot than would a single original test sample.
2. *Bulk compositing and testing*: The second approach combines all 10 samples at the pre-enrichment stage; that is, 250 g of combined sample is inoculated directly into 2250 mL of pre-enrichment medium. After incubation, the subsequent stages are done on the same basis as for a single sample (ie, 10-mL pre-enrichment medium is inoculated into 100 mL of enrichment medium, etc.).

The key issue here is the level of sensitivity of the test method. If only 1 of the 10 samples contains the target organism, what is the likelihood that a method capable of detecting, say, 1 viable cell in a 25-g sample of food material will detect 1 viable cell in 250 g of the same food? A positive result would indicate that at least one of the original samples was contaminated, but a negative result (ie, failure to detect the target organism) might suggest wrongly that none of the test samples was contaminated. It may be that the sensitivity of the test method can be improved by, for example, extending the pre-enrichment incubation period, but this would work only if growth of competitive organisms can be adequately controlled. Only if it has been demonstrated that the level of sensitivity of the method is at least 1 organism in 250 g is it acceptable to test such composite samples.

3. *Wet compositing*: The only fully acceptable method of compositing is to inoculate 10 pre-enrichment cultures (ie, 1 for each 25-g sample) and, after incubation, to inoculate equal volumes of each pre-enrichment culture (ie, 10 × 1 mL) into 100-mL enrichment broth. By retaining the original pre-enrichment samples, it is possible also to re-culture from these if the composite enrichment broth gives a positive result, so that the prevalence of positive samples can be determined.

Example 8.7 illustrates the statistical aspects that must be taken into account in compositing of samples for microbiological analyses. For a more detailed appraisal of compositing see Jarvis (2007).

EXAMPLE 8.7 STATISTICAL ASPECTS OF SAMPLE COMPOSITING

A microbiological food safety criterion requires the testing of 30 samples, each of 25 g, for the presence of salmonellae using defined methods (eg, EU, 2007, for infant formulae and dried dietary foods). Can we use the same test method if we combine (composit) a number of food samples prior to testing?

The defined test procedure is sufficiently sensitive to detect at least one salmonella cell in a quantity of 25 g. Provided that the test procedure can detect a single cell in a larger quantity of sample (eg, 250 g or even 750 g) from a statistical perspective there is no difference in testing 1 × 750, 3 × 250 or 30 × 25 g samples. But if there is *any doubt* as to the level of sensitivity of the test procedure, then compositing should not be done, except using a 'wet' compositing approach where each individual sample unit is pre-enriched separately and then composited for enrichment (for details see text or Jarvis, 2007).

Assume n sample units, each unit (i) being of size k_i g with a salmonella cell density of δ_i cells/g. Provided the test procedure is sufficiently sensitive to detect the organism if present, the probability of finding a negative result [$P_{(x=0)}$] for any sample i is given by

$$P_{(x_i=0)} = e^{-\delta_i k_i}$$

If $\delta_1 = 1$ viable cell/10 g and $k_1 = 25$ g, then the probability of not detecting any organism in the 25-g sample (assuming the test protocol is 100% effective) is given by

$$P_{(x_1=0)} = e^{-0.1\times 25} = e^{-2.5} = 0.082$$

and the probability of finding one or more salmonellae is given by

$$P_{(x\geq 1)} = 1 - P_{(x=0)} = 1 - 0.082 = 0.918$$

The probability of finding a negative result on all n units is the product of the individual probabilities for a negative result in each unit:

$$P_{(x_1=0)(x_2=0)(x_3=0)\cdots(x_n=0)} = e^{-\delta k_1} \times e^{-\delta k_2} \times e^{-\delta k_3} \times \cdots \times e^{-\delta_n k_n} = e^{-\Sigma \delta k_i}$$

To illustrate this effect, suppose that we have 3 × 250 g sample units with levels of 0.01, 0.00 and 0.02 viable cell/g, respectively; then the total cell count in the three sample units will be 2.5 + 0.0 + 5.0 = 7.5 cells per 750 g, which is equivalent to an average prevalence of 0.01 cell/g. Assuming that the method can detect such a low level of salmonellae, the probability of obtaining a negative (ie, of not detecting a positive) result in the individual sample units will be 0.082, 1.00 and 0.007, respectively, and the combined probability of a negative result on the three sample units will be given by

$$P_{(x_1=0)(x_2=0)(x_3=0)} = e^{-\Sigma d_i k_i} = e^{-\Sigma 2.5+0+5.0} = e^{-7.5} = 0.00055$$

This is the same probability that would be found by using the mean cell count (0.01 cell/g) and multiplying by a k value of 750 g, that is, $P_{(x=0)} = e^{-0.01\times 750} = e^{-7.5} = 0.00055$.

However, the validity of the statistical calculation is totally dependent on the technical efficiency of the method. A method capable of detecting contamination at a level of (say) 1 salmonella cell in a 25-g sample may not be able to detect 1 salmonella cell in a combined sample of 250 g, if only 1 of the 10 × 25 g sample units were contaminated. It is therefore essential to carry out trials using different levels of contamination before making any assumption that samples can be composited.

Based in part on Jarvis (2007) with permission of Blackwell Publishing.

SELECTION OF COLONIES FOR IDENTIFICATION

After development of isolated colonies on a diagnostic medium, it is necessary to select colonies showing typical morphology for further identification. The chance of isolation of specific organisms from an almost pure culture will of course be high. However, when mixed cultures are streaked onto agar plates, different types of organism may have apparently similar colonial morphologies. It is essential to test several apparently typical colonies from such plates. In some laboratory manuals no reference is made to the number of colonies to be tested; ICMSF (1978) recommends testing at least two typical colonies whilst Anon (2002) recommends testing up to five typical colonies.

The chance of picking any one colony of the type sought is dependent on the number (n) of the specific type of target colonies relative to the total number (N) of colonies present. Since N is not large, the relationship is governed by the hypergeometric distribution. This is illustrated in Example 8.8. Table 8.10 shows the number of typical colonies of the target organism that must be present in order to select, with a given statistical probability, at least one colony of the target organism from amongst a population of otherwise apparently typical colonies. For instance, if the true prevalence of *E. coli* colonies is 50% of the colonies having 'typical' coliform morphology, then the

EXAMPLE 8.8 PROBABILITY FOR SELECTION OF A SPECIFIC ORGANISM FROM A DIAGNOSTIC PLATE CULTURE

Assume that we wish to confirm the presence of E. coli colonies on a culture plate that also contains non–E. coli organisms that may give typical E. coli reactions. Suppose that the actual number of non–E. coli colonies on a diagnostic plate with 30 apparently typical colonies is 10 or 25. If five colonies are picked, what is the probability that at least one colony will be an E. coli?

Since the population is finite and (relatively) small, removal of any one colony will reduce the residual number from which to select another colony. The probability of selecting a specific number of *E. coli* colonies is described by the hypergeometric equation:

$$P_{(X=x)} = h(x, n, M, N) = \frac{\binom{M}{x}\binom{N-M}{n-x}}{\binom{N}{n}}$$

where N is the total number of typical colonies on the plate, M is the expected number of *E. coli* colonies on the plate, n is the number of colonies picked for identification and x is the number of successes required.

Assuming that the number colonies on the plate is $N = 30$ and m (=10) is the number of non–*E. coli* typical colonies, the number of potential *E. coli* colonies = $M = N - m = 20$; the number of colonies to be picked is $n = 5$ and $x \geq 1$, since we are looking for *at least* 1 *E. coli* colony. The probability of *not* picking an *E. coli* colony is $P_{(x=0)} < 0.002$; so the probability of picking one or more *E. coli* colonies is almost certain [$P_{(x\geq 1)} = 0.998$].

However, if $m = 25$ (ie, the number of non–*E. coli* typical colonies), then $M = 5$ and the chance of picking no *E. coli* colonies increases to $P_{(x=0)} = 0.37$, so the chance of picking 1 or more *E. coli* colonies in a random sample of 5 reduces to $P_{(x\geq 1)} = 0.627$. If only one colony was picked, the chances of finding an *E. coli* would be only $P_{(x=1)} = 0.17$. Table 8.10 provides other examples and shows the importance of having a high proportion of target colonies to ensure picking at least one target organism.

Although this calculation can be done by hand using an expansion of the equation, Excel™ has a function that can be used for the hypergeometric calculation and a 'free to use' on-line hypergeometric calculator can be accessed on http://stattrek.com/online-calculator/hypergeometric.aspx (accessed 13.11.15).

Table 8.10 Probability of Picking at Least One Target Organism From Amongst a Number of 'Typical' Colonies on a Diagnostic Culture Plate (Based on the Hypergeometric Distribution)

No. of 'Typical' Colonies (*N*)	No. of Colonies Picked (*n*)	The Approximate Probability (*P*) that At Least One Target Colony Will Be Picked From Amongst the Colonies Present if the Proportion of Target Organisms Is At Least		
		10%	20%	50%
10	5	0.50	0.77	>0.99
20	5	0.45	0.72	0.98
30	5	0.43	0.70	0.98
40	5	0.43	0.69	0.98
50	5	0.42	0.69	0.97

probability of selecting at least 1 *E. coli* by testing 5 colonies out of a population of 30 colonies is better than 95% (P = 0.98). If, however, the true prevalence is only 10% (ie, 3 *E. coli* in 30 colonies), then the chance that at least 1 *E. coli* colony is included in the sample of 5 colonies is only 43% (P = 0.43), that is, a less than 1 in 2 chance of finding 1 or more *E. coli* colonies. A similar situation would apply to the testing of individual colonies using PCR or other methods of identification.

Hedges et al. (1977) published a theoretical study of the expected distribution of *E. coli* serotypes amongst samples drawn randomly from populations containing different numbers of serotypes. From these studies it was possible to recommend guidelines for planning efficient sampling programmes to determine the serotypes present in various animal species. This is of equal importance in the testing of typical colonies from culture plates of mixed culture isolates of other organisms from natural sources.

REFERENCES

Anon, 2002. Microbiology of Food and Animal Feeding Stuffs – Horizontal Method for the Detection of *salmonellae* Species. ISO 6579:2002. International Organization for Standardization, Geneva.

Anon, 2005a. Microbiology of Food and Animal Feeding Stuffs – Polymerase Chain Reaction (PCR) for the Detection of Food-Borne Pathogens – Performance Testing for Thermal Cyclers. ISO/TS 20836:2005. International Organization for Standardization, Geneva.

Anon, 2005b. Microbiology of Food and Animal Feeding Stuffs – Polymerase Chain Reaction (PCR) for the Detection of Food-Borne Pathogens – General Requirements and Definitions. ISO 22174:2005. International Organization for Standardization, Geneva.

Anon, 2005c. Microbiology of Food and Animal Feeding Stuffs – Horizontal Method for the Detection and Enumeration of Presumptive *Escherichia coli* – Most Probable Number Technique. ISO 7251:2005. International Organization for Standardization, Geneva.

Anon, 2006a. Microbiology of Food and Animal Feeding Stuffs – Polymerase Chain Reaction (PCR) for the Detection of Food-Borne Pathogens – Requirements for Sample Preparation for Qualitative Detection. ISO 20837:2006. International Organization for Standardization, Geneva.

Anon, 2006b. Microbiology of Food and Animal Feeding Stuffs – Polymerase Chain Reaction (PCR) for the Detection of Food-Borne Pathogens – Requirements for Amplification and Detection for Qualitative Methods. ISO 20838:2006. International Organization for Standardization, Geneva.

Anon, 2007. Microbiology of Food and Animal Feeding Stuffs – Horizontal Method for the Detection of *Salmonella* Species. ISO 6579:2002/Amd 1:2007. International Organization for Standardization, Geneva.

AOAC, 2012. Official Methods of Analysis, 19th ed. Association of Official Analytical Chemists, Washington, DC (revision 1).

Aspinall, L.J., Kilsby, D.C., 1979. A microbiological quality control procedure based on tube counts. J. Appl. Bacteriol. 46, 325–330.

Blodgett, R., 2002. Measuring improbability of outcomes from a serial dilution test. Comm. Stat. Theory M 31, 2209–2223.

Cochran, W.G., 1950. Estimation of bacterial densities by means of the 'most probable number'. Biometrics 6, 105–116.

De Man, J.C., 1975. The probability of most probable numbers. Eur. J. Appl. Microbiol. 1, 67–78.

De Man, J.C., 1983. MPN tables corrected. Eur. J. Appl. Microbiol. 17, 301–305.

Eisenhart, C., Wilson, P.W., 1943. Statistical methods and control in bacteriology. Bacteriol. Rev. 7, 57–137.

EU, 2007. Commission Regulation (EC) No. 1441/2007 of 5 December 2007 Amending Regulation (EC) No. 2073/2005 on Microbiological Criteria for Foodstuffs.

Finney, D.J., 1947. The principles of biological assay. J. R. Stat. Soc. Ser. B 9, 46–91.

Fisher, R.A., Yates, F., 1963. Statistical Tables for Biological, Agricultural and Medical Research, sixth ed. Longman Group, London.

Halvorson, H.O., Ziegler, N.A., 1933. Application of statistics to problems in bacteriology. I. A means of determining bacterial population by the dilution method. J. Bacteriol. 25, 101–121.

Hedges, A.J., Howe, K., Linton, A.H., 1977. Statistical considerations in the sampling of *Escherichia coli* from intestinal sources for serotyping. J. Appl. Bacteriol. 43, 271–280.

Hurley, M.A., Roscoe, M.E., 1983. Automated statistical analysis of microbial enumeration by dilution series. J. Appl. Bacteriol. 55, 159–164.

ICMSF, 1978. Micro-Organisms in Foods. Vol. 1. Their Significance and Methods of Enumeration, second ed. University of Toronto Press, Toronto.

Jarvis, B., 2007. On the compositing of samples for qualitative microbiological testing. Lett. Appl. Microbiol. 45, 592–598.

Jarvis, B., Wilrich, C., Wilrich, P.Th., 2010. Reconsideration of the derivation of most probable numbers, their standard deviations, confidence bounds and rarity values. J. Appl. Microbiol. 109, 1660–1667.

Meynell, G.G., Meynell, E., second ed.,1970. Theory and Practice in Experimental Bacteriology. Cambridge University Press, London.

Moran, P.A.P., 1954a. The dilution assay of viruses. I. J. Hyg. Cambridge 52, 189–193.

Moran, P.A.P., 1954b. The dilution assay of viruses. II. J. Hyg. Cambridge 52, 444–446.

Moran, P.A.P., 1958. Another test for heterogeneity of host resistance in dilution assays. J. Hyg. Cambridge 56, 319.

Myers, L.E., McQuay, L.J.M., Hollinger, F.B., 1994. Dilution assay statistics. J. Clin. Microbiol. 32, 732–739.

Pearson. E.S., Hartley, H.O. 1976. Biometrika Tables for Statisticians, third edn. Griffin, High Wycombe, for The Biometrika Trustees.

Safarik, I., Safarikova, M., Forsythe, S.M., 1995. The application of magnetic separations in applied microbiology. J. Appl. Bacteriol. 78, 575–585.

Stevens, K.A., Jaykus, L.A., 2004. Bacterial separation and concentration from complex sample matrices: a review. Crit. Rev. Microbiol. 30, 7–24.

Woodward, R.L., 1957. How probable is the most probable number? J. Am. Water Works. Assoc. 49, 1060–1068.

CHAPTER

Statistical considerations of other methods in quantitative microbiology

9

In addition to those methods considered previously, alternative methods to estimate microbial numbers in foods include those often referred to as 'alternative methods'. These methods fall into two major categories: direct counting of cells, for example, by microscopy or by electronic cell counting; and indirect methods of analysis, which are dependent on physical or chemical means of indirectly estimating microbial cell numbers.

DIRECT MICROSCOPIC METHODS

Much has been written in the past about direct microscopic counts using techniques such as the haemocytometer, or other types of counting chamber, the Breed smear (Breed and Brew, 1925), the proportional (or ratio) count technique (Thornton and Gray, 1934) and the membrane filter technique. Technical details for these procedures are given in standard laboratory texts such as Meynell and Meynell (1965) and Harrigan and McCance (1998).

The average number of microbial cells counted on each microscopic field will be related to the initial level present in the sample, or in a macerate of a solid food sample, and will be dependent on the dilution factor, the area over which the sample is distributed and the area of the field of view of the microscope. The distribution of microbial cells is usually considered to follow either a negative binomial or a Poisson series (Chapters 3 and 4; see also Examples 3.4, 4.3 and 4.4), although other more complex distributions may occur. Takahasi et al. (1964) used the ratio method to enumerate microbial cells blended with a suspension of particles of known density and similar size to bacterial cells in a haemocytometer chamber. The numbers of microbial cells and particles were counted for each of 1000 haemocytometer squares and the results collated. A bivariate frequency distribution was obtained (Table 9.1) showing the distribution of both the microbial cells and the reference particles. Using the first data row as an example, 268 squares contained either 3 or 4 microbial cells; 138 of these contained 0–4 reference particles, 114 contained 5–9 particles, 15 contained 10–14 particles and 1 square contained a single particle. Using a goodness of fit (χ^2) test they showed that the distributions of cells and particles conformed to a negative binomial distribution.

Statistical Aspects of the Microbiological Examination of Foods. http://dx.doi.org/10.1016/B978-0-12-803973-1.00009-7

Table 9.1 Bivariate Frequency Distribution of Reference and Test Cells in Squares of a Haemocytometer Counting Chamber

Number of Microbial Cells/Square	Number of Reference Particles/Square						
	0–4	5–9	10–14	15–19	20–24	30–34	Total
3–4	138 (146.2)	114 (113.5)	15 (10.9)	1*	—	—	268
5–9	172 (171.5)	268 (287.5)	66 (60.6)	4*	—	—	528
10–14	39 (27.9)	98 (100)	30 (44.2)	7 (6.8)	13** (9.2)	—	174
15–19	1*	8 (11.7)	14 (10)	2*		—	25
20–24	—	1*	3*	—	—	—	4
25–29	—	—	—	—	—	1*	1
Total	350	507	128	14	1	1	1000

Values in ( ) are expected frequencies based on bivariate negative binomial distribution.
Values with single asterisk (*) were combined to make value marked with double asterisk (**).
Goodness-of-fit test to a negative binomial distribution is $\chi^2 = 15.86$*; the critical limit for compliance with* $\upsilon = 9$ *is* $\chi^2_{(P=0.05)} = 16.92$*; hence, 0.10 > P > 0.05, so the null hypothesis of compliance is not rejected.*
Reproduced from Takahasi et al. (1964) by permission of the Japanese Journal of Infectious Diseases.

For a distribution conforming to Poisson, it has already been shown (Table 7.2) that precision [as measured by the coefficient of variation (CV)] increases with an increase in the number of individuals counted; the CV of the mean count $= \sigma/\mu = \sqrt{\mu}/\mu = 1/\sqrt{\mu}$. In the negative binomial series the variance and mean are related (Chapter 3) by the expression $\sigma_2 = \mu + (\mu^2/k)$ and $1/k$ provides a measure of the excess variance due to clumping. Therefore, greatest precision will result from counting a large number of cells but for maximum precision these must be distributed over a large number of fields since the variance is related to both the overall mean and the negative binomial parameter k. Table 9.2 clearly shows that greater precision, presented as the 95% CI around the mean count/field, is obtained by counting, for example, 50 (or fewer) cells/field than by counting 100 or more cells/field if evidence of cell clumping is found.

Hence, it is essential to count as many organisms as is technically feasible, over a large number of fields, if the clumping effect is pronounced. Where ratio counts are made, greatest precision results from use of approximately equal numbers of microbial cells and reference particles in each field. Standardised latex particles supplied for use with the Coulter Electronic Particle Counter™ provide a suitable source of reference materials.

Apart from distribution errors, the errors of direct counts are related mainly to the inaccuracies of pipetting very small volumes of sample (Table 6.3; Brew, 1914), uneven distribution of the sample on the slide or in the counting chamber, inaccuracies in the dimensions or use of counting chambers, the massive multiplication factor involved (see Example. 9.1), inadequate mixing of the original sample and the worker error associated with counting many microbial cells and reference particles.

Table 9.2 Confidence Intervals for Selected Numbers of Cells Counted Per Microscopic Field and Values of the Negative Binomial Factor $\hat{k}$ When Contagion Is Apparent

Mean Cell Count/Field	Number of Fields Counted[a]	95% Confidence Intervals for Mean Cell Count/Field With k =			
		2	3	4	5
500	2	2–82,100	7–27,101	15–14,735	23–9,912
330	3	105–1,052	134–819	153–713	167–651
250	4	131–473	150–411	162–380	169–361
200	5	127–311	140–282	147–267	152–257
100	10	82–119	85–114	87–111	87–109
50	20	45–53	45–52	45–51	45–51
25	40	20–29	19–29	19–28	18–27

For details of calculation see Example 7.2.
[a]Assumes a total of 1000 cells counted.

EXAMPLE 9.1 DILUTION EFFECT FOR MICROSCOPIC COUNTS ON A BACTERIAL SMEAR

Assume that 0.01-mL sample suspension is spread over an area of 1 cm^2. After fixing and staining, the slide is examined by microscopy using a field diameter of 16 μm. From 400 fields counted the mean cell count/field is 2.42 with s^2 = 4.06. What is the estimate of cell numbers per millilitre of suspension?

Calculate:

1. Area of field = $\pi r^2 = \pi(16/2)^2 = 201.09\ \mu m^2 = A$.
2. The number of fields/cm^2 = $(1/A) \times 10^8 \approx 10^8/201 = 4.97 \times 10^5$.
3. The mean number of cells/field = 2.42.
4. The mean number of cells/cm^2 = $4.97 \times 10^5 \times 2.42 \approx 1.2 \times 10^6$.
5. Since 0.01 mL of sample was tested, the average direct count of cells in the sample = $1.2 \times 10^6 \times 10^2 = 1.2 \times 10^8$ cells/mL.

Note that as the standard deviation for the average cell count/field is ±2.01, the 95% confidence interval (based on 400 fields) ranges from 0 (2.42 − 4.02 = −1.60) to 6.44 (2.42 + 4.02) so the true count would lie between 0 and 3.2×10^8 ($4.97 \times 10^5 \times 6.44 \approx 3.2 \times 10^6 \times 10^2$)! Note also that the negative value for the lower 95% CL suggests that the cell counts are not 'Normally' distributed.

Furthermore, the overall dilution factor of 4.96×10^7 is much greater than the mean cell count/field or the 95% CL of that count. An error of 10% in pipetting and spreading the sample would affect the count relatively little, for example, if 0.009 mL had been pipetted instead of 0.010 mL, then the apparent mean count would have been 1.33×10^8. Because of the large dilution factors involved, microscopic counts can be used only when high total cell numbers are anticipated and the precision of such counts is low.

Data from Example 3.4.

HOWARD MOULD COUNT

A special form of microscopic count of fungal hyphae, developed to improve the quality of tomato-based products, is dependent for quantification on scoring fields as positive, that is, containing at least a defined 'critical' amount of fungal mycelium, or negative (ie, less than the critical level). Technical details of methodology are given by AOAC (2012).

The method is highly subjective. Even skilled analysts obtain a CV of the order of 55% or more (Vas et al., 1959; Dakin et al., 1964; Jarvis, 1977). Since the analyst scores the number of fields with or without fungal mycelium, each series of replicate tests can be expected to follow the binomial distribution, but as the number of fields to be counted is large (not less than 100/sample) the Normal approximation may be accepted. With a CV of about ±7.5%, we can put 95% confidence limits of ±15% on the observed value from a single Howard mould count (HMC) estimate. For a manufacturer to supply tomato-based products to a specification of (say) not greater than 50% positive fields, control analyses must make due allowance for the variation in the method. If the CV is ±7.5%, then to avoid producing more than 1 in 20 lots with a HMC $\geq$ 50%, the manufacturer's control average must not exceed 39% when duplicate samples are analysed (95% tolerance limits given by $\bar{x} \pm 1.96s/\sqrt{n}$). For further consideration on lot tolerances see Davies (1954) or Duncan (1965).

The HMC is affected by processes such as homogenisation that disrupt and disperse mould mycelium (Eisenberg, 1968). Such processes totally negate the use of the method, which is, at best, a quality control method, although it is used by regulatory authorities in the USA and elsewhere.

ELECTRICAL COUNTING OF CELLS (FLOW CYTOMETRY)

Flow cytometry is a powerful technique that can be used to enumerate microbial cells in pure culture suspension but is of limited applicability in food examination unless the microbial cells can be quantitatively separated from food and other natural substrates (Wood et al., 1976). The technique is based on measurement of the number of cells (particles) passing through an electronic sensing gate that enumerates the cells in relation to their cell size, or some other chosen parameter. The procedure is often linked with use of fluorochromes that react with cell constituents. Robinson (2009) provides a fascinating account of the history of the development of the technique and Diaz et al. (2010) described its applications in monitoring fermentation processes, such as those used for beer and cider.

A directly relevant application is that used in preparing frozen reference suspensions of microbial cells for use in microbiological quality control by the BioBall™ process. Reconstitution of a BioBall in a suitable volume of diluent provides a reference suspension of a defined organism at a known average concentration. Generally the CV of the colony counts on replicate balls of the same batch will be <10% of the stated cell content; for example, the mean count of *Staphylococcus aureus* cells in batch B1882 was 28.5 cfu with $s = 2.2$ giving a 95% predicted CI ranging from 24.1 to 32.9 cells. The reference standard can be used for culture media control (Jarvis, 2012) and, using products at higher cell concentrations, for studies of lethality, growth rates, etc.

INDIRECT METHODS

Although many indirect methods of enumeration have been developed, few are useful for food analysis without considerable modification. Indirect methods are based on measurements of physical (eg, turbidity, packed cell volume) or chemical (eg, protein, nucleic acid, ATP, chitin) properties of cell constituents, or of cell metabolism (eg, electrical measurements, oxygen utilisation, carbon dioxide production, acid production). Food constituents affect many of these techniques significantly, unless the microbial cells are separated from the food matrix before analysis (Wood et al., 1976, 1977).

Since all indirect estimation systems are based on measurements other than counts, one may suppose that the results from any estimate will follow a Normal distribution. Whilst this is true for estimation of chemicals in solution (eg, pH, acidity, proteins, nucleic acids, ATP, chitin, oxygen uptake), measurements of such compounds in microbial cells rarely conform to a Normal distribution, even when pure cultures are tested.

For instance, a plot of mean values against the variance estimates for chitin, as a measure of fungal mycelium in tomato juice (Jarvis, 1977), showed a high degree of correlation between the $\log_{10}$ of variance and the $\log_{10}$ of mean chitin content ($r_{xy} = 0.91$, $SE_r = 0.354$) indicating a lack of independence (Fig. 9.1). Six of the nine sets of data, all associated with high mould count estimates, conformed to a negative binomial distribution with individual values for $\hat{k}$ ranging from 9 to 42. Consequently, although the results of estimations of pure chitin are Normally distributed around the mean (Potts et al., 2001), the variance associated with estimation of chitin in food is affected by the contagious distribution of mould in the food.

IMPEDANCE METHODS

Estimation of microbial numbers by electrical impedance measurement is dependent on electrical conductance and capacitance changes in culture media resulting from microbial growth. The time to detection of a specific impedance change (in μ-siemens) can be related to the initial number of microorganisms in the medium. For a pure culture growing synchronously, we should expect normal growth kinetics to apply (Wood et al., 1977; Richards et al., 1978), so that an inverse linear relationship is found between the log initial cell density and the time to detection of a specific impedance change (Fig. 9.2; Example 9.2). For pure cultures, the variance of the detection time is largely independent of the detection time so that, as expected, the detection times are distributed Normally. However, when food samples

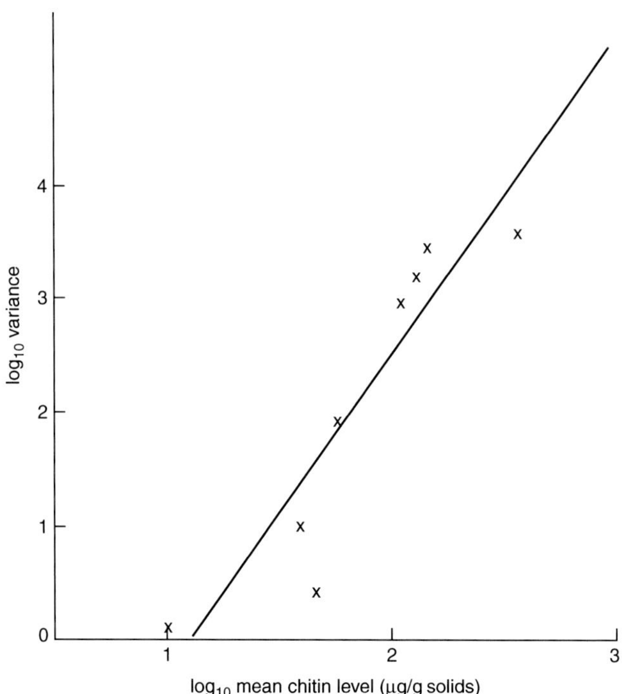

FIGURE 9.1 Plot of $\log_{10}$ Variance Against $\log_{10}$ Mean Level of Fungal Chitin in Tomato Products (Jarvis, 1977)

Regression equation: $y = 2.7867x - 3.04$; $r = 0.91$; standard error of $r = 0.354$

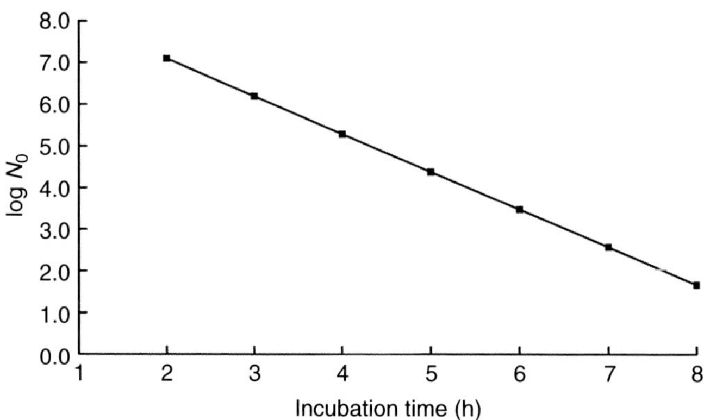

FIGURE 9.2 Plot of 'Time to Detection' Using Impedance With an Assumed Final Cell Density of 10^8 cells/mL, Plotted Against Initial Cell Numbers (N_0)

Data from Example 9.2

EXAMPLE 9.2 IMPEDANCE MEASUREMENTS – RELATIONSHIP BETWEEN INITIAL NUMBERS AND TIME TO DETECTION OF SPECIFIC IMPEDANCE CHANGE

How many organisms must be inoculated for detection to be achieved after incubation for various times at a defined temperature?

Synchronous growth is described by the following equation: $N = N_0 \times 2^{(T-L)/Z_d}$, where N is the number of organisms after time T, N_0 is the number of organisms at time T_0, Z_d is the doubling time (h) and L is the lag period (h). We assume that the culture has an average population doubling time of 0.33 h (20 min) and a lag period of 1 h, and that 10^8 organisms/mL are needed to show a specific impedance change.

We assume synchronous growth and maintenance of the average population doubling time, so the initial contamination level (N_0) can be determined by rearranging the equation as follows: $N_0 = N/2^{(T-L)/Z_d}$; we also assume that $N = 10^8$, $Z_d = 0.333$ and $L = 1$. So for growth to be detected after 6 h ($T = 6$), $N_0 = 10^8/2^{15} = 3052$. Calculated results for various detection times are as follows:

Detection Time (h), T	Number of Generations, $(T - L)/Z_d$	Population Increase, $2^{(T-L)/Z_d}$	Initial Number (Organisms/mL[a]), $N_0 = 10^8/2^{(T-L)/Z_d}$	$\log_{10} N_0$
2	3	8	1.3×10^7	7.1
3	6	64	1.6×10^6	6.2
4	9	512	2.0×10^5	5.3
5	12	4,096	2.4×10^4	4.4
6	15	32,768	3.1×10^3	3.5
7	18	262,144	3.8×10^2	2.6
8	21	2,097,152	4.8×10^1	1.7

[a]Rounded to one decimal place.

Hence, an initial cell concentration of 10^6 cells/mL should be detected after about 3 h but a concentration of 10^3 cells/mL would be detected only after about 6.5 h (Fig. 9.2).

are examined, the variance is often related to the detection time so the data do not conform to the Normal distribution (Fig. 9.3).

It may be supposed that this association reflects the distribution of organisms in the replicate test samples. Naturally-contaminated samples will have a distribution not only of numbers of organisms capable of growth in the test system but some will have different specific growth rates, lag times, nutritional requirements, etc. Although physical and chemical analytical methods would be expected to yield Normally distributed results, the types, numbers, condition and distribution of the microorganisms in the sample will affect estimation of microorganisms. This should not be taken to imply that such alternative methods are unsuitable for the purposes intended. Rather, its purpose is to make readers aware of misleading statements and claims about the precision of such methods by some who market new methodology concepts. It is for such reasons that it is necessary for all new methods to be validated using approved statistical procedures (see Chapter 13).

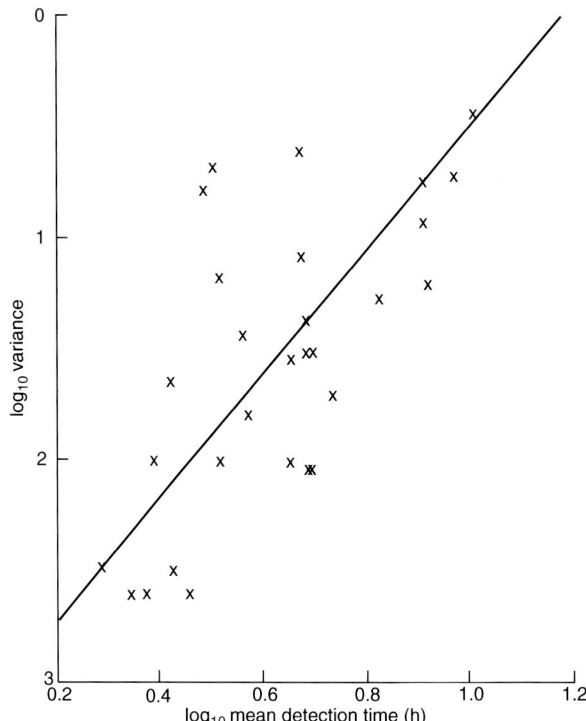

FIGURE 9.3 Plot of $\log_{10}$ Variance Against $\log_{10}$ Mean Detection Time for Food Samples Analysed by Impediometry

Regression equation: $y = 2.59x - 1.638$; $r = 0.72$; standard error of $r = 0.189$

REAL-TIME POLYMERASE CHAIN REACTION METHODS

The terminology associated with polymerase chain reaction (PCR) methods has been widely confused in the past; to seek to remove the confusion an international working party has defined preferred terminology for PCR methods (Bustin et al., 2009). Abbreviations used here conform to those recommendations.

The PCR is a method for amplifying, that is, synthesising multiple copies of, specific pieces of DNA or RNA. Since its introduction in the 1980s, the application of PCR has changed the face of qualitative microbiology in many laboratories and has the potential to revolutionise quantitative practices. A number of commercial kits are available for the purpose. The primary use of the procedure to date in food microbiology has been in the detection of specific microorganisms, especially pathogens such as *Salmonella*, *Listeria*, *E. coli* O157:H7, *Campylobacter* and *Cronobacter* in foods. It is also very useful for the speciation of some genera – for example, *Campylobacter* spp. However, current developments seek increasingly to quantify the levels of defined organisms, or groups of organisms, in foods using quantitative PCR (qPCR; Postollec et al., 2011; Filion, 2012).

For PCR to take place four basic components are required:

- a DNA template containing the target sequence that is to be amplified (for pathogen detection this sequence must be highly specific to the organism concerned, often a single gene);
- primers – a pair of short single-stranded oligonucleotides that complement exactly specific parts of the target sequence;
- a heat-stable DNA-polymerase enzyme to catalyse the reaction, usually a *Taq* polymerase from a thermophilic bacterium;
- free nucleotides that are used as the building blocks for multiple copies of the DNA template.

For qualitative tests on a food matrix, the procedure requires extraction of the DNA from target organisms separated from the food matrix, for example, by growing the organism in enrichment culture and destruction of DNA from dead cells. However, the technique will also detect DNA from viable but non-culturable cells and may detect DNA from non-viable cells. Quantitative tests require maceration of the sample in, for example, a Stomacher™ filter bag to retain as much as possible of the food substrate before chemical extraction of the DNA into a suitable buffer. Examples of procedures that have been used include those described by Martín et al. (2006), Hong et al. (2007), Malorny et al. (2008), Postollec et al. (2011) and Garcia et al. (2013).

After extraction, the first stage of the analytical process is to raise the temperature of the reaction mixture to 90–95°C. This causes the double-stranded DNA to denature into single strands. The temperature is then reduced to about 50–60°C to allow the two primers to bind (anneal) at specific points on the single-stranded DNA of the target sequence. Finally, the temperature is raised to 68–74°C to enable the DNA-polymerase enzyme to catalyse the duplication of the target sequence, starting at the annealed primers on each single strand, in a process known as extension. The exact temperature chosen for these two steps is dependent on the particular assay being used. The process results in two double-stranded DNA fragments that are identical copies of the original target sequence. The temperature cycling process is then repeated, typically 30–40 times, creating a theoretical doubling of the number of copies of the target sequence at each cycle. This gives an exponential increase in target DNA concentration and produces sufficient DNA for reliable detection from a single target sequence in a few hours. The exact conditions depend on the reagents used and the annealing temperatures of the primers. When used as a presence or absence test, the product of the PCR is visualised after a specific number of cycles using gel electrophoresis – checking for DNA of the expected size, and confirming, if desired, by sequencing the nucleotides in the DNA fragment.

'Real-time Quantitative PCR' (qPCR) requires the use of a fluorescent reporter dye or DNA probe and measurement of the fluorescence during each cycle of the PCR. The probe is an oligonucleotide that binds specifically to the target DNA sequence during annealing. Probes typically have a fluorescent reporter dye at the 5′

end and a quencher dye, which inhibits fluorescence, at the 3′ end. During the extension stage the probe is hydrolysed by the DNA-polymerase enzyme and the released dye fluoresces more strongly. The fluorescence emitted is measured at each cycle and increases in proportion to the number of target sequence copies produced. To quantify the assay, the cycle number (C_T), or time, at which the fluorescence intensity rises above a pre-set threshold level, is recorded for each sample (Fig. 9.4); in absolute copy number qPCR a set of standards is always run at the same time. Fewer cycles are required to detect threshold fluorescence with high levels of target DNA. Since the probe binds specifically to the target DNA, non-specific PCR products are not detected. A standard curve can be constructed based on the relationship between the cell numbers and the C_T (Fig. 9.5), which then permits calculation of the amount of target DNA present in the sample in terms of the expected cell density. Results on unknown sample extracts can be interpolated against the data on the standard curve. With this technique it is essential to run control and reference samples on each occasion to ensure validation of the standard curve. These usually include a 'blank' and a sample known to include a specified concentration of the target DNA. In addition, it is strongly advised to spike the template DNA with an 'internal amplification control' (IAC), which consists of an artificial DNA template sequence and corresponding qPCR assay, using a different fluorescent reporter, which has a known C_T for detection. If the C_T for the IAC significantly exceeds the expected C_T, the presence of substances inhibitory to the qPCR reaction is indicated, and therefore the result of the main assay is in doubt – that is, it could be a false-negative, or a higher level than

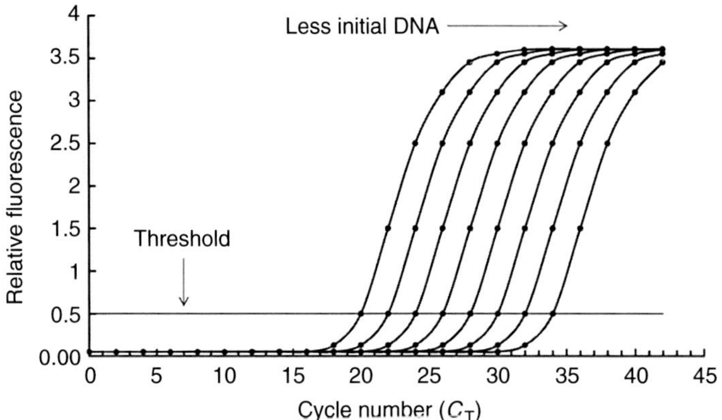

FIGURE 9.4 Hypothetical Example of Increasing Fluorescence Response With Increasing Levels of DNA in a qPCR Test

Higher levels of extracted DNA require fewer cycles (C_T) to develop a quantifiable level of fluorescence and vice versa. A low level of fluorescence may occur before sufficient DNA replication is detected and a threshold level of fluorescence is usually pre-set for recording the effective C_T value for detection (illustrated here by a horizontal line set above the point of inflexion of the curve)

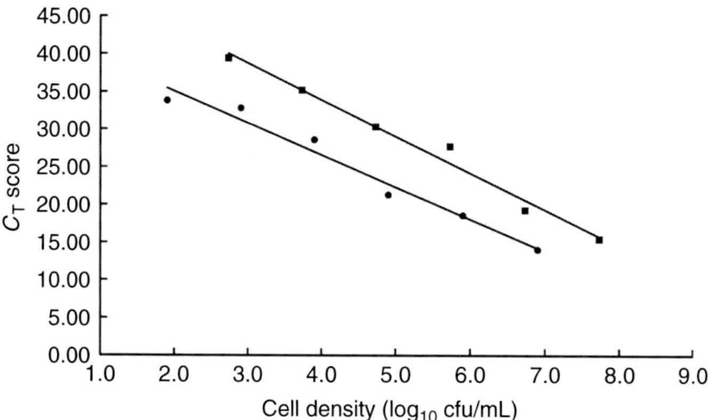

FIGURE 9.5

Relationship between C_T values and mean counts (as log cfu/mL) of cells from dilutions of pure cultures of *Campylobacter jejuni* (■) and *C. coli* (●). Regression equation: *C. jejuni*: $y = 53.34 - 4.86x$ with $R^2 = 0.98$; *C. coli*: $y = 43.57 - 4.25x$ with $R^2 = 0.97$. For the sake of clarity the 95% CIs and prediction bands are not shown

Unpublished data reproduced with permission of Dr J.E.L. Corry (2015)

indicated. In such a situation, it is necessary to re-extract the DNA from the sample in order to try to obtain a valid result.

The major benefit of the qPCR technique lies in its rapidity, giving estimates of contamination within hours, as opposed to the days often required for traditional cultural methods. It is also both quantitative and more specific than PCR followed by gel electrophoresis for which the PCR products need to be confirmed by DNA sequencing. However, the technique requires specialist hardware, available from several manufacturers. Commercial test kits, which are quite expensive when compared with the capital and running costs of traditional methods of detection and enumeration, are not always necessary provided the laboratory has the time and expertise to isolate and prepare its own molecular markers and other tools. As with other alternative methods, from a statistical perspective the relationship between the observed C_T and the level of DNA present in the reference sample (as a marker of the cell density) generally conforms well to the Normal distribution, but often shows overdispersion in relation to occurrence of organisms in a natural matrix, for example, a food. Hence, results often indicate higher apparent levels of target organisms than those found by traditional colony count methods (Postollec et al., 2011). This may be due to the presence of non-viable cells with detectable DNA or to free DNA in the samples.

RT-qPCR

A further development in PCR technology is that of reverse transcription PCR (RT-qPCR) that is used widely in genetic studies since it detects (and estimates) RNA rather than DNA. Whilst it is used to detect only viable cells, it suffers from the potential problem that the half-life of extracted RNA is very limited. However, RT-qPCR has been used to detect food-borne viruses in shellfish (eg, Häfliger et al., 1997), *Listeria monocytogenes* in ground meats (Klein and Juneja, 1997), yeasts and moulds in yoghurts (Bleve et al., 2003) and the toxigenic potential of *Clostridium botulinum* strains (McGrath et al., 2000).

Increasingly, it is being used for quantitative applications. For instance, Zhang et al. (2011) were able to detect as few as 2 cfu *Salmonella*/25 g in various raw plant foods and Falentin et al. (2010) have used RT-qPCR to follow the quantitative development of propionibacteria and lactobacilli during ripening of Emmental cheese.

There is no doubt about the potential for various forms of PCR analysis to revolutionise food microbiology but we must recognise that whilst it is essentially both a very sensitive and rapid analytical tool, the statistics associated with analysis of its findings do not differ from those of other quantal or quantitative methods in food microbiology.

REFERENCES

AOAC, 2012. Official Methods of Analysis, 19th ed. Association of Official Analytical Chemists, Washington, DC (updated 2015).

Bleve, G., Rizzotti, L., Dellaglio, F., Torriani, S., 2003. Development of reverse transcription (RT)-PCR and real-time RT-PCR assays for rapid detection and quantification of viable yeasts and molds contaminating yogurts and pasteurized food products. Appl. Environ. Microbiol. 69, 4116–4122.

Breed, R.S., Brew, J.D., 1925. Counting bacteria by means of the microscope, second reprint. New York State Agricultural Experiment Station, Geneva, Technical Circular 58, pp. 1–12.

Brew, J.D., 1914. A comparison of the microscopical method and the plate method of counting bacteria in milk. New York Agricultural Experiment Station, Geneva Technical Bulletin, p. 373.

Bustin, S.A., et al., 2009. The MIQE guidelines: minimum information for publication of quantitative real-time pcr experiments. Clin. Chem. 55, 611–624.

Dakin, J.C.Y., Smith, D.J., Taylor, A.Mc.M., 1964. Variance in the Howard mould count arising from observer and distribution differences. Leatherhead Food Research Association Technical Circular No. 259.

Davies, O.L., 1954. The Design and Analysis of Industrial Experiments. Oliver & Boyd, London.

Diaz, M., Herrero, M., Garcia, L.A., Quirōs, C., 2010. Application of flow cytometry to industrial microbiological processes. Biochem. Eng. J. 48 (3), 385–407.

Duncan, A.J., 1965. Quality Control and Industrial Statistics, third ed. R.D. Irwin, Chicago.

Eisenberg, W.V., 1968. Mould counts of tomato products as influenced by different degrees of comminution. Q. Bull. Assoc. Food Drug Off. U.S. 32, 173–179.

Falentin, H., Postollec, F., Parayre, S., Henaff, N., Le Bivic, P., Richoux, P.R., Thierry, A., Sohier, D., 2010. Specific metabolic activity of ripening bacteria quantified by real-time reverse transcription PCR throughout Emmental cheese manufacture. Int. J. Food Microbiol. 144, 10–19.

Filion, M., 2012. Quantitative Real-Time PCR in Applied Microbiology. Caister Academic Press, Poole, UK.

Garcia, A.B., Kamara, J.N., Vigre, H., Hoorfar, J., Josefsen, M.H., 2013. Direct quantification of *Campylobacter jejuni* in chicken fecal samples using real time PCR: evaluation of six rapid DNA extraction methods. Food Anal. Methods 6, 1728–1738.

Häfliger, D., Gilgen, M., Lüthy, J., Hübner, P., 1997. Semi-nested RT-PCR systems for small round structured viruses and detection of enteric viruses in seafood. Int. J. Food Microbiol. 37, 27–36.

Harrigan, W.F., McCance, M.E., 1998. Laboratory Methods in Food Microbiology, third ed. Academic Press, London.

Hong, J., Jung, W.K., Kim, J.M., Kim, S.H., Koo, H.C., Ser, J., Park, Y.H., 2007. Quantification and differentiation of *Campylobacter jejuni* and *Campylobacter coli* in raw chicken meats using a real-time PCR method. J. Food Prot. 70, 2015–2022.

Jarvis, B., 1977. A chemical method for the estimation of mould in tomato products. J. Food Technol. 12, 581–591.

Jarvis, B., 2012. Some practical and statistical aspects of the comparative evaluation of microbiological culture media. In: Corry, J.E.L., Curtis, G.D.W., Baird, R.M. (Eds.), Handbook of Culture Media for Food and Water Microbiology. third ed. Royal Society of Chemistry, Cambridge, UK, pp. 3–38.

Klein, P.G., Juneja, V.K., 1997. Sensitive detection of viable *Listeria monocytogenes* by reverse transcription-PCR. Appl. Environ. Microbiol. 63, 4441–4448.

Malorny, B., Löfström, C., Wagner, M., Kramer, N., Hoorfar, J., 2008. Enumeration of *Salmonella* bacteria in food and feed samples by real-time PCR for quantitative microbial risk assessment. Appl. Environ. Microbiol. 74, 1299–1304.

Martín, B., Jofré, A., Garriga, M., Pla, M., Aymerich, T., 2006. Rapid quantitative detection of *Lactobacillus sakei* in meat and fermented sausages by real-time PCR. Appl. Environ. Microbiol. 72, 6040–6048.

McGrath, S., Dooley, J.S.G., Haylock, R.W., 2000. Quantification of *Clostridium botulinum* toxin gene expression by competitive reverse transcription-PCR. Appl. Environ. Microbiol. 66, 1423–1428.

Meynell, G.G., Meynell, E., 1965. Theory and Practice in Experimental Bacteriology. Cambridge University Press, London.

Postollec, F., Falentin, H., Pavan, S., Combrisson, J., Sohier, D., 2011. Recent advances in quantitative PCR (qPCR) applications in food microbiology. Food Microbiol. 28, 848–861.

Potts, S.J., Slaughter, D.C., Thompson, J.F., 2001. Measuring mold infestation in raw tomato juice. J Food Sci. 67, 321–325.

Richards, J.C.S., Jason, A.C., Hobbs, G., Gibson, D.M., Christie, R.H., 1978. Electronic measurement of bacterial growth. J. Phys. E 11, 560–568.

Robinson, J.P., 2009. Cytometry – a definitive history of the early days. In: Sack, U., Tárnok, A., Rothe, G. (Eds.), Cellular Diagnostics. Basics, Methods and Clinical Applications of Flow. Karger Medical and Scientific Publishers, Basel, Switzerland, pp. 1–28.

Takahasi, K., Ishida, S., Kurokawa, M., 1964. Statistical consideration on sampling errors in total bacteria cell count. J. Med. Sci. Biol. 17, 73–86.

Thornton, H.G., Gray, P.H.H., 1934. The numbers of bacterial cells in field soils as estimated by the ratio method. Proc. R. Soc. Lond. B Biol. Sci. 115, 522–543.

Vas, K., Fabri, I., Kutz, N., Lang, A., Orbanyi, T., Szabo, G., 1959. Factors involved in the interpretation of mold counts of tomato products. Food Technol. 13, 318–322.

Wood, J.M., Jarvis, B., Wiseman, A., 1976. The separation of microorganisms from food. Chem. Ind. (Lond.) 27, 783–784.

Wood, J.M., Lach, V.H., Jarvis, B., 1977. Detection of food-associated microbes using electrical impedance measurements. J. Appl. Bacteriol. 43, xiv–xiv10.

Zhang, G., Brown, E.W., González-Escalona, N., 2011. Comparison of real-time PCR, reverse transcriptase real-time PCR, loop-mediated isothermal amplification, and the FDA conventional microbiological method for the detection of *Salmonella* spp. in produce. Appl. Environ. Microbiol. 77, 6495–6501.

CHAPTER

Measurement uncertainty in microbiological analysis 10

There are only two absolute certainties in life: death and taxes! Whatever task we undertake, no matter how menial or how sophisticated, we are usually faced with some lack of certainty in the outcome! It is therefore essential to have a basic understanding of what is meant by uncertainty in relation to microbiological data. In this chapter we shall consider briefly the definitions of uncertainty, its causes and general aspects of uncertainty measurement. In the next chapter we shall consider details of the various approaches to its determination. Corry et al. (2006) have published a detailed critical review of measurement uncertainty in quantitative microbiological analysis.

We can identify many causes of variability in microbiological laboratory practices. For instance, the inability of an isolate to give typical reactions on a diagnostic medium; the use of the incorrect ingredients in a culture medium; the consequence of changing brands of commercial media; use of non-standard conditions in the preparation, sterilisation and use of a culture medium or diluent; equipment errors; the tolerance applied to the shelf life of reagents; human errors in weighing, dispensing, pipetting and other laboratory activities; the relative skill levels of different analysts; the relative well-being of anyone who is doing the analyses; and so on – *ad infinitum*!

These are trite examples of biological, instrumental and personal bias that affect the accuracy and precision and, hence, the uncertainty of microbiological tests, a situation that constantly faces scientists involved in laboratory management. To interpret properly the results obtained using any analytical procedure, whether physical, chemical or biological, requires careful consideration of the diverse sources of actual or potential error associated with the results obtained. It is important to differentiate between the two meanings of the term 'error': in common parlance, error means mistakes, for instance, failure properly to undertake a defined test protocol; in statistics 'error' means statistical variation, although much variation can be caused by common errors!

Three groups of statistical errors influence any analytical result:

1. *random errors*, associated with the distribution of analytical targets in the primary sample matrix and in the analytical (test) portion, errors in the composition of the culture media, etc.;
2. *systematic* errors associated with analytical equipment and procedures; and
3. *modification of the systematic errors* in a laboratory carrying out test procedures due to environmental, equipment and individual analysts' personal traits.

Statistical Aspects of the Microbiological Examination of Foods. http://dx.doi.org/10.1016/B978-0-12-803973-1.00010-3

ACCURACY, TRUENESS AND PRECISION

Accuracy is a qualitative concept (VIM; *International Vocabulary of Metrology*: JCGM, 2012). In simple terms, accuracy can be defined as the correctness of a result relative to an expected outcome, whilst *precision* is a measure of the variability of test results. Accuracy is defined (Anon, 2006a) as 'the closeness of agreement between a measurement result and the true value'. Accuracy is a combination of trueness and precision (a combination of random components and systematic error or bias components), although in biostatistics terminology 'trueness' is often referred to as 'accuracy'! This differs somewhat from the definition given by VIM (JCGM, 2012): 'closeness of agreement between a measured quantity value and a true quantity value of a measurand'. Accuracy is essentially the 'absence of error'; the more accurate a result, the lower is the associated error of the test. It is important to note that this definition of the term 'accuracy' applies only to results; in common parlance, people often misuse the term accuracy when they refer to 'the accuracy of a method' or 'the accuracy of a piece of equipment' such as a pipette.

Trueness is defined (Anon, 2006a) as 'the closeness of agreement between the average value obtained from a large series of test results and an accepted reference value'. Trueness is equivalent to an absence of 'bias', which is the difference between the expectation of the test results and an accepted reference value and is a measure of systematic error. Trueness, sometimes referred to as accuracy, may correctly be contrasted with precision.

Precision is defined as the 'closeness of agreement between independent test results obtained under stipulated conditions'. Precision depends only on the distribution of random errors and does not relate to a true or specified value. The *measure of precision* is expressed usually in terms of imprecision and computed as a standard deviation of the test results. Lack of precision is reflected by a large standard deviation.

Independent test results are results obtained in a manner that is not influenced by any previous or subsequent results on the same or similar test material. Quantitative measures of precision depend critically on the stipulated conditions. Repeatability and reproducibility conditions are particular sets of extreme stipulated conditions (Anon, 2006a).

Relationships between trueness, accuracy, precision and uncertainty are illustrated schematically in Fig. 10.1 (AMC, 2003). The concepts of accuracy and trueness must take account of error and precision. Uncertainty estimates (*qv*) provide a simple way to quantify such parameters.

However, since in a real-life situation we can never know the 'true' or 'correct' answer, trueness can be assessed only in a validation-type trial using an accepted reference material for which a 'true' concentration is known. This is more complex in microbiology than it is in physics and chemistry. Of course, repeating the test using one or more fundamentally different methods might also provide an indication of the trueness of the original estimate!

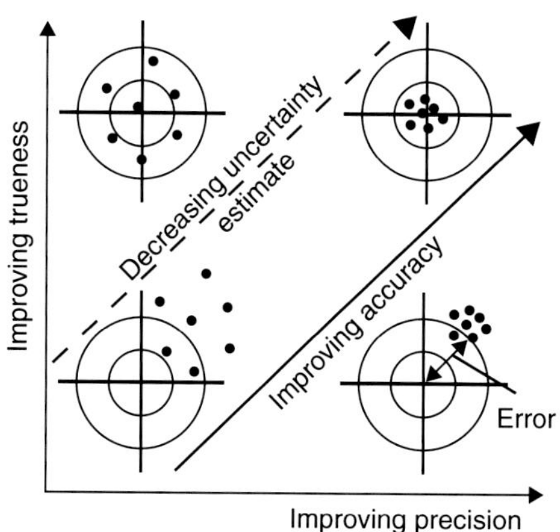

FIGURE 10.1 Relationships Between Trueness, Accuracy, Precision and Uncertainty in Analytical Results (AMC, 2003)

The schematic illustrates a series of shots at a target. At the bottom left the grouping of shots is neither true nor precise. Improving trueness without improving precision moves the grouping onto the target (top left), whilst improving precision but not trueness results in a closer grouping of shots that are still some distance (the error) from the 'bull's eye' (bottom right). Improving both trueness and precision (ie, improving accuracy and decreasing uncertainty) gives a close grouping of shots around the bull's eye.

Reproduced by permission of the Royal Society of Chemistry, London

MEASUREMENT UNCERTAINTY

The Eurachem definition of uncertainty of a measurement (Ellison and Williams, 2012) is: 'A parameter associated with the result of a measurement that characterises the dispersion of the values that could reasonably be attributed to the measurand'. The term 'measurand' is a bureaucratic way of saying 'analyte'. Translated into simple English this definition can be rewritten as 'Uncertainty is a measure of the likely range of values that is indicated by an analytical result'. For quantitative data [eg, colony counts or Most Probable Number (MPN) values] a measure of uncertainty may be any appropriate statistical parameter associated with the test result. Such parameters include the standard deviation, the standard error of the mean or a confidence interval around that mean.

Measures of repeatability (r) and reproducibility (R) are the cornerstones of the estimation of analytical uncertainty. Anon (2010) defines repeatability as 'a measure of variability derived under specified repeatability conditions', that is, independent test results obtained with the same method on identical test items in the same laboratory by the same analyst using the same equipment, the same batches of culture media and diluents, and tested within a short interval of time. By contrast, reproducibility is 'a measure of variability derived under reproducibility conditions', that is, test results are obtained with the same method on identical test items in different laboratories with different operators using different equipment and media at different times. Valid statements of repeatability and reproducibility require specification of the conditions used.

Intermediate reproducibility (Anon, 1994) is defined as 'a measure of reproducibility derived under reproducibility conditions within a single laboratory', that is, test results obtained with the same method on identical test items in one laboratory with different operators using different equipment and/or media at different times.

Standard uncertainty (u_y) of a measurement (y) is defined (Anon, 2009) as 'the result obtained from the values of a number of other quantities, equal to the positive square root of a sum of terms, the terms being the variances or covariances of these other quantities weighted according to how the measurement result varies with changes in these quantities'. More simply, this means that standard uncertainty is the square root of the sum of those variances that are likely to have influenced the result, that is, the overall standard deviation.

Expanded uncertainty (U) is 'the quantity defining an interval about the result of a measurement expected to encompass a large fraction of the distribution of values that could reasonably be attributed to the measurand' (Anon, 2009). Expanded uncertainty values are derived by multiplying the standard uncertainty by a 'coverage factor' to provide approximate confidence intervals for repeatability and reproducibility around the mean value. Routinely, a coverage factor of 2 is used to give approximate 95% distribution limits (95% confidence interval) around a 'Normalised' mean value.

For quantal data (eg, presence or absence tests) uncertainty measures cannot be derived in the same way. However, procedures that derive quantitative estimates from quantal responses can be used to provide a measure of variability and hence of uncertainty, for example, the standard error associated with derived values for a level of detection of 50% positive results (LOD_{50}) (*qv*) or for the relative proportions of positive and negative results that conform to a binomial distribution in a comparative evaluation of methods (see later).

HOW IS UNCERTAINTY ESTIMATED?

There are two totally different approaches to the estimation of uncertainty (Corry et al., 2006):

> The 'bottom-up' approach uses the estimates of all errors associated with all the relevant steps undertaken during an analysis to derive a value for a 'combined standard uncertainty' associated with a method (Ellison and Williams, 2012; Niemelä, 2002, 2003; NMKL, 2002). Essentially this approach provides a broad indication of the possible level of uncertainty associated with a method rather than a measurement. In practice this approach will always under-estimate the extent of variation since it cannot take into account either sample matrix-associated errors or the actual day-to-day

variation seen in a laboratory. For these and other reasons, this approach is not considered to be appropriate for microbiological analyses (Corry et al., 2006; Anon, 2006b).

The 'top-down' approach (Anon, 2006b) is based on statistical analysis of data generated in intra- or inter-laboratory collaborative studies of a method used to analyse a diversity of matrices. It therefore provides an estimate of the uncertainty of a measurement associated with the result obtained using a specific method (see, eg, Corry et al., 2006, 2007; Jarvis et al., 2007a,b, 2012). This approach is also used when estimating microbiological Intermediate Reproducibility.

Quantitative tests

For quantitative data (eg, colony counts and MPN estimates), measures of 'repeatability' and 'reproducibility' are derived as the standard deviations of repeatability (s_r) and reproducibility (s_R). However, since microbiological data do not usually conform to a 'Normal' distribution, the estimates require mathematical transformation prior to statistical analysis. For most purposes, a $\log_{10}$ transformation is used to 'Normalise' the data but in cases of significant over-dispersion the use of a negative binomial transformation may be necessary (Niemelä, 2002). If there is reason to believe that data conform to a Poisson distribution, then a square root transformation is required, since the population variance (σ^2) is numerically equal to the population mean (λ) value.

Statistical analyses of collaborative trial data are generally done by analysis of variance (ANOVA) after removing any outlying values, as described by Youden and Steiner (1975) and by Horwitz (1995). However, it has been argued (eg, AMC, 1989a,b, 2001) that it is wrong to eliminate outlier data and that application of robust methods of analysis is preferable. One approach to robust analysis is a 'robusticised' ANOVA procedure based on Huber's H15 estimators for the robust mean and standard deviation of the data (AMC, 1989a,b, 2001; Anon, 1998). An alternative approach uses Rousseeuw's Scale Estimator (see Chapter 11). A major drawback to use of published robust techniques for analysis of data from inter-laboratory trials is that they do not permit simple derivation of components of variance. The components of variance are the variances of the fixed or random factors associated with an ANOVA and are identified within the 'expected mean square'. They might, for instance, include the variances within and between test samples, analysts, laboratories, etc. A novel approach to overcome this disadvantage is by the use of a stepwise robust analysis for 'nested' trial data, described by Hedges and Jarvis (2006) and Hedges (2008). Examples of the use of traditional and robust ANOVA and related procedures are given in Chapter 11.

Intermediate reproducibility of quantitative data can be done using similar procedures to estimate intermediate (ie, within-laboratory) reproducibility associated with the use of an analytical procedure in a single laboratory. Even data obtained, for instance, in laboratory quality monitoring can be used to provide an estimate of intra-laboratory reproducibility. Anon (2006b) describes a statistical procedure for analysis of paired data. A worked example is provided in Chapter 11.

Qualitative tests

Estimation of uncertainty associated with quantal methods (eg, presence or absence tests) is currently the subject of much discussion. Many of the potential errors that affect quantitative methods also affect qualitative methods, but there are also additional errors that are associated with the analytical procedures.

By definition, the output from a series of quantal tests is a number of either positive or negative responses (see Chapter 8). There is an intrinsic need to ensure effective growth of the target organism to appropriate levels during all the cultural stages of such tests – so culture medium composition, incubation times and temperatures, etc., are critical to the success of the test. It is essential also to ensure that the confirmatory stages of a test protocol can actually identify the target organism. Knowledge of matrix effects, including the potential effect of competitive organisms, is also of major importance for both cultural and confirmatory stages of a test protocol.

One approach to quantifying such data was the derivation of the Accordance and Concordance concept (Langton et al., 2002) that sought to provide measures 'equivalent to the conceptual aspects of repeatability and reproducibility'. However, experience suggests that this approach is not sufficiently robust to be used in the manner proposed since it merely reinterprets the original data.

Provided that a sufficient number of parallel tests has been undertaken at each of several levels of potential contamination, it is possible to quantify test responses in terms of an estimated level of detection for (eg) 50% positives (LOD_{50}). Note that the term LOD refers to the level of detection, *not* to the limit of detection.

The statistical approach of Hitchins and Burney (2004) essentially estimated the MPN of organisms at each test level and then analysed the relative MPN values using the non-parametric Spearman-Kärber approach. Subsequent work has shown that other approaches based on a generalised linear model for logit or loglog data are more appropriate (Wilrich and Wilrich, 2009; Wilrich, 2010; Mărgăritescu and Wilrich, 2013).

The common theme of these approaches is to transform purely qualitative data into a quantitative format for which error values can be assigned in order to derive an estimate of the uncertainty of the test result. An extrapolation of the approach is to determine the LOD_0 and LOD_{90} values such that a dose–response curve can be derived. This may be of importance in differentiating between methods capable of detecting specific organisms at a similar LOD_{50} level but for which the absolute limit of detection (LOD_0) and a selected higher level of detection (eg, LOD_{90}) differ.

A simpler alternative approach is to estimate the uncertainty associated with the proportions of test samples giving a positive response, based on the binomial or other distribution. Examples of the way in which such approaches to analysis of qualitative data can be used are illustrated in worked examples in Chapter 11.

REPORTING OF UNCERTAINTY

The expression of uncertainty is of some importance. Strictly, a transformed value such as a log count of colony numbers is dimensionless, but for simplicity I will ignore this nicety!

Assuming a mean aerobic colony count = 5.00 $\log_{10}$ cfu/g, a reproducibility standard deviation of ±0.25 $\log_{10}$ cfu/g and a coverage factor of 2, then the 95% expanded uncertainty value is ±0.50 $\log_{10}$ cfu/g. The results might be reported as, for instance, "the aerobic colony count on product X is:

5.00 ± 0.50 $\log_{10}$ cfu/g, with a 95% probability; or
5.00 $\log_{10}$ cfu/g ± 10% with a 95% probability; or
between $10^{4.5}$ and $10^{5.5}$ cfu/g with a 95% probability; or
between 4.5 and 5.5 $\log_{10}$ cfu/g with a 95% probability".

If log-transformed data are back-transformed, that is, 5.0 $\log_{10}$ cfu/g = 100,000 cfu/g, then the associated uncertainty estimate limits will not be symmetrical. For instance, back-transformation of a log colony count of 5.0 ± 0.50 $\log_{10}$ cfu/g is equivalent to a count of 100,000 cfu/g with 95% uncertainty limits of approximately 31,600 and 316,200 cfu/g.

It is important not to refer to analytical methods as having a precision of, for example, ±10% based on uncertainty estimates. Uncertainty is a measure of variability in a test result and is therefore a measure of the lack of precision in the result rather than an estimate of the precision of the method itself.

SAMPLING UNCERTAINTY

Estimates of measurement uncertainty do not take account of the distribution of organisms within a lot, within a primary sample or within a test portion. Individual samples of foodstuff may be drawn either from a bulk lot or from individual primary and secondary packaged units within a lot, as discussed in Chapters 5 and 6. Estimates of measurement uncertainty are based on the preparation of an initial suspension of a test portion from which serial dilutions have been prepared for testing. It is assumed that the primary samples have been drawn randomly and with care to ensure that they are representative of the entire lot of product and of the primary sample unit. In practice this may not be so, no matter how carefully the samples are drawn. It is also generally assumed that microbial cells are randomly distributed within the lot, the sample unit and the test portion but, as stated earlier, there are many cases of heterogeneous dispersion of microbial cells within a food matrix. Consequently, it is not unreasonable to suppose that any estimation of measurement uncertainty will underestimate the total level of uncertainty associated with an analysis. The food matrix may itself affect the estimate of measurement uncertainty (Anon, 2006b) because of its effect on the recovery of organisms and/or interactions between food particles and the test system. Such effects will tend to widen the estimate of measurement uncertainty *per se* but do not provide estimates of sampling uncertainty.

Investigation of the chemical composition of diverse foodstuffs (Ramsey et al., 2001; Lyn et al., 2002, 2003) has demonstrated that the sampling uncertainty is at least as great, and is often greater than, the estimate of measurement uncertainty. Jarvis et al. (2007b, 2012) showed that a similar situation applies to the microbiological examinations of foods. In an examination of imported prawns by a UK Health Protection Agency laboratory, sufficient replicate samples and analyses were taken to estimate the extent of both sampling uncertainty and measurement uncertainty (Jarvis et al., 2007b). The estimate of total uncertainty was ±18.6%. The contribution of sampling uncertainty to the total was ±15.3%, whilst the estimated measurement uncertainty was ±10.8%. A more detailed study of sampling uncertainty (Jarvis et al., 2012) demonstrated that the variance associated with sampling *per se* accounted for 50–95% of the total uncertainty of reproducibility for a range of microbes in a number of foodstuffs. It is clear from these studies that the extent of sampling uncertainty for microbiological examination of foods is likely to exceed the estimate of the actual measurement uncertainty. In other words, the measurement uncertainty of a microbiological examination of a sample of product will generally underestimate the extent of the overall dispersion of microorganisms within a lot. As is the case with chemical analyses, such underestimates may be critical in assessing the compliance of food materials with legislative and commercial microbiological criteria for foods (Chapter 14).

REFERENCES

AMC, 1989a. Robust statistics – how not to reject outliers. Part 1: basic concepts. Analyst 114, 1693–1697.

AMC, 1989b. Robust statistics – how not to reject outliers. Part 2: inter-laboratory trials. Analyst 114, 1699–1702.

AMC, 2001. Robust statistics: a method of coping with outliers. AMC Brief No. 6. Analytical Methods Committee, Royal Society of Chemistry, London. <http://www.rsc.org/images/robust-statistics-technical-brief-6_tcm18-214850.pdf> (accessed 13.07.15).

AMC, 2003. Terminology – the key to understanding analytical science. Part 1: accuracy, precision and uncertainty. AMC Brief No. 13. Analytical Methods Committee, Royal Society of Chemistry, London. <http://www.rsc.org/images/terminology-part-1-technical-brief-13_tcm18-214863.pdf> (accessed 13.07.15).

Anon, 1994. Accuracy (Trueness and Precision) of Measurement Methods and Results – Part 2: Basic Method for the Determination of Repeatability and Reproducibility of a Standard Measurement Method. ISO 5725-2:1994. International Standards Organisation, Geneva.

Anon, 1998. Accuracy (Trueness and Precision) of Measurement Methods and Results – Part 5: Alternative Methods for the Determination of the Precision of a Standard Measurement Method. ISO 5725-5:1998/COR 1:2005. International Standards Organisation, Geneva.

Anon, 2006a. Statistics – Vocabulary and Symbols – Part 1: General Statistical Terms and Terms Used in Probability. ISO 3534-1:2006. International Standards Organisation, Geneva.

Anon, 2006b. Microbiology of Food and Animal Feeding Stuffs – Guide on Estimation of Measurement Uncertainty for Quantitative Determinations. ISO TS 19036:2006/Amd 1:2009. International Standards Organisation, Geneva.

Anon, 2009. Guide to the Expression of Uncertainty in Measurement (GUM). ISO/IEC Guide 98-1:2009. International Standards Organisation, Geneva.

Anon, 2010. Guidance for the Use of Repeatability, Reproducibility and Trueness Estimates in Measurement Uncertainty Estimation. ISO TS 21748:2010. International Standards Organisation, Geneva.

Corry, J.E.L., Jarvis, B., Passmore, S., Hedges, A.J., 2006. A critical review of measurement uncertainty in the enumeration of food microorganisms. Food Microbiol. 24, 230–253.

Corry, J.E.L., Jarvis, B., Hedges, A.J., 2007. Measurement uncertainty of the EU methods for microbiological examination of red meat. Food Microbiol. 24, 652–657.

Ellison, S.L.R., Williams, A. (Eds.), 2012. Eurachem/CITAC Guide: Quantifying Uncertainty in Analytical Measurement, third ed. Available from: <https://www.eurachem.org/index.php/publications/guides/quam> (accessed 13.07.15).

Hedges, A.J., 2008. A method to apply the robust estimator of dispersion, *Qn*, to fully-nested designs in the analysis of variance of microbiological count data. J. Microbiol. Methods 72, 206–208.

Hedges, A.J., Jarvis, B., 2006. Application of robust methods to the analysis of collaborative trial data using bacterial colony counts. J Microbiol. Methods 66, 504–511.

Hitchins, A.D., Burney, A.A., 2004. Determination of the limits of detection of AOAC validated qualitative microbiology methods. In: AOAC International 118th Annual Meeting Program, Poster Abstract #P-1021, p. 153.

Horwitz, W., 1995. Protocol for the design, conduct and interpretation of method performance studies. Pure Appl. Chem. 67, 331–343.

Jarvis, B., Corry, J.E.L., Hedges, A.J., 2007a. Estimates of measurement uncertainty from proficiency testing schemes, internal laboratory quality monitoring and during routine enforcement examination of foods. J. Appl. Microbiol. 103, 462–467.

Jarvis, B., Hedges, A.J., Corry, J.E.L., 2007b. Assessment of measurement uncertainty for quantitative methods of analysis: comparative assessment of the precision (uncertainty) of bacterial colony counts. Int. J. Food Microbiol. 116, 44–51.

Jarvis, B., Hedges, A.J., Corry, J.E.L., 2012. The contribution of sampling uncertainty to total measurement uncertainty in the enumeration of microorganisms in foods. Food Microbiol. 30, 362–371.

JCGM (Joint Committee for Guides in Metrology), 2012. International Vocabulary of Metrology – Basic and General Concepts and Associated Terms (VIM). Report 200, third ed. Bureau International des Poids et Mesures, Paris, France. <http://www.bipm.org/en/publications/guides/> (accessed 11.07.15).

Langton, S.D., Chevennement, R., Nagelkeke, N., Lombard, B., 2002. Analysing collaborative trials for qualitative microbiological methods: accordance and concordance. Int. J. Food Microbiol. 79, 171–181.

Lyn, J.A., Ramsey, M.H., Wood, R., 2002. Optimised uncertainty in food analysis: application and comparison between four contrasting 'analyte–commodity' combinations. Analyst 127, 1252–1260.

Lyn, J.A., Ramsey, M.H., Wood, R., 2003. Multi-analyte optimisation of uncertainty in infant food analysis. Analyst 128, 379–388.

Mărgăritescu, I., Wilrich, P.Th., 2013. Determination of the RLOD of a qualitative microbiological measurement method with respect to a reference measurement method. J. AOAC Int. 96, 1086–1091.

Niemelä, S.I., 2002. Uncertainty of Quantitative Determinations Derived by Cultivation of Microorganisms, second ed. Centre for Metrology and Accreditation, Advisory Commission for Metrology, Chemistry Section, Expert Group for Microbiology, Helsinki, Finland, Publication J3/2002.

Niemelä, S.I., 2003. Measurement uncertainty of microbiological viable counts. Accred. Qual. Assur. 8, 559–563.

NMKL, 2002. Measurement of Uncertainty in Microbiological Examination of Foods. NMKL Procedure No. 8, second ed. Nordic Committee on Food Analysis.

Ramsey, M.H., Lyn, J.A., Wood, R., 2001. Optimised uncertainty at minimum overall cost to achieve fitness-for-purpose in food analysis. Analyst 126, 1777–1783.

Wilrich, P.Th., 2010. The determination of precision of qualitative measurement methods by interlaboratory experiments. Accred. Qual. Assur. 15, 439–444.

Wilrich, C., Wilrich, P.Th., 2009. Estimation of the POD function and the LOD of a qualitative microbiological measurement method. J. AOAC Int. 92, 1763–1772.

Youden, W.J., Steiner, E.H., 1975. Statistical Manual of the AOAC. AOAC, Washington.

CHAPTER

Estimation of measurement uncertainty 11

The previous chapter considered the definitions and general approaches to estimation of measurement uncertainty. This chapter considers the procedures in more detail and provides worked examples of different methods for estimation of uncertainty.

THE 'GENERALISED UNCERTAINTY METHOD' OR 'BOTTOM-UP' PROCEDURE

The basis of the generalised uncertainty method (GUM) or 'bottom-up' approach, described by Eurachem (Ellison and Williams, 2012), is to identify and take into account the cumulative variance associated with all stages of an analytical method. In order to estimate a generic level of uncertainty for a method, the variance associated with an individual stage is combined with the variances and covariances of all the other stages that make up an analytical procedure. This is illustrated diagrammatically in the following schematic:

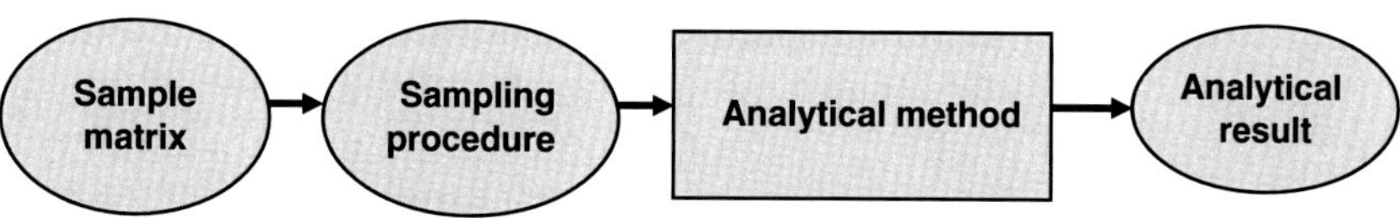

ERRORS ASSOCIATED WITH THE MICROBIAL DISTRIBUTION IN THE SAMPLE MATRIX

The largest potential error sources associated with the sample matrix are as follows: the spatial distribution of microorganisms (eg, random, under- or over-dispersion); the condition of the microorganisms (viable, sub-lethally damaged, non-cultivatable); the presence of competitive organisms that might affect recoverability of the target organisms; and the location of the organisms on or within the matrix (ie, primarily on the surface or distributed throughout the matrix). In addition, factors associated with the intrinsic composition of the matrix may also affect the results of an analysis.

Statistical Aspects of the Microbiological Examination of Foods. http://dx.doi.org/10.1016/B978-0-12-803973-1.00011-5

ERRORS ASSOCIATED WITH THE SAMPLING PROCESS

It is assumed that any sample has been drawn totally at random and without any deliberate bias. How representative of the lot is the analytical sample? Should the analytical sample be representative of the whole matrix, or should it relate only to a specific part, for example, the surface of a meat carcase? If the former, should the matrix be homogenised prior to taking a sample? If the latter, will the technique used to sample the surface layer (excision, swabbing, rinsing or use of a replica plating technique) affect the results obtained? Are the microflora in the analytical sample numerically and typically representative of the microorganisms in the original matrix? What size of sample should be tested? Increasing the size of an analytical sample frequently results in an apparent increase in the colony count whilst reducing the variance (Brown, 1977; Kilsby and Pugh, 1981; Corry et al., 2010).

Most importantly, it is essential to recognise that what is often called a 'sample' is better referred to as a 'laboratory sample', that is, a quantity of material submitted for analysis that purports to have been drawn randomly from a 'lot' or 'sampling target'; there may be more than one laboratory sample for a 'lot'. The laboratory sample is itself sampled to produce an analytical (or test) sample that should be representative of the laboratory sample and that will eventually yield a test portion from which results are obtained (AMC, 2005).

ERRORS ASSOCIATED WITH USE OF A MICROBIOLOGICAL METHOD

At its simplest, microbiological analysis consists of:

- taking (by weight or volume) a test sample from a laboratory sample;
- macerating the test sample in a defined volume of a suitable primary diluent (the test portion);
- preparing serial dilutions of the test portion;
- transferring measured volumes of the dilutions onto or into a culture medium;
- incubating the culture plates (or tubes);
- counting and recording the numbers of colonies; and
- deriving a final estimate of colony-forming units (cfu) in the original matrix.

As discussed previously, some errors, for example, those associated with the accuracies of weighing, pipette volumes and colony counting, can be quantified and measures of the variance can be derived. Other errors can be assessed, but not necessarily quantified: for instance, the extent to which a batch of culture medium supports the growth of specific organisms. It has been suggested that errors in the ability of a particular culture medium to support growth of specific organisms should be quantified and used to provide a correction factor for the yield of such organisms. In most circumstances, it is debatable whether correction factors should ever be used in microbiological practice, although it is commonplace in chemical analysis!

Errors associated with individual technical performance on a day cannot be quantified. Some analytical errors associated with microbiological practices are probably

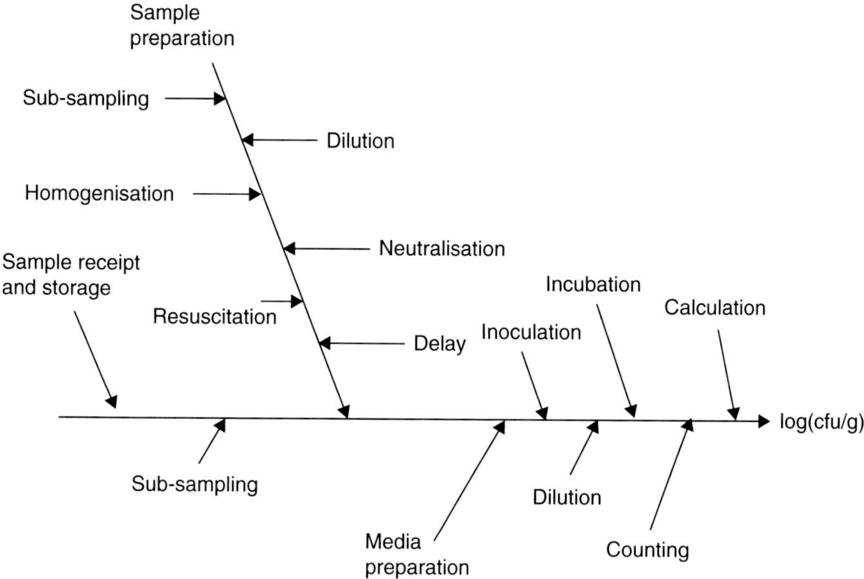

FIGURE 11.1 A Simple Cause and Effect (Fishtail) Diagram Illustrating Some of the Many Sources of Variance in a Standard Aerobic Colony Count

Reproduced from Jewell (2004) by permission of the Campden and Chorleywood Food Research Association

less significant than others, but how do you know which these are if the errors cannot be quantified? To assess the uncertainty of an analytical microbiological procedure from the 'bottom-up' requires a full evaluation of all potential sources of error for each and every stage of an analytical procedure. Fig. 11.1 is a simple, yet incomplete, 'cause and effect' diagram to illustrate some of the major sources of errors likely to affect a colony count procedure.

Once a reliable schedule of quantifiable errors has been produced, the combined variance is obtained by summing the individual contributory variances obtained in several laboratories:

$$s_R^2 = s_a^2 + s_b^2 + \cdots + s_x^2 + s_y^2 + s_z^2$$

where s_R^2 is the reproducibility variance of the method and $s_a^2 \ldots s_z^2$ is the variance of stages $a \ldots z$ within the overall method. By definition, the standard uncertainty is the reproducibility standard deviation (s_R) derived from the square root of the combined variance:

$$s_R = \sqrt{s_a^2 + s_b^2 + \cdots + s_x^2 + s_y^2 + s_z^2}$$

Covariances may also need to be taken into account – but many covariances are not immediately obvious.

The expanded uncertainty is determined by multiplying the standard uncertainty by a coverage factor k, which has a value from 2 to 3. A value of 2 is normally used to give approximate 95% confidence limits (CLs); hence:

$$U = k \cdot s_R = 2 \times s_R \text{ for an approximate 95\% confidence interval}$$

Niemelä (2002) provides a more detailed explanation of this approach to assessment of measurement uncertainty in microbiological analysis.

This stepwise approach is not satisfactory for microbiological examination of foods because of the difficulty of building a fully comprehensive model of the measurement process. It is difficult to quantify the variance contribution of many individual steps in the analytical process not least because the analyte is a living organism, whose physiological state can be variable and the analytical target may include different strains, species and genera of microbes. In other words, microbiological methods are generally unsuitable for a rigorous and statistically valid metrological procedure for estimation of measurement uncertainty (Anon, 2009), although its use has been widely recommended in Scandinavia (Niemelä, 2002, 2003).

THE 'TOP-DOWN' APPROACH TO ESTIMATION OF UNCERTAINTY

In this approach, the parameters used to derive uncertainty measures are estimated from the pooled results of valid inter-laboratory collaborative studies, or, in the case of intermediate reproducibility, from an intra-laboratory study (see, eg, Corry et al., 2007; Jarvis et al., 2007a,b, 2012). Appropriate procedures to ensure that a study design is valid have been described by Youden and Steiner (1975), Anon (1994, 1998, 2005a) and Horwitz (1995).

Quantitative microbiological data [eg, colony counts and Most Probable Numbers (MPNs)] do not conform to a Normal distribution and require transformation to 'Normalise' the data and 'stabilise' the variance. The choice of transformation is dependent on the distribution of the original data. Routinely, most microbiologists transform their data by converting each data value (x_i) into the $\log_{10}$ value (y_i), where $y_i = \log_{10} x_i$. For low-level counts (typically <100 cfu/g) that conform to the Poisson distribution [mean value (m) = variance (s^2)], the data should be transformed by taking the square root of each data value (ie, $y_i = \sqrt{x_i}$). Even at low count levels the sampling variance will still be inflated by the variances of the counting method so the log transformation is usually preferable. However, because of problems of overdispersion frequently associated with microbial distributions, it may be preferable to test for conformance with a negative binomial distribution. Some statistical packages (eg, Genstat) include a facility to assess conformance with a negative binomial, using the maximum likelihood method programme RNEGBINOMIAL, but such procedures are not universally available and it is very time consuming to calculate manually (see Chapter 3; Niemelä, 2002).

ANALYSIS OF VARIANCE

The 'Normalised' data from all participating laboratories are subjected to analysis of variance (ANOVA) after first checking for conformance to a Normal distribution (ND; Chapter 3) and identification and removal of 'outliers' (*qv*) followed, if necessary, by repeating the tests for conformance to 'Normality'.

TESTS FOR 'NORMALITY'

Before doing any tests for 'Normality' it is essential first to undertake a descriptive analysis of the distribution of the data by plotting histograms, box -plots, etc., in order to inspect the distribution. The key element is to obtain approximate symmetry (see Fig. 11.2) and to stabilise the variance in order to make it independent of the

Graphical and descriptive analysis of data

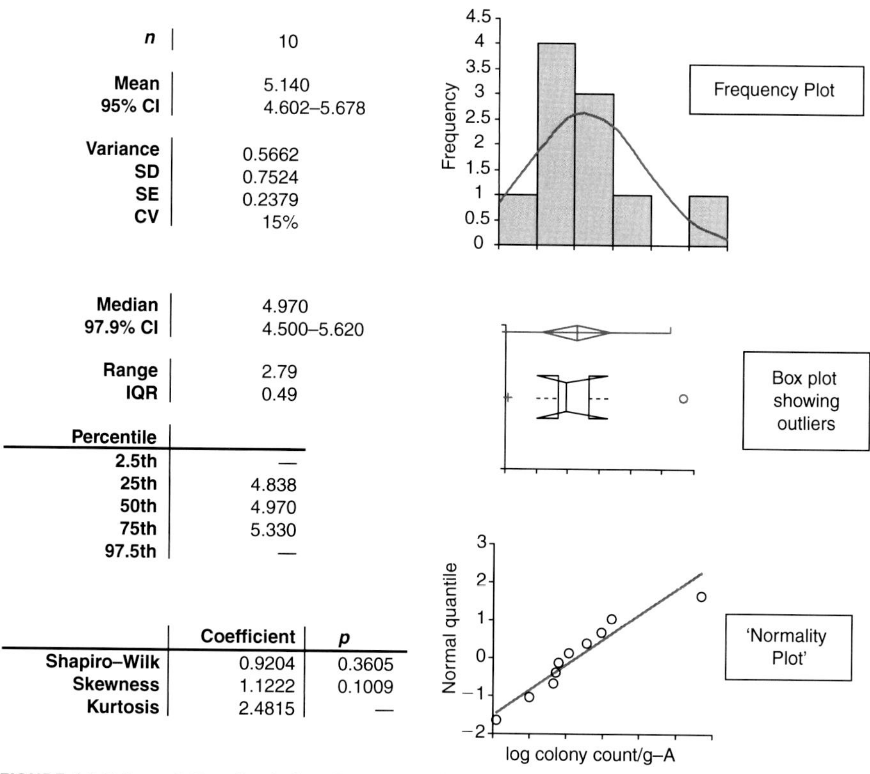

n	10
Mean	5.140
95% CI	4.602–5.678
Variance	0.5662
SD	0.7524
SE	0.2379
CV	15%
Median	4.970
97.9% CI	4.500–5.620
Range	2.79
IQR	0.49

Percentile	
2.5th	—
25th	4.838
50th	4.970
75th	5.330
97.5th	—

	Coefficient	*p*
Shapiro–Wilk	0.9204	0.3605
Skewness	1.1222	0.1009
Kurtosis	2.4815	—

FIGURE 11.2 Descriptive Statistics, Together With Frequency, Box and Normality Plots of One Set of Inter-Laboratory Study Data

Although there is evidence of kurtosis and positive skewness, the log-transformed data conform reasonably to a 'Normal' distribution. The box plot shows the presence of a potential low-level outlier (+) and a significant high-level outlier (ˆ)

mean value. Various tests can be used to assess the 'Normality' of data including the Box–Cox Normality plot (Box and Cox, 1964), the Shapiro–Wilk test (Shapiro and Wilk, 1965), the Anderson–Darling test (Stephens, 1974), the D'Agostino–Pearson test (D'Agostino, 1986) and the Kolmogorov–Smirnov test (Chakravarti et al., 1967). Details of such tests are given in standard statistical texts (eg, D'Agostino, 1986) and are included also in many computer software packages of statistical methods. How useful are the tests?

Normality tests have little power to tell whether or not a small sample of data comes from a Normal distribution such that small data sets almost always pass a Normality test. With large samples, minor deviations from Normality may be flagged as statistically significant, even though such deviations would not affect the results of subsequent statistical tests. If you decide to use Normality tests, first consider whether the chosen test will provide useful information.

D'Agostino (1986) says, 'The Kolmogorov–Smirnov test is only a historical curiosity. It should never be used'. It is, however, one of the standard tests included in many statistical software packages. The Shapiro–Wilk Normality test is complex and does not work well when several items in a data set have the same value. In contrast, the D'Agostino–Pearson omnibus test is easy to understand. It first analyses the data to determine skewness (ie, to quantify the distribution asymmetry) and kurtosis (to quantify the shape of the distribution), calculates the extent of the difference of each of these values from the value expected for a Normal distribution and computes a single probability value from the sum of the squares of these discrepancies. Unlike the Shapiro–Wilk test, this test is not affected if the data contain identical values. For further considerations of Normality tests see Granato et al. (2014).

A simple test for conformance to ND (Tabachnick and Fidell, 1996) is to determine the standard error (SE) of skewness, which approximates to $\sqrt{6/n}$, and the SE of kurtosis, which approximates to $\sqrt{24/n}$, where n is the number of samples tested. For a perfect ND, the SE for skewness should equal 0 (acceptable range 0–2) and the SE for kurtosis should equal 3. The descriptive statistics illustrated in Fig. 11.2, with $n = 10$, has a SE for skewness of $\sqrt{6/10} = 0.775$ and a SE for kurtosis of $\sqrt{24/10} = 1.549$. The skewness of the distribution is larger than ideal, due to the presence of two potential outlier values, and its kurtosis is smaller than the ideal. However, both lie within the range of acceptable values so the hypothesis of conformance to a ND is not rejected, in agreement with the result of the Shapiro–Wilk test.

TESTS FOR OUTLIERS

In a set of replicate measurements one or more values may differ considerably from the majority. In such a case there is always a strong motivation to eliminate the deviant values because of their effect on subsequent calculations, such as variance measurements. This is permissible only if the suspect values can be 'legitimately' characterised as outliers. An outlier is defined as 'a value that arises from a distribution that differs from the main body of data'. Although this definition implies that an outlier may be found anywhere within the range of observations, it is natural to suspect and examine as possible outliers only the extreme values. Before taking a decision, it is essential to check the raw data for evidence of calculation or transcription errors, or identifiable technical errors in the test undertaken. In the absence of any valid explanation for an apparent outlier, such data are considered to be 'outliers'. The rejection of suspect observations should not be based exclusively on objective criteria if there is a risk that the data are not Normally distributed, or that there is extensive kurtosis or skewness in the distribution. Statistically sound tests for 'the detection of outliers' provide evidence that one or more data values lie outside the expected ND of data, but not that the data values are intrinsically wrong.

Youden and Steiner (1975), Anon (1994) and Horwitz (1995) provide details of the most commonly used statistical tests for outliers. Youden and Steiner (1975) describe procedures for deciding whether or not one or more values should be excluded from a data set, including tests to assess whether sub-sets of data from individual laboratories conform to the general data set using a ranking procedure. They also discuss tests for ruggedness and the effect of missing values on the use of an ANOVA procedure. Anon (1994) and Horwitz (1995) provide details of the Dixon's (1953) and Grubbs' (1969) tests that are used to identify which individual data values should be described as 'outliers'. Details of the methods and illustration of their use to examine data for outliers are given in Examples 11.1 and 11.2.

EXAMPLE 11.1 THE YOUDEN AND STEINER (1975) PROCEDURE TO IDENTIFY OUTLIER LABORATORIES IN A COLLABORATIVE STUDY

Seven laboratories carried out tests for total aerobic and Enterobacteriaceae colony counts on a number of samples (Corry et al., 2007). Are all the data valid?

A preliminary evaluation was made to assess whether the data from any one laboratory might be statistically different to any of the others. The mean log colony counts were tabulated and ranked across laboratories from 1 to 7 for the mean colony count on each sample, rank 1 being the lowest count and 7 the highest on each individual test, as illustrated in Table 11.1. If two laboratories had submitted identical counts for one of the samples such that each could have been ranked as, for example, '4', then both laboratories would have been given the rank 4.5; the laboratory with the next higher count would then be scored as 6.

The total rank score for each laboratory is compared with the 95% critical limit values in Table B of Youden and Steiner (1975). For 7 laboratories, each testing 4 samples, the upper and lower limit values are 27 and 5, respectively. In this example, the counts from laboratory 5 are consistently lower than those from the other laboratories, and have a total rank score of 4 (below the lower 95% critical limit). Before further analysis the data from laboratory 5 were investigated to ensure that that there was no acceptable technical explanation for the low results, such as a calculation error that could have been corrected. For instance, if the results from laboratory 5 had been calculated wrongly, such that the results should have been 10-fold higher (ie, about 4.5 log cfu/g rather than 3.5 log cfu/g), then none of the laboratories would have been scored as an 'outlier'. In the absence of any such error the data were eliminated. It is worthy of note that the total rank scores for laboratories 3 and 6 are close to the upper critical limits but are not significantly different to those of the other laboratory data sets.

Table 11.1 Ranking of Enterobacteriaceae Colony Counts from an Inter-Laboratory Study of Microbiological Methods for Analysis of Meat Samples

	Colony Count (log_{10} cfu/g) and Rank (by Laboratory)								Total Rank Score
	A1S1[a]		A1S2		A2S1		A2S2		
Laboratory	Count	Rank	Count	Rank	Count	Rank	Count	Rank	Score
L1	3.86	3	3.82	3	4.06	3	3.96	4	13
L2	4.07	4	3.99	4	3.77	2	4.11	5	15
L3	4.95	7	4.99	7	4.60	6	4.63	6	26
L4	4.40	6	4.40	5	4.09	4	3.86	2	17
L5[b]	3.58	1	3.53	1	3.47	1	3.47	1	4
L6	4.29	5	4.41	6	4.95	7	4.79	7	25
L7	3.81	2	3.78	2	4.18	5	3.95	3	12

[a]*A1S1: analyst 1, sample 1, etc.*
[b]*Colony counts by laboratory L5 were significantly lower than counts by other laboratories and were rejected as being outliers.*
Data of Corry et al. (2007).

Table 11.2 Ranking of Enterobacteriaceae Colony Counts After Removal of Data for Laboratory 5

	Colony Count (log_{10} cfu/g) and Rank (by Laboratory)								Total Rank Score
	A1S1		A1S2		A2S1		A2S2		
Laboratory	Count (*x*)	Rank	Count (*x*)	Rank	Count (*x*)	Rank	Count (*x*)	Rank	Score
L1	3.86	2	3.82	2	4.06	2	3.96	3	9
L2	4.07	3	3.99	3	3.77	1	4.11	4	11
L3	4.95	6	4.99	6	4.60	5	4.63	5	22
L4	4.40	5	4.40	4	4.09	3	3.86	1	13
L6	4.29	4	4.41	5	4.95	6	4.79	6	21
L7	3.81	1	3.78	1	4.18	4	3.95	2	8

Note that the Youden and Steiner (1975) lower and upper limits of total rank scores for 6 laboratories each examining 4 samples are 5 and 23, respectively.
Data of Corry et al. (2007).

EXAMPLE 11.2 USE OF DIXON'S AND GRUBB'S TESTS TO IDENTIFY OUTLIER DATA

Dixon's test compares the differences between the highest two (or the lowest two) data values in a set with the difference between the highest and the lowest values in that set. In a set with n values, where n is 8 or less, the highest value (x_n) should be rejected if

$$\frac{x_n - x_{n-1}}{x_n - x_1} > r$$

where r is the critical value for a 95% probability, given by Dixon (1953) and Youden and Steiner (1975). Similarly, the lowest value (x_1) should be rejected if

$$\frac{x_2 - x_1}{x_n - x_1} > r$$

In Example 11.1, we eliminated laboratory 5, so we now apply Dixon's test for individual outliers to the remaining data.

For a set of 6 measurements, rejection at $P = 0.05$ of the highest or lowest values is given by the critical value of 0.56 (Table C1, Youden and Steiner, 1975). So, if either

$$\frac{x_6 - x_5}{x_6 - x_1} \text{ or } \frac{x_2 - x_1}{x_6 - x_1} > 0.56$$

the data would be eliminated.

All of the calculated values in Table 11.3 were smaller than Dixon's criteria, so no data values were eliminated as outliers.

Table 11.3 Dixon's Test Values for Data in Table 11.2

	A1S1	A1S2	A2S1	A2S2
Highest value	$\frac{4.95-4.40}{4.95-3.81}=0.48$	$\frac{4.99-4.40}{4.99-3.78}=0.49$	$\frac{4.95-4.60}{4.95-3.77}=0.30$	$\frac{4.79-4.63}{4.79-3.95}=0.19$
Lowest value	$\frac{3.86-3.81}{4.95-3.81}=0.04$	$\frac{3.82-3.78}{4.99-3.78}=0.03$	$\frac{4.06-3.77}{4.95-3.77}=0.15$	$\frac{3.96-3.95}{4.79-3.95}=0.01$

Grubbs' test (Grubbs, 1969) is an alternative method for data that conform reasonably to a Normal distribution. The test detects one outlier at a time; after that outlier has been removed from a data set, the test is repeated iteratively, but it should not be used for sample sizes of less than 6. The null hypothesis (H_0) is that there are no outliers in the data set; the alternative hypothesis (H_a) is that there is at least one outlier in the data set. The Grubb's test statistic is a two-sided test that measures the largest absolute deviation from the sample mean in units of the sample standard deviation and is derived using the following equation:

$$G = \frac{\max|y_i - \bar{y}|}{s}$$

where $\bar{y}$ and s denote the sample mean and standard deviation, respectively. Alternative versions to test for minimum (y_{min}) and maximum (y_{max}) value outliers are, respectively:

$$G_{min} = \frac{\bar{y} - y_{min}}{s} \text{ and } G_{max} = \frac{y_{max} - \bar{y}}{s}$$

For the data in Table 11.2, the overall mean value ($\bar{y}$) of the 24 colony counts is 4.23 log cfu/g and $s = 0.398$. The maximum absolute difference in the $\log_{10}$ colony counts is $4.99 - 4.23 = 0.76$ $\log_{10}$ units, so the G value = 0.76/0.398 = 1.91, which is less than the 2-sided critical value of 2.80 for n = 24. So we accept the null hypothesis that none of the values are outliers. Critical values for G can be obtained from Table 6 of Grubbs (1969); a simple calculator for determining the Grubb's G value is available at http://graphpad.com/quickcalcs/Grubbs1.cfm (accessed 03.06.15).

IUPAC recommends an iterative procedure (Horwitz, 1995). Cochran's test for within-laboratory variance (stage 1) is used first to eliminate any outlier laboratory and then Grubb's two-tailed test for between-laboratory variance is used (stage 2), again removing any laboratory that fails; if no laboratory fails, then Grubb's pair-value test is applied (stage 3) to identify any failure. The procedure is then repeated sequentially either until no more outliers are found or until the maximum permitted number of data values (22%; 2/9) has been removed. Details of the procedure and critical values for the tests are given in Grubbs (1969) and Horwitz (1995).

THE STANDARD ANOVA PROCEDURE

Analysis of Variance (ANOVA) is a statistical method that is used to uncover the main and interacting effects of independent variables on a dependent variable. The procedure estimates the extent to which an observed overall variance can be partitioned into components each of which causes a specific effect. In the case of an inter-laboratory collaborative trial, independent variables might include replication of samples, analysts, laboratories, test methods or other factors. ANOVA can be done on the assumption of fixed-effect, random-effect or mixed-effect models – the random-effects model generally applies in the context of the analysis of collaborative trial data. Depending on the experimental design, ANOVA can be used either as a simple one-way or as a nested design (eg, 2 × 2, 3 × 3). The procedure provides an estimate of the significance of an effect through derivation of the Fisher's F-ratio (see Example 11.3) but it is also important to assess the confidence intervals of the distribution for each set of data to ensure that data from, for example, individual laboratories or analysts are consistent (see Hector, 2015).

The estimate of variance is determined from the sum of squares of the differences between each mean data value and the overall mean value. In a simple model, the sum of the squares of the variance is equal to the sum of the squares of both the treatments and the 'error' (ie, the between-replicate variance):

$$\text{Variance} = \text{SS}_{\text{Total}} = \text{SS}_{\text{Treatments}} + \text{SS}_{\text{Error}}$$

The contribution of each effect to the total degrees of freedom (df) of the estimated sum of squares can be partitioned similarly: $\text{df}_{\text{Total}} = \text{df}_{\text{Treatments}} + \text{df}_{\text{Error}}$.

Assuming a fully 'nested' experimental design (eg, duplicate testing of S samples by A analysts in each of L laboratories), the residual error, that is, the residual mean square, provides an estimate of the variance of replicated analyses on a single sample and gives the repeatability variance (s_r^2). The estimate of reproducibility variance (s_R^2) first requires computation of the contributions to variance of the samples, analysts and laboratories (see below). The repeatability standard deviation (s_r) and the reproducibility standard deviation (s_R), being the square root values of the respective variances, are the measures of standard uncertainty from which the expanded uncertainty estimates are derived. These derivations are illustrated in Example 11.4.

EXAMPLE 11.3 USE OF THE ANOVA PROCEDURE TO DERIVE THE COMPONENT VARIANCES FROM A COLLABORATIVE ANALYSIS

An inter-laboratory trial has been done in 10 laboratories (p = 10) in each of which 2 analysts each tested 2 replicate samples in duplicate. Hence, each laboratory carried out 8 replicate analyses and the total number of analyses was 8p = 80. Each data value (y_{pijk}) is allocated to a cell in Table 11.4 (data table) in the sequence laboratory (p), analyst (i), sample (j) and replicate (k), and the data are then analysed by nested Analysis of Variance.

Table 11.4 Layout of Data in a Cell Format[a] for ANOVA

	Analyst (i = 1)				Analyst (i = 2)			
	Sample (j = 1)		Sample (j = 2)		Sample (j = 1)		Sample (j = 2)	
Laboratory (p = 1–10)	Replicate (k = 1)	Replicate (k = 2)	Replicate (k = 1)	Replicate (k = 2)	Replicate (k = 1)	Replicate (k = 2)	Replicate (k = 1)	Replicate (k = 2)
1	y_{1111}	y_{1112}	y_{1121}	y_{1122}	y_{1211}	y_{1212}	y_{1221}	y_{1222}
2	y_{2111}	y_{2112}	y_{2121}	y_{2122}	y_{2211}	y_{2212}	y_{2221}	y_{2222}
3	y_{3111}	y_{3112}	y_{3121}	y_{3122}	y_{3211}	y_{3212}	y_{3221}	y_{3222}
4	y_{4111}	y_{4112}	y_{4121}	…	…	…	…	…
…	…	…	…	…				
…	…	…	…	…				
…	…	…	…	…				
10	y_{10111}	y_{10112}	y_{10121}	y_{10122}	y_{10211}	y_{10212}	y_{10221}	y_{10222}

[a]*Cell formats are labelled in the sequence, for example, y_{1111}: laboratory 1, analyst 1, sample 1, replicate 1, etc.*

Table 11.5 is the ANOVA table for a four-factor fully nested experiment.

The residual mean square (MS_{res}) provides the repeatability variance (s_r^2) between duplicate analyses done on the same replicate sample and the repeatability standard deviation is $\sqrt{s_r^2}$.

Table 11.5 ANOVA Table for a Four-Factor Fully Nested Experiment for 10 Laboratories, 2 Analysts and 2 Samples Each Tested in Duplicate

Source of Variation	Sum of Squares	Degrees of Freedom[a]	Mean Square	Expected Mean Square Components[b]
Laboratories	SS_{lab}	$p - 1 = 9$	$SS_{lab}/9 = MS_{lab}$	$\sigma_r^2 + 2\sigma_{sam}^2 + 4\sigma_{ana}^2 + 8\sigma_{lab}^2$
Analysts	SS_{ana}	$p = 10$	$SS_{ana}/10 = MS_{ana}$	$\sigma_r^2 + 2\sigma_{sam}^2 + 4\sigma_{ana}^2$
Samples	SS_{sam}	$2p = 20$	$SS_{sam}/20 = MS_{sam}$	$\sigma_r^2 + 2\sigma_{sam}^2$
Residual	SS_{res}	$4p = 40$	$SS_{res}/40 = MS_{res}$	σ_r^2
Total	Total SS	$8p - 1 = 79$		

[a]*p, the number of laboratories in each of which one analyst examined two samples in duplicate.*
[b]*The components of variance are shown as population variances since this is an expectation table.*

The variance due to samples (s_{sam}^2) is given by $(MS_{sam} - MS_{res})/2$.
The variance due to analysts (s_{ana}^2) is given by $(MS_{ana} - MS_{sam})/4$.
The variance due to laboratories (s_{lab}^2) is given by $(MS_{lab} - MS_{ana})/8$.
The reproducibility variance (s_R^2) is given by $\left(s_{sam}^2 + s_{ana}^2 + s_{lab}^2 + s_r^2\right)$.
The reproducibility standard deviation is given by $\sqrt{s_{sam}^2 + s_{ana}^2 + s_{lab}^2 + s_r^2}$.

EXAMPLE 11.4 DERIVATION OF THE COMPONENT VARIANCES FROM AN ANOVA ANALYSIS OF A COLLABORATIVE TRIAL

An inter-laboratory trial was run in 10 laboratories; in each laboratory 2 analysts each tested 2 samples in duplicate (ie, 2 replicate analyses per sample) for aerobic colony counts. How do we derive component variances?

The colony counts (as $\log_{10}$ cfu/g) are tabulated in Table 11.6.

Table 11.6 Example of Layout of Data for Aerobic Colony Counts ($\log_{10}$ cfu/g) on Replicate Samples Examined in Duplicate by 2 Analysts in Each of 10 Laboratories

	Analyst (i = 1)				Analyst (i = 2)			
	Sample (j = 1)		Sample (j = 2)		Sample (j = 1)		Sample (j = 2)	
Laboratory (p = 10)	Replicate (k = 1)	Replicate (k = 2)	Replicate (k = 1)	Replicate (k = 2)	Replicate (k = 1)	Replicate (k = 2)	Replicate (k = 1)	Replicate (k = 2)
1	5.56	5.73	5.76	5.59	6.08	5.96	6.07	5.99
2	6.02	5.88	5.87	5.80	5.54	5.63	5.92	5.79
3	6.26	6.30	6.46	6.54	6.42	6.49	6.11	6.42
4	5.07	5.11	4.90	4.61	4.63	4.81	4.42	4.56
5	5.39	5.25	5.28	5.52	5.34	5.46	5.47	5.49
6	5.98	5.88	6.02	5.64	5.96	6.06	5.70	5.57
7	5.43	5.18	5.16	5.08	6.15	5.76	5.44	5.43
8	5.94	5.73	5.28	5.47	5.99	6.01	5.92	6.13
9	5.45	5.35	5.49	5.42	5.68	5.57	5.74	5.69
10	5.51	5.74	6.18	6.13	5.83	5.91	5.76	5.60

A test for Normality (eg, Shapiro–Wilk, $W = 0.9830$, $p = 0.0885$) did not reject the hypothesis that the $\log_{10}$ colony counts conform reasonably (although not perfectly) to a Normal distribution. The variance of the laboratory 7 data was larger than for the other laboratories but the Cochran test (Horwitz, 1995) did not reject the null hypothesis that the individual variances did not differ. Subsequent evaluation using the Dixon and Grubbs' tests also failed to demonstrate that individual data sets or laboratories were outliers. The results of the multivariate analysis of variance (ANOVA) are given in Table 11.7.

Table 11.7 ANOVA Table for the Four-Factor Fully Nested Experiment (No Data Excluded)

Source of Variation	Sum of Squares	Degrees of Freedom	Mean Square (to 4 Decimal Places)	Component Contributions to the Mean Square
Laboratories	12.636	9	1.4040	$s_r^2 + 2s_{sam}^2 + 4s_{ana}^2 + 8s_{lab}^2$
Analysts	1.4906	10	0.1491	$s_r^2 + 2s_{sam}^2 + 4s_{ana}^2$
Samples	1.346	20	0.0673	$s_r^2 + 2s_{sam}^2$
Residual	0.5554	40	0.0139	s_r^2
Total	16.0272	79		

The component variances are derived as follows:
repeatability variance $(s_r^2) = \text{MS}_{\text{rep}} = 0.0139$;
sample variance $(s_{\text{sam}}^2) = (\text{MS}_{\text{sam}} - s_r^2)/2 = (0.0673 - 0.0139)/2 = 0.0267$;
analyst variance $(s_{\text{ana}}^2) = (\text{MS}_{\text{ana}} - \text{MS}_{\text{sam}})/4 = (0.1494 - 0.0267)/4 = 0.03068$;
laboratory variance $(s_{\text{lab}}^2) = (\text{MS}_{\text{lab}} - \text{MS}_{\text{ana}})/8 = (1.4040 - 0.03068)/8 = 0.17167$;
reproducibility variance $(s_R^2) = 0.0139 + 0.0267 + 0.03068 + 0.17167 = 0.2426$;
reproducibility standard deviation = $s_R = \sqrt{0.2426} = \pm 0.4926$;
repeatability standard deviation = $s_r = \sqrt{0.0139} = \pm 0.1179$;
the mean colony count = 5.6921 ≈ 5.69($\log_{10}$) cfu/g.

Hence, the relative standard deviation of reproducibility (RSD_R) = 100 × 0.4926/5.69 = 8.65% and the relative standard deviation of repeatability (RSD_r) = 100 × 0.1179/5.69 = 2.07%.

The 95% expanded uncertainty of reproducibility is given by

$$U = 2s_R = 2 \times 0.4926 = 0.9852 \approx 0.99 \log_{10} \text{cfu/g}$$

The upper and lower bounds of the 95% confidence interval on the mean colony count are as follows:

$$U_{\text{CI}} = 5.69 + 0.99 = 6.68 \log_{10} \text{cfu/g}$$
$$L_{\text{CI}} = 5.69 - 0.99 = 4.70 \log_{10} \text{cfu/g}$$

Hence, repeat tests in different laboratories on samples drawn from this particular lot could be expected, with 95% probability, to lie within the range 4.70–6.68 $\log_{10}$ cfu/g. Whilst this range may, at first sight, appear to be very wide, it is not an unusual result. It is also noteworthy that the expanded repeatability uncertainty is ±2 × 0.1179 ≈ ±0.24 $\log_{10}$ cfu/g, a value that would be found typically for many microbiological colony count procedures.

ROBUST METHODS OF ANOVA

Because of problems with the occurrence of outlier data, several alternative approaches to the ANOVA have been developed, based on robust methods of statistical analysis. Robust procedures estimate variances around the median values and do not require identification and removal of outlying data (which might be valid results).

Mean values are affected significantly by one or more outlier values within a data set, whereas the median value is less affected. For series A in Table 11.8, the mean and median values are the same, but the presence of one or more high values (as in series B, C, D) significantly increases the value of the mean but has no effect on the median value. Removal of the high outliers (series E) reduces both the mean and the median values but with the loss of original data that may be valid. An equivalent effect would be seen with low-value outliers, although the occurrence of both high and low outliers could balance out the effect on the mean, but not on the variances. Before deciding whether or not to accept data with obvious outlier values it is essential to check that the values are not due to a simple transcription, calculation or technical error.

Table 11.8 Effect of Outlier Values on Mean and Median Values

Series	Data Values (x)	Number of Values, n	Sum of Values, Σ	Mean Value, $\bar{x}$	Median Value[a]
A	1, 3, 3, 3, 4, 4, 5, 5, 6, 6	10	40	4	4
B	1, 3, 3, 3, 4, 4, 5, 5, 6, **26**	10	60	6	4
C	1, 3, 3, 3, 4, 4, 5, 6, **15**, **26**	10	70	7	4
D	1, 3, 3, 3, 4, 4, 5, 6, **15**, **126**	10	170	17	4
E	1, 3, 3, 3, 4, 4, 5, 6	8	29	3.6	3.5

[a] *The median is the mid-value of a series having an odd number of values, or the average of the two mid-values for a series having an even number of values.*

There are two primary approaches to robust analysis. One approach (RobStat; AMC, 1989a,b, 2001) calculates the median absolute difference (MAD) between the results and their median value and then applies Hüber's method of winsorisation, a technique for reducing the effect of outlying observations on data sets (for detail see Smith and Kokic, 1997). The procedure can be used with data that conform approximately to a ND but having heavy tails and/or outliers. The procedure is illustrated in Example 11.5 for the set of descriptive statistical data shown in Fig. 11.2. However, the procedure is not suitable for multimodal or heavily skewed data sets. The AMC website (http://www.rsc.org/Membership/Networking/InterestGroups/Analytical/AMC/Software/RobustStatistics.asp; accessed 31.10.15) provides downloadable software for use in either Minitab or Excel 97 and later versions.

Alternative approaches are based on an extrapolation of the work of Rousseeuw and Croux (1993). One version of this approach used Rousseeuw's recursive median S_n but, following the work of Wilrich (2007), an alternative approach to the robust estimation of repeatability and reproducibility standard deviations is described in a revised edition of the ISO standard (Anon, 2011). Wilrich demonstrated that estimates of S_r and S_R based on the procedures described in Anon (1994, 1998, 2013), including identification and removals of outliers where appropriate, are inefficient (maximum efficiency of the estimates range from 25% to 58%). His alternative procedure, based on that of Rousseeuw and Croux (1993), uses a derived value Q_n, which has an efficiency of 82.3%. The procedure is illustrated in Example 11.6.

A problem with robust methods of analysis is that they do not directly provide a means of assessing component variances. Hedges and Jarvis (2006) have described a stepwise procedure for determination of component variances. All results are first analysed to determine the between- and within-group variances associated with duplicate tests. This is then followed by sequential analysis of the mean values to determine the between and within variances for samples, analysts and laboratories. The procedure is illustrated in Example 11.7. Hedges (2008) used the Wilrich (2010) approach to revise the procedure described by Hedges and Jarvis (2006).

EXAMPLE 11.5 ROBUST ANALYSIS OF VARIANCE (ANOVA) USING THE ROBSTAT PROCEDURE (AMC, 2001)

The data examined by ANOVA in Example 11.4 were analysed using the RobStat robust procedure. The replicate pairs of data were tabulated (Table 11.9) in an Excel spreadsheet to estimate the repeatability between replicates and the overall reproducibility.

The RobStat menu is set up to analyse the data in rows (ie, as replicate values) by an iterative method, with a convergence criterion of 0.0001. The output (Table 11.10) gives a repeatability SD (ie, between replicates) of 0.12 and a reproducibility SD of 0.42. The reproducibility SD is calculated from $\sqrt{\text{between-group SD}^2 + \text{within-group SD}^2}$ (Table 11.10).

The relative SD of repeatability is $100 \times 0.12/5.69 = 2.11\%$ and the relative SD of reproducibility is $100 \times 0.42/5.69 = 7.38\%$. The expanded uncertainties of repeatability and reproducibility are ±0.24 and ±0.84 $\log_{10}$ cfu/g, respectively. Note that the estimate of expanded uncertainty of repeatability is the same as that determined by ANOVA but that of the expanded reproducibility estimate is smaller. This is not unexpected since the data set only conformed approximately to Normal.

The RobStat procedure can be repeated on values for each pair of samples to determine the between- and within-analyst SDs, and similarly to determine the between- and within-laboratory SDs. For these data the results are shown in Table 11.11.

Unfortunately, these tabulated standard deviations do not directly provide estimates of the component variances from the collaborative trial. The next example demonstrates a way in which component variances can be derived from such data.

Table 11.9 Data Organised Sequentially for Analysis of Within- and Between-Sample Standard Deviation Using the RobStat Procedure

			Replicate		
Laboratory	**Analyst**	**Sample**	**1**	**2**	**Mean**
1	1	1	5.56	5.73	5.65
		2	5.76	5.59	5.68
	2	1	6.08	5.96	6.02
		2	6.07	5.99	6.03
2	1	1	6.02	5.88	5.95
		2	5.87	5.80	5.84
	2	1	5.54	5.63	5.59
		2	5.92	5.79	5.86
3	1	1	6.26	6.30	6.28
		2	6.46	6.54	6.50
…	…	…	…	…	…
…	…	…	…	…	…
10	1	1	5.51	5.74	5.63
		2	6.18	6.13	6.16
	2	1	5.83	5.91	5.87
		2	5.76	5.60	5.68

Table 11.10 RobStat Output Table Showing the Within-Group (Repeatability) SD, the Between-Group SD and the Reproducibility SD for the Data From Table 11.9

Parameter	
Grand mean	5.691514
Within-group/repeatability SD	0.119667
Between-group SD	0.400970
Reproducibility SD	0.418446

The groups are the results obtained on samples 1 and 2, respectively.

Table 11.11 RobStat Robust Estimates of the Between- and Within-Group SD for Samples, Analysts and Laboratories

Parameter	Samples	Analysts	Laboratories
Grand mean	5.691514	5.692971	5.692971
Within-group/repeatability SD	0.119667	0.181677	0.215163
Between-group SD	0.40097	0.341481	0.251358
Reproducibility SD	0.418446	0.386802	0.330871

Note that the reproducibility SD is greatest for the samples and that the reproducibility for laboratories is lower than that for the analysts.

EXAMPLE 11.6 DETERMINATION OF COMPONENT VARIANCES FROM ROBUST ANOVA USING THE METHOD OF HEDGES AND JARVIS (2006)

In the previous example we determined the between and within standard deviations for a set of nested data involving 10 laboratories, in each of which 2 analysts each examined 2 samples in duplicate – a total of 80 sets of aerobic colony counts. The robust analyses (Table 11.11) provide the starting point to determine the component variances for samples, analysts and laboratories.

We use the notation that the variance is the square of the relevant group SD, that is, 'within-group' variance is shown by W^2_{group} and 'between-group' variance by B^2_{group}.

The residual variance within the samples (ie, the error of the analysis) is given by

$$s_r^2 = W^2_{\text{sam}} = (0.119667)^2 = 0.014320191 \approx 0.0143$$

The component variance due to the samples is given by

$$s^2_{\text{sam}} = W^2_{\text{ana}} - W^2_{\text{sam}} = W^2_{\text{ana}} - s_r^2/2 = (0.181677)^2 - (0.119667)^2/2 \approx 0.025846$$

where W^2_{ana} is the variance within analysts and W^2_{sam} is the variance within samples, that is, the residual variance s_r^2 divided by 2.

Similarly, the component variance due to analysts is given by

$$s^2_{\text{ana}} = W^2_{\text{lab}} - W^2_{\text{ana}}/2 = (0.215163)^2 - (0.181677)^2/2 = 0.029791 \approx \mathbf{0.0298},$$

where W^2_{lab} is the within-laboratory variance.

The component variance due to laboratories is given by

$$s_{\text{lab}}^2 = B_{\text{lab}}^2 = (0.251358)^2 = 0.0631808 \approx 0.0632$$

where B_{lab}^2 is the between-laboratory variance.

Then the reproducibility SD is given by

$$\text{Reproducibility SD} = \sqrt{s_{\text{sam}}^2 + s_{\text{ana}}^2 + s_{\text{lab}}^2 + s_r^2}$$
$$= \sqrt{0.0632 + 0.0298 + 0.0258 + 0.0143} = \sqrt{0.1331} = \pm 0.3648$$

Hence, for a mean colony count of 5.69 $\log_{10}$ cfu/g, the relative SDs of repeatability and reproducibility determined by the Hedges and Jarvis (2006) modification of the RobStat procedure are 2.10% and 6.41%, respectively. In Example 11.4, the traditional ANOVA of the same data gave relative standard deviations of repeatability and reproducibility of 2.07% and 8.69%, respectively. As was noted in that example, the data set contained some potential outliers and did not conform well to a Normal distribution, showing evidence of kurtosis and skewness. It is not surprising therefore that the traditional ANOVA gave larger values for inter-laboratory reproducibility than did the robust method.

EXAMPLE 11.7 ROBUST ANALYSIS OF VARIANCE (ANOVA) USING THE *Qn* PROCEDURE (WILRICH, 2007; ANON 2011)

The second edition of this book contained details of an ANOVA using the recursive median procedure described originally in 2003. That procedure has been replaced by the Qn procedure described by Wilrich (2007) and Anon (2011). For the sake of simplicity, a reduced data set has been chosen for purposes of illustration of the procedure. Assume that one analyst in each of five laboratories tested duplicate samples of a product.

The colony counts, as $\log_{10}$ cfu/g, are shown in Table 11.12 together with their mean values and the deviance of each count from the mean count. Note that the deviances within each laboratory will be numerically identical but one will be positive and the other negative.

Table 11.12 Duplicate Colony Counts (as log cfu/g) Determined in Each of Five Laboratories, Together With Their Mean Values and the Deviances of Each Value From the Mean Count on the Duplicate Samples

Laboratory	S1[a]	S2[a]	Mean	Dev 1[b]	Dev 2[b]
1	5.560	5.732	5.6460	−0.0860	0.0860
2	6.024	5.880	5.9520	0.0720	−0.0720
3	6.260	6.309	6.2845	−0.0245	0.0245
4	5.078	5.110	5.0940	−0.0160	0.0160
5	5.395	5.250	5.3225	0.0725	−0.0725
Median			5.6460		

[a]*log colony counts on the two analytical samples S1 and S2.*
[b]*Dev 1 and Dev 2 refer to the deviances between each value S1 and S2, from their mean value; hence, one deviance is negative and the other is positive.*

In accordance with the method described by Wilrich (2007), the absolute differences of each deviance (ie, the difference between the duplicate colony counts) from each of the other deviances are set out as a matrix (Table 11.13) from which the lower half has been deleted for this procedure.

Table 11.13 Determination of Absolute Differences in the Deviances

Laboratory			1	2	3	4	5	1	2	3	4	5
	Sample		S1	S1	S1	S1	S1	S2	S2	S2	S2	S2
		Deviance	−0.086	0.072	−0.024	−0.016	0.072	0.086	−0.072	0.025	0.016	−0.072
1	S1	−0.086		0.1580	0.0615	0.0700	0.1585	0.1720	0.0140	0.1105	0.1020	0.0135
2	S1	0.072			0.0965	0.0880	0.0005	0.0140	0.1440	0.0475	0.0560	0.1445
3	S1	−0.024				0.0085	0.0970	0.1105	0.0475	0.0490	0.0405	0.0480
4	S1	−0.016					0.0885	0.1020	0.0560	0.0405	0.0320	0.0565
5	S1	0.072						0.0135	0.1445	0.0480	0.0565	0.1450
1	S2	0.086							0.1580	0.0615	0.0700	0.1585
2	S2	−0.072								0.0965	0.0880	0.0005
3	S2	0.025									0.0085	0.0970
4	S2	0.016										0.0885
5	S2	−0.072										

The value in each deviance column has been subtracted from the deviance in each row and the absolute value of the difference is listed in the matrix, for example, $|(-0.086) - (0.072)| = |-0.1580| = 0.1580$. All zero differences have been removed.

Although these data are shown to four significant places, as in all statistical work the actual data used for the calculations were not rounded down.

The $n(n-1)/2$ (= 45) absolute differenvces in the deviations are sorted in ascending order, and the ith smallest difference is identified as Q_n, using the procedure:

$$Q_n = (|x_i - x_j|_{i<j})_l$$

where $|x_i - x_j|_{i<j}$ is the ith difference, with

$$l = \binom{f}{2}$$

and

$$f = \begin{cases} \frac{n}{2}+1 & \text{if } n \text{ is even} \\ \frac{n+1}{2} & \text{if } n \text{ is odd} \end{cases} \quad \text{and} \quad l = \frac{f(f-1)}{2}$$

For the absolute differences in Table 11.13, where $n = 10$, $f = (10/2) + 1 = 6$ and $l = (6 \times 5)/2 = 15$, Q_n is the 15th smallest ranked value = 0.0480.

To obtain an unbiased estimate of the standard deviation, Q_n must be multiplied by a bias correction factor (C_n), determined as follows:

$$C_n = \begin{cases} 2.2219\left(\frac{n}{n+1.4}\right) & \text{if } n \text{ is odd} \\ 2.2219\left(\frac{n}{n+3.8}\right) & \text{if } n \text{ is even} \end{cases}$$

For $n = 10$, $C_n = 2.2219 \times 10/(10 + 3.8) = 1.610$, so $Q_{\text{intra}} = C_n Q_n = 1.610 \times 0.0480 = 0.07728$.

We need now to determine the Q_n of the differences in the laboratory mean values. We again construct a matrix and delete the lower half (Table 11.14).

Table 11.14 Determination of Absolute Differences in the Mean Values

		1	2	3	4	5
Laboratory	**Mean**	**5.646**	**5.9520**	**6.2845**	**5.094**	**5.3225**
1	5.6460		0.30600	0.63850	0.55200	0.32350
2	5.9520			0.33250	0.85800	0.62950
3	6.2845				1.19050	0.96200
4	5.0940					0.22850
5	5.3225					

As before we sort the differences into ascending order and identify the 'ith' value. Since $n = 5$, $f = (n + 1)/2 = 3$ and $l = f(f - 1)/2 = 3(3 - 1)/2 = 3$; the third smallest difference (= Qn) is 0.3235.

Bias correction (for an odd number) is given by $C_n = 2.2219 \times 5/(5 + 1.4) = 1.7359$ and $Q_{\text{inter}} = 1.736 \times Q_n = 1.736 \times 0.3235 = 0.56155$.

The repeatability SD = $S_r = \sqrt{2} \times Q_{\text{intra}} = 1.4142 \times 0.07728 = 0.109289$.

The between-laboratories $\text{SD} = S_L = \sqrt{Q_{\text{inter}}^2 - Q_{\text{intra}}^2} = \sqrt{0.56155^2 - 0.108289^2} = 0.556207$.

The reproducibility $SD = S_R = \sqrt{S_L^2 + S_r^2} = \sqrt{0.556207^2 + 0.109289^2} = 0.566842$.

The median of the mean values = 5.646 $\log_{10}$ cfu/g; the estimate of uncertainty of repeatability SD is ±0.1093 and that of reproducibility is ±0.567 $\log_{10}$ cfu/g. Results obtained by the MAD procedure (AMC, 1989a,b) for the same data set give the overall mean count = 5.660 with repeatability SD = 0.098, between-laboratory SD = 0.536 and reproducibility SD = 0.545, so giving an estimate of the expanded uncertainty of reproducibility = ±1.09. For this data set the two robust methods give almost identical outputs.

The component variances for the *Qn* procedure can be derived in a similar manner to those used for the RobStat procedure (Example 11.6) but the working is somewhat more complex (for details see Hedges and Jarvis, 2006). As can be seen from the worked example, the *Qn* procedure is much more complicated and time consuming than is the RobStat approach to robust ANOVA.

MEASUREMENT OF INTERMEDIATE REPRODUCIBILITY

Intermediate reproducibility is the term used to describe estimates of reproducibility done in a single laboratory. Confusion sometimes arises between repeatability and intermediate reproducibility. Estimates of repeatability require all stages of the replicated testing to be done *only* by a single analyst working in one laboratory and carrying out repeat determinations on a single laboratory sample, using identical culture media, diluents, etc., within a short time period, for example, a few hours. If more than one analyst undertakes the analyses and/or tests are done on different samples and/or on different days, then the calculation derives a measure of intermediate reproducibility. The procedure described in the following can be used to determine average repeatability estimates for individual analysts provided all the repeatability criteria are met, or for estimates of intermediate reproducibility. Such estimates provide a source of data amenable to statistical process control (Chapter 12), which can be valuable in laboratory quality management.

Intra-laboratory uncertainty estimates can be made by carrying out an internal collaborative trial, with different analysts testing the same samples over a number of days, using different batches or even different brands of commercial culture media, etc. In such a case the statistical procedures of choice are those described under ANOVA or robust ANOVA. However, if a laboratory undertakes routine quality monitoring tests, it is possible to estimate reproducibility from such data also.

A simple procedure, described in Anon (2009), can be used to determine the variance for each set of transformed replicate data values. The reproducibility standard deviation is derived from the square root of the sum of the variances on duplicate, or replicate, test samples divided by the number of data sets. For duplicate tests the equation is as follows:

$$S_R = \sqrt{\frac{\sum_{i=1}^{n} (y_{i1} - y_{i2})^2 / 2}{n}}$$

where y_{i1} and y_{i2} are the log-transformed values of duplicate counts (x_{i1} and x_{i2}) and n is the number of pairs of counts. For more than 2 replicate tests a one-way ANOVA is required to determine the variance for each set of tests. Calculation of intermediate reproducibility values is shown in Example 11.8.

MATRIX AND OTHER EFFECTS

Current thinking requires incorporation of 'corrections' to allow for the matrix and other effects, such as counts of low numbers of organisms, on the estimates of uncertainty. I include no details on this because the approaches have still to be agreed internationally, although Anon (2009) provides limited information.

EXAMPLE 11.8 DERIVATION OF INTERMEDIATE REPRODUCIBILITY (MODIFIED FROM ANON, 2009)

Intermediate reproducibility is an estimate of reproducibility obtained in a single laboratory. It is usually based on results from duplicate test portions on a number of samples, all of which are tested by different analysts using the same test procedure. The data in Table 11.15 were derived from enumeration of aerobic mesophilic flora in mixed poultry meat samples.

The duplicate data values (x_{iA} and x_{iB}) are $\log_{10}$ transformed to give y_{iA} and y_{iB}, respectively. The mean $\log_{10}$ counts ($\bar{y}_i$) are determined as $(y_{iA} + y_{iB})/2$ and the variances (s_i^2) from $(y_{iA} - y_{iB})^2/2$. The relative standard deviation (RSD, %) for each pair of counts is given by $100\sqrt{s_i^2}/\bar{y}$.

Table 11.15 Procedure for Calculation of Intermediate Reproducibility Based on 10 Sets of Paired Colony Counts (Anon, 2006a)

Test Sample	Colony Count (cfu/g) A	Colony Count (cfu/g) B	Log Colony Count ($\log_{10}$ cfu/g) A	Log Colony Count ($\log_{10}$ cfu/g) B	Mean	Absolute Difference	Variance	Relative S D (%)
i	x_{iA}	x_{iB}	y_{iA}	y_{iB}	$\bar{y}$	$\lvert y_{iA} - y_{iB}\rvert$	S^2_{Ri}	RSD_{Ri}
1	6.70E+04	8.70E+04	4.83	4.94	4.89	0.11	0.00605	1.59
2	7.10E+06	6.20E+06	6.85	6.79	6.82	0.06	0.00180	0.62
3	3.50E+05	4.40E+05	5.54	5.64	5.59	0.10	0.00500	1.26
4	1.00E+07	4.30E+06	7.00	6.63	6.82	0.37	0.06845	3.84
5	1.90E+07	1.70E+07	7.28	7.23	7.26	0.05	0.00125	0.49
6	2.30E+05	1.50E+05	5.36	5.18	5.27	0.18	0.01620	2.42
7	5.30E+08	4.10E+08	8.72	8.61	8.67	0.11	0.00605	0.90
8	1.00E+04	1.20E+04	4.00	4.08	4.04	0.08	0.00320	1.40
9	3.00E+04	1.30E+04	4.48	4.11	4.30	0.37	0.06845	6.09
10	1.10E+08	2.20E+08	8.04	8.34	8.19	0.30	0.04500	2.59
Average					**6.18**		**0.022145**	

The reproducibility standard deviation is derived as the square root of the mean variance for the individual data sets, determined by summating the variances and dividing by n:

$$S_R = \sqrt{\frac{\sum_{i=1}^{n}(y_{i1} - y_{i2})^2/2}{n}} = \sqrt{\frac{0.22145}{10}} = \sqrt{0.022145} = \pm 0.1488 \approx \pm 0.15$$

Note that the average of the individual RSD values (2.12%) must not be used.

The overall intermediate RSD is given by the ratio between the overall SD and the overall mean log count $= 100 \cdot S_R/\bar{y} = 100(0.1488/6.18) = 2.41\%$. Note that the RSD values on individual duplicate analyses (i = 1–10) range from 0.62% to 6.09%.

This method of analysis can be extended to analyse results from three or more replicates by using a simple one-way ANOVA of the log counts on each occasion. It can also be used to estimate method repeatability if analyses on a set of samples are all done by a single analyst on a single day. This approach can then be used to assess analyst performance within a laboratory if all analyses are totally replicated by one or more additional analysts.

ESTIMATION OF UNCERTAINTY ASSOCIATED WITH QUANTAL METHODS

By definition, a quantal method is non-quantitative and merely provides an empirical answer regarding the presence or absence of a specific index organism or a group of related organisms in a given quantity of a representative sample. However, provided that multiple samples are analysed, and on the assumption that the test method is 'perfect', the number of tests giving a positive response provides an indication of the overall prevalence of defective samples (see Chapter 8). For instance, if a test on 10 parallel samples gave 4 positive and 6 negative results, then the apparent prevalence of defectives would be 40% (of the samples analysed) for which the 'exact' 95% CLs are 12.2–73.8%. However, if no positive samples were found, it would be common practice to refer, incorrectly, to a prevalence of defectives of 0% when in fact the true prevalence would actually lie between 0% and 31% (95% CLs).

ESTIMATION OF VARIANCE BASED ON BINOMIAL DISTRIBUTION

Determination of repeatability and reproducibility estimates for quantal test procedures can be based on the binomial probability of detection (POD) of positive and negative results. As described in Chapter 8, the probability of an event occurring (P) can be derived from the binomial response and an error estimate made provided that an appropriate number of samples have been analysed. For events that conform to the binomial distribution, the probability of obtaining k successes out of n trials is given by

$$P_{(k \text{ out of } n)} = \frac{n!}{k!(n-k)!} \cdot (p^k)(q^{n-k})$$

where k is the observed number of positive events from n independent trials, p is the probability of success (ie, the proportion of positive events) in a single trial and q $(=1 - p)$ is the probability of failure (ie, the proportion of negative events) in a single trial. The mean of a binomial distribution is derived from the Normal approximation and is given by $\mu = np$; the variance is given by $\sigma^2 = npq = np(1 - p)$.

Suppose that in a series of 10 trials, each of 10 tests, on a reference sample a total of 60 positive results are obtained out of the 100 tests. Then the overall proportion of positive results is $p = 60/100 = 0.6$ and that of the negative tests is $q = 1 - p = 0.4$. The variance (s^2) around the mean ($np = 60$) is given by $np(1 - p) = 60 \times 0.4 = 24$; so the standard deviation (s) $= \sqrt{24} \approx 4.90$. Hence, the approximate expanded uncertainty around the mean is $\pm 2 \times 4.9 = \pm 9.8 \approx \pm 10$. So the expected proportion of positive results would be contained within a 95% inter-percentile range of 0.6 ± 0.1. Clearly, such limits are very wide but the approach does derive an estimate for a parameter (s) that could be used as a measure of the expanded uncertainty.

An approach, suggested by Wilrich (2010), uses the Anon (1994) procedure to determine the variances of repeatability and reproducibility for continuous characteristics in inter-laboratory experiments in which a qualitative characteristic is investigated. A measurement series y_{ij} is made up of either 1's (positive results) or 0's (negative results); the laboratory is i ($i = 1...k$) and the measurement result is j ($j = 1...n$). Its average, given by $\bar{y}_i = \frac{1}{n}\sum_{j=1}^{n} y_{ij}$, is the proportion of positive results obtained in laboratory i and the average proportion of positive results over all laboratories by $\bar{\bar{y}} = \frac{1}{k_n}\sum_{i=1}^{k}\sum_{j=1}^{n} y_{ij} = \frac{1}{k}\sum_{i=1}^{k} \bar{y}_i$. Applying these concepts to the ISO Anon (1994, 2006) statistical model permits the estimation of variance both within and between laboratories (ie, repeatability and reproducibility variances) by use of ANOVA. The between- and within-laboratory standard deviations can be obtained from the ANOVA and hence estimates of measurement uncertainty can be derived.

In recent years much attention has been paid to the provision of confidence intervals for the mean POD (Probability of Detection) of a target substance (eg, chemical or microbiological contaminants). Wilrich (2010, 2015), Wehling et al. (2011), Macarthur and von Holst (2012), Uhlig et al. (2013) and Schneeweiß and Wilrich (2014) all deal with estimation of CLs for collaborative (inter-laboratory) studies where each of several laboratories performs binomial measurement studies. Schneeweiß and Wilrich (2014) discuss these approaches, which vary considerably both in approach and in difficulty, and propose a simplified approach. Wilrich (2015) provides details of a slightly different procedure that is illustrated in Examples 11.9 and 11.10. Further discussion of this topic is provided in Chapter 13.

EXAMPLE 11.9 USE OF THE WILRICH SPREADSHEET PROCEDURE TO DETERMINE THE LOD_{50} VALUE FOR A QUANTAL ASSAY

Suppose that we wish to compare the performance of a method for detection of a target organism in samples of three different foods. Tests are carried out using samples inoculated at five levels using a culture of the target organism, at least one level being intended to give a graduated response, that is, some positive and some negative results. After incubation under appropriate conditions, the numbers of positive and negative results at each level of each test are recorded and the results are entered into the spreadsheet. The spreadsheet can be downloaded from http://www.wiwiss.fu-berlin.de/fachbereich/vwl/iso/ehemalige/wilrich/index.html.

Open the spreadsheet (by clicking to accept the macros) and enter the following details in the yellow box on the left-hand side of the spreadsheet:

- name of experiment;
- date;
- sample size (as gram or millilitre), for example, 25 g;
- number of sample matrices, for example, 3; and
- number of contamination levels for each matrix, for example, 5.

For three sample types, three new yellow boxes are generated. For each sample type enter:

- sample matrix name;
- the data for the nominal levels of inoculation;
- the number of parallel tests at each inoculum level; and
- the results obtained as the number of positive tests at each level.

This is illustrated in Table 11.16.

Once all the data have been input, press *Ctrl* + *b* to generate the results which appear in green boxes to the right of the screen as shown in Table 11.17; a set of individual boxes also appears alongside each input sample data.

For each matrix, the results show the following:

- The matrix effect (F_i), the standard deviation of the $\log_{10}$ matrix effect and, at the end of the row, the absolute test statistic ($|z_i|$) for that effect. F_i estimates the extent to which the data deviate from an expected perfect response and $|z_i|$ is the Normalised range – so in the case of the meat balls the observed matrix effect is small ($|z| < 0.3$) but that for the frozen vegetables is almost at the limit of acceptable tolerance ($|z| = 2$).
- The level of detection at 50% positive results (LOD_{50}) and the lower and upper confidence limits of that estimate. In the case of the pasteurised milk the $LOD_{50} = 0.033$ organisms/g (0.825 organisms/25 g sample) with confidence limits of 0.019–0.057 organisms/g (0.475–1.425 organisms/25 g).
- The level of detection at 95% positive results (LOD_{95}) and the lower and upper confidence limits of that estimate.

The spreadsheet output also calculates values for the combined data inputs – this is particularly useful if parallel replicate tests of the same foodstuff have been done or if, as in this case, different foods are being evaluated.

The procedure can be used also to compare two or more different analytical methods. The purpose in providing an estimate for the LOD_{95} is that sometimes the responses obtained from two different methods may be similar at the LOD_{50} level but differ widely at the LOD_{95} level, or vice versa.

The spreadsheet plots the data for each individual set of results to show the data points against the calculated line of best fit and the lines of confidence around the best fit. Fig. 11.3 shows the data for the individual matrices and Fig. 11.4 compares these with the 'perfect' response and its 95% confidence limits. Note that the curves for pasteurised milk and meat balls are very close to the 'perfect' line but the line for the frozen vegetables is close to the upper 95% CL.

Table 11.16 Screenshot of the Use of the Excel Spreadsheet of Wilrich and Wilrich (2009) reproduced by kind permission of Prof P Th Wilrich

20	Name of experiment:	Examples	
21	Date of experiment:	16/06/2015	
22	Sample size A0 in g or ml:	25	
23	Total no. of matrices:	3	
24	Max. no. of contamination levels:	5	
25			
26	No. of matrix:	1	
27	Name of matrix/sample reference:	Pateurized Milk	
28	Level of inoculum in cfu/g or cfu/ml	No. of inoculated tubes	No. of postive tubes
29	d	n	y
30	0.0112	6	1
31	0.0224	6	2
32	0.0448	6	4
33	0.0672	6	4
34	0.1416	6	6
35			
36	No. of matrix:	2	
37	Name of matrix/sample reference:	Meat Balls	
38	Level of inoculum in cfu/g or cfu/ml	No. of inoculated tubes	No. of postive tubes
39	d	n	y
40	0.0156	6	1
41	0.0308	6	4
42	0.062	6	4
43	0.0928	6	5
44	0.098	6	6
45			
46	No. of matrix:	3	
47	Name of matrix/sample reference:	Frozen vegetables	
48	Level of inoculum in cfu/g or cfu/ml	No. of inoculated tubes	No. of postive tubes
49	d	n	y
50	0.0108	6	2
51	0.0216	6	4
52	0.0432	6	4
53	0.0648	6	6
54	0.102	6	6
55			
56			
57			

Table 11.17 Screenshot of Results for LOD Determinations on Three Sample Matrices

No. of matrix	Name of matrix	Matrix effect	SD of log matrix effect	LOD50% [1] Detection limit	Lower conf. limit	Upper conf. limit	LOD95% [2] Detection limit	Lower conf. limit	Upper conf. limit	Test statistic matrix effect
i	$matrix_i$	F_i	s_{fi}	$d_{0.5,i}$	$d_{0.5,i,L}$	$d_{0.5,i,U}$	$d_{0.95,i}$	$d_{0.95,i,L}$	$d_{0.95,i,U}$	$\lvert z_i \rvert$
1	Pateurized Milk	0.833	0.272	0.033	0.019	0.057	0.144	0.084	0.248	0.679
2	Meat Balls	0.932	0.251	0.030	0.018	0.049	0.129	0.078	0.213	0.279
3	Frozen vegetable	1.644	0.266	0.017	0.010	0.029	0.073	0.043	0.124	1.903
Combined data		1.065	0.150	0.026	0.019	0.035	0.113	0.083	0.152	0.417

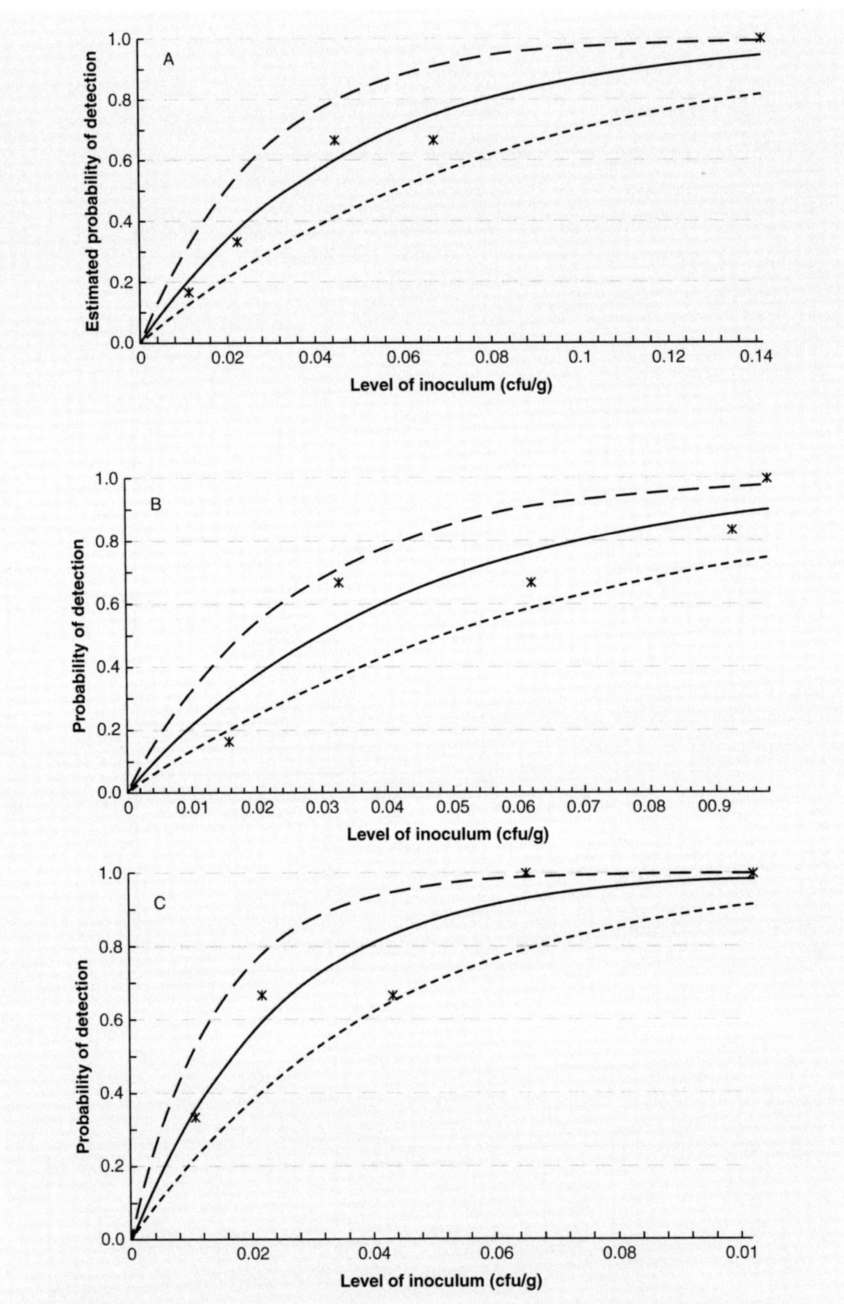

FIGURE 11.3 Probability of Detection (POD) Response Curves for the Matrices Examined in Example 11.9 Using the Spreadsheet Method of Wilrich and Wilrich (2009)

(A) Pasteurised milk; (B) meat balls; (C) frozen vegetables. Note that the legend to the graphs refers to the POD, not to the limit of detection at a specific probability (eg, the LOD_{50})

Reproduced with permission of AOAC International and the authors

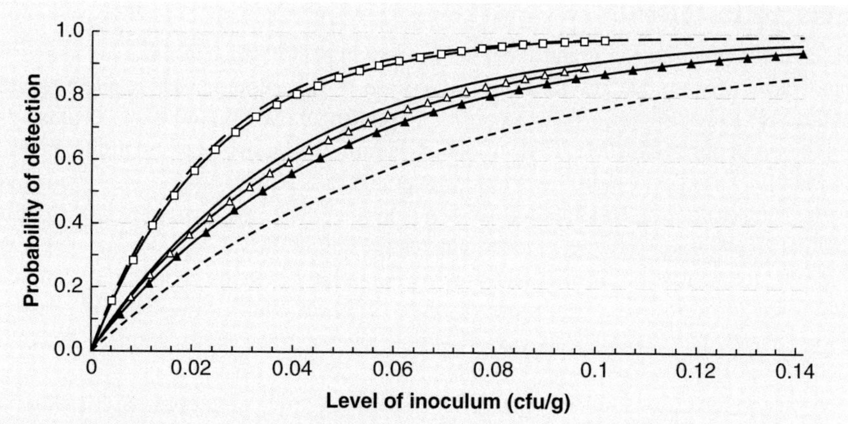

FIGURE 11.4 Combined POD curves for the frozen vegetables (□), meat balls (△) and pasteurised milk (▲) matrices shown in Fig. 11.3 together with the 'perfect' curve and the bounds of its confidence limits.

Reproduced by permission of AOAC International and the authors

MPN ESTIMATES

When a proportion of parallel test results are positive, an estimate of population density can be derived for multiple tests at a single dilution level, using the following equation:

$$\hat{m} = -\frac{1}{V} \cdot \ln\left(\frac{S}{n}\right)$$

where $\hat{m}$ is the Poisson estimate of the mean, V is the quantity of sample and S is the number of sterile tests out of n tests inoculated. Assuming 10 tests are set up on replicate 25-g samples of product, 3 tests are positive and 7 are negative, then the value of $\hat{m}$ is given by:

$$\hat{m} = -\frac{1000}{25}\ln\left(\frac{7}{10}\right) = -40(-0.3567) = 14.27 \approx 14 \text{ organisms/kg}$$

Since, for a Poisson estimate, the mean equals the variance, the standard error will be $\sqrt{14.27} = 3.78$. The uncertainty, as the limits of the 95% CI, will be given by $1.96 \times 3.78 \approx \pm 7.4$ so the estimate of the mean contamination level and its uncertainty is given by 14.27 ± 7.4 organisms/kg, that is, approximately from 6.9 to 21.7 organisms/kg. Note that calculated 'exact' CIs are 7.7 to 23.5 organisms/kg.

When tests are done at several levels of inoculation, it is possible to derive estimates of MPNs together with their upper and lower CLs from either tables or computer spreadsheet (Chapter 8; Jarvis et al., 2010). The CLs define the relative uncertainty of the estimate.

LEVEL OF DETECTION ESTIMATES

The equation given earlier is also the basis for deriving MPN values for use in the Spearman–Kärber (SK) procedure to estimate the level of detection at 50% positive results (LOD_{50}) for a test. Finney (1971) describes the statistical background to the SK test.

The procedure (BPMM, 2006) uses non-parametric statistics to calculate the microbial concentration (and CLs) that corresponds to a 50% probability of a positive result using a defined test method. It is not applicable to routine test data since it requires an estimate of the level of organisms present in the inoculum but it had been proposed for validation of quantal methods (Chapter 13) using inoculated samples. The procedure requires a minimum of three different concentrations of organisms at log-spaced intervals, at least one of which gives a fractional response (ie, some positive and some negative values) and one gives a 100% response. In addition, it is necessary to have one totally negative set of data (ie, a negative control) for which a low, but artificial, contamination level is assumed, for example, 0.01% response. This method of quantification enables the derivation of a value for the standard error, and hence of the uncertainty, of the LOD_{50} estimate.

Alternative approaches including probit and logit procedures have been evaluated for analysis of microbiological quantal data. Van der Voet (personal communication, 2007) considered that a generalised linear model of the complimentary loglog response provided a better option. This procedure was developed further (Wilrich and Wilrich, 2009; Wilrich, 2010; Mărgăritescu and Wilrich, 2013) and is described in Anon (2011). Wilrich and Wilrich (2009) noted major deficiencies in the SK procedure: the need for monotonically increasing contamination levels; use of a weighted average of the data for each concentration; the need to include a 'low-level pseudo-measurement' result; and confidence levels that provide coverage of only ca. 90% around the calculated LOD_{50} estimate. An Excel spreadsheet system for the Wilrich and Wilrich (2009) method (PODLOD ver6.1) can be downloaded from http://www.wiwiss.fu-berlin.de/fachbereich/vwl/iso/ehemalige/wilrich/index.html (accessed 15.11.15).

EXAMPLE 11.10 ESTIMATION OF THE CLS FOR A BINOMIAL DISTRIBUTION

Assume that 10 laboratories carry out tests to detect the presence of a target organism in each of 12 replicate 25-g samples of a food using a standard method and that the following results are obtained:

	Laboratory (*k*)									
	1	2	3	4	5	6	7	8	9	10
No. of tests (n)	12	12	12	12	12	12	12	12	12	12
No. of positive results	7	9	6	10	5	7	5	7	11	9
Proportion positive, $\hat{p}_i$	0.583	0.750	0.500	0.833	0.417	0.583	0.417	0.583	0.917	0.750

For $k = 10$ laboratories each testing $n = 12$ parallel samples and, for the data shown, the mean proportion of positives $= \bar{p} = (1/k)\Sigma_{i=1}^{k} \hat{p}_i = (1/10) \times 6.33 = 0.6\dot{3}\dot{3}$.

The repeatability variance is given as follows:

$$\begin{aligned}\overline{s}_r^2 &= \left(\frac{n}{n-1}\right)\frac{1}{k}\sum_{i=1}^{k}[\hat{p}_i(1-\hat{p}_i)] \\ &= \left(\frac{12}{12-1}\right)\frac{1}{10}[(0.583\times 0.417)+(0.750\times 0.250)+\cdots+(0.750\times 0.250)] \\ &= \left(\frac{12}{11\times 10}\right)2.056 = 0.2242\end{aligned}$$

and

$$\overline{s}_r = \sqrt{0.2242} = 0.4735$$

An unbiased estimate of between-laboratory variance (s_L^2) is given by

$$s_L^2 = s_{\hat{p}}^2\left\{1-\frac{\exp[1-c(n-1)]}{c(n-1)}\right\}$$

where $s_{\hat{p}}^2$ is the variance of the proportions of positive results ($\hat{p}_i$) in the laboratories:

$$s_{\hat{p}}^2 = \frac{1}{k-1}\sum_{i=1}^{k}(\hat{p}_i-\overline{p})^2 \quad \text{and} \quad c=\frac{s_{\hat{p}}^2}{\overline{s}_r^2}$$

For these data, $\overline{p}=0.633$, $s_{\hat{p}}^2=[1/(10-1)]0.2667=0.0296$ and $s_r^2=0.2242$ (see earlier text), so

$$c=\frac{s_{\hat{p}}^2}{\overline{s}_r^2}=\frac{0.0296}{0.2242}=0.1321$$

Hence:

$$s_L^2 = s_{\hat{p}}^2\left\{1-\frac{\exp[1-c(n-1)]}{c(n-1)}\right\}=0.0296\left\{1-\frac{\exp[1-0.1321(12-1)]}{0.1321(12-1)}\right\}=0.01665 \text{ and}$$

$s_L=\sqrt{0.01665}=0.129$.

The reproducibility variance $=s_R^2=s_L^2+s_r^2=0.01665+0.2242=0.2409$ and the reproducibility standard deviation $s_R=\sqrt{0.2409}=\pm 0.4908$.

The ratio $L_R=s_L^2/s_R^2=0.01665/0.2409=0.069$ will be 0 if there is no inter-laboratory variation and will tend to 1 if that variation is large.

The confidence interval that covers the variation in laboratory outputs can be determined as a parametric (1 − γ)-expectation tolerance interval, that is, the interval calculated with the results of the experiment that is expected to cover the fraction (1 − γ) of the probability of detection (POD) of the population of laboratories. Wilrich (2015) uses $\overline{p}$ and s_L for the calculation of the 90%-expectation tolerance interval with lower limit $\overline{p}-2s_L$ and upper limit $\overline{p}+2s_L$. If the lower limit is negative we substitute it by 0, and if the upper limit is larger than 1 we substitute it by 1. Hence, the lower limit of the 90%-expectation tolerance interval for the POD across the laboratories is given by the maximum of 0 or $\overline{p}-2s_L$ and the upper limit by the minimum of 1 or $\overline{p}+2s_L$. In this example, $\overline{p}=0.633$, $s_L=0.129$ and the 90%-expectation limits are 0.633 ± 2 × 0.129, ie, 0.38 and 0.89.

The mean result can be stated as a POD = 0.63 with a 90% uncertainty of ±0.26.

Based on the procedure of Wilrich (2015).

USE OF REFERENCE MATERIALS IN QUANTAL TESTING

One of the inherent problems with presence or absence tests relates to the likelihood that a method may give either a false-positive (Type I error) or a false-negative (Type II error) (Chapter 8). A false-positive is a result that indicates the apparent presence of the target organism when it is *not* present in the sample; a false-negative result is one where the presence of a target organism is not detected when it *is* present in a sample. Such errors create specific problems in the interpretation of quantal test results. If tests are done on natural food matrices, it is impossible to estimate the likely occurrence of false results, other than by comparison of results in one laboratory with those in another. During development, evaluation and routine use of any method it is essential to ensure that the likelihood of such errors occurring is minimised. Laboratory proficiency schemes (Jarvis, 2014a) provide a way to monitor the efficiency of test procedures in any individual laboratory and are now mandatory in many countries for food laboratories accredited to ISO 17025 (Anon, 2005a,b, 2010a,b).

Minimising the occurrence of false results requires the use of standardised reference materials that can be relied on to contain the target organism at a defined level. For high-level contamination (eg, 1000 cfu/mL) that is not a major problem, but problems arise when the actual level of detection (LOD) is intended to be close to the minimum LOD. For instance, to detect 1 cfu of a specific organism in (say) 25 g of sample implies that the organism is evenly distributed throughout a lot of test material such that each 25-g sample unit is likely to contain 1 viable cell of the target organism. Unfortunately it is not possible to add a single cell of a test organism to each individual 25-g sample to guarantee with 100% probability that each sample would contain 1, but neither more nor less than 1, viable organism. Even if it were possible, there is a distinct probability that the organism might not survive the preparation and storage process. There is little guarantee that low levels of inoculated organisms will be distributed randomly throughout a batch of food matrix such that each of a number of randomly drawn samples will contain the test organism at the same concentration.

Standard reference materials for use in quantal microbiological tests are now available either for direct use in a test system or for dilution into a batch of material. Certified reference materials, that is, those conforming to the requirements of *ISO Guide 35* (Anon, 2006), have defined levels for specific types of organism (Jarvis, 2014b). However, no matter how well a commercial- or laboratory-standardised preparation has been produced, the indicated level of target organism can *never* be absolute, not least since it has to be estimated by a viable counting method. With such high inherent distribution variability it is not surprising that in microbiological testing there can be no 'absolute' reference values.

REFERENCES

AMC, 1989a. Robust statistics – how not to reject outliers. Part 1: basic concepts. Analyst 114, 1693–1697.

AMC, 1989b. Robust statistics – how not to reject outliers. Part 2: inter-laboratory trials. Analyst 114, 1699–1702.

AMC, 2001. Robust statistics: a method of coping with outliers. AMC Brief No. 6. Royal Society of Chemistry, London. <http://www.rsc.org/images/robust-statistics-technical-brief-6_tcm18-214850.pdf> (accessed 14.07.15).

AMC, 2005. Terminology – the key to understanding analytical science. Part 2: sampling and sample preparation. AMC Brief No. 19. Royal Society of Chemistry, London. <http://www.rsc.org/images/sampling-sample-preparation-technical-brief-19_tcm18-214856.pdf> (accessed 14.07.15).

Anon, 1994. Accuracy (Trueness and Precision) of Measurement Methods and Results – Part 2: Basic Methods for the Determination of Repeatability and Reproducibility of a Standard Measurement Method. ISO 5725-2:1994. International Standards Organization, Geneva.

Anon, 1998. Accuracy (Trueness and Precision) of Measurement Methods and Results – Part 5: Alternative Methods for the Determination of the Precision of a Standard Measurement Method. ISO 5725-5:1998. International Standards Organization, Geneva.

Anon, 2005a. Statistical Methods for Use in Proficiency Testing by Interlaboratory Comparisons. ISO 13528:2005. International Standards Organization, Geneva.

Anon, 2005b. General Requirements for the Competence of Testing and Calibration Laboratories. ISO/IEC 17025:2005. International Standards Organization, Geneva.

Anon, 2006. Reference Materials – General and Statistical Principles for Certification. ISO Guide 35: 2006. International Standards Organization, Geneva.

Anon, 2009. Microbiology of Food and Animal Feeding Stuffs – Guide on Estimation of Measurement Uncertainty for Quantitative Determinations. ISO TS 19036:2006/Amd 1:2009. International Standards Organization, Geneva (under revision).

Anon, 2010a. Conformity Assessment – General Requirements for Proficiency Testing. ISO 17043:2010. International Standards Organization, Geneva.

Anon, 2010b. Microbiology of Food and Animal Feeding Stuffs – Specific Requirements and Guidance for Proficiency Testing by Interlaboratory Comparison. ISO/TS 22117: 2010. International Standards Organization, Geneva.

Anon, 2011. Microbiology of Food and Animal Feeding Stuffs – Protocol for the Validation of Alternative Methods. ISO 16140:2003+A1. International Standards Organization, Geneva (under review).

Anon, 2013. Microbiology of Food and Animal Feed – Method Validation – Part 2: Protocol for the Validation of Alternative (Proprietary) Methods Against a Reference Method. Committee Draft ISO 16140-2. International Standards Organization, Geneva (under revision).

Box, G.E.P., Cox, D.R., 1964. An analysis of transformations. J. R. Stat. Soc. B26, 211–243, (discussion 244–252).

BPMM, 2006. Proposed use of a 50% limit of detection value in defining uncertainty limits in the validation of presence–absence microbial detection methods. In: Best Practices in Microbiological Methodology, AOAC Final BPMM Task Force Report, Statistics WG Reports 4a and 4b, Annexes K & L. <http://www.fda.gov/Food/FoodScienceResearch/LaboratoryMethods/ucm124900.htm> (accessed 31.10.15).

Brown, M.H., 1977. Microbiology of the British fresh sausage. PhD Thesis. University of Bath, UK.

Chakravarti, I.M., Laha, R.G., Roy, J., 1967. Handbook of Methods of Applied Statistics, Volume I, Techniques of Computation, Descriptive Methods, and Statistical Inference. John Wiley and Sons, New York, pp. 392–394.

Corry, J.E.L., Hedges, A.J., Jarvis, B., 2007. Measurement uncertainty of the EU methods for microbiological examination of red meat. Food Microbiol. 24, 652–657.

Corry, J.E.L., Hedges, A.J., Jarvis, B., 2010. Minimising the between-sample variance in colony counts on foods. Food Microbiol. 27, 598–603.

D'Agostino, R.B., 1986. Tests for Normal distribution. In: D'Agostino, R.B., Stephens, M.A. (Eds.), Goodness-of-Fit Techniques. Marcel Decker, New York, pp. 36–41.

Dixon, W.J., 1953. Processing data for outliers. Biometrics 9, 74–89.

Ellison, S.L.R., Williams, A. (Eds.), 2012. Eurachem/CITAC Guide: Quantifying Uncertainty in Analytical Measurement, third ed. Available from: <https://www.eurachem.org/index.php/publications/guides/quam> (accessed 13.07.15).

Finney, D.J., 1971. Probit Analysis, third Ed. University Press, Cambridge, UK.

Granato, D., Calado, V.M.A., Jarvis, B., 2014. Observations on the use of statistical methods in food science and technology. Food Res. Int. 55, 137–149.

Grubbs, F., 1969. Procedures for detecting outlying observations in samples. Technometrics 11, 1–21.

Hector, A., 2015. The New Statistics With R: An Introduction for Biologists. Oxford University Press, Oxford, UK.

Hedges, A.J., 2008. A method to apply the robust estimator of dispersion, Qn, to fully-nested designs in the analysis of variance of microbiological count data. J. Microbiol. Methods 72, 206–207.

Hedges, A.J., Jarvis, B., 2006. Application of 'robust' methods to the analysis of collaborative trial data using bacterial colony counts. J. Microbiol. Methods 66, 504–511.

Horwitz, W., 1995. Protocol for the design, conduct and interpretation of method performance studies. Pure Appl. Chem. 67, 331–343.

Jarvis, B., 2014a. Proficiency testing schemes – a European perspective. Batt, C.A., Tortorello, M.L. (Eds.), Encyclopaedia of Food Microbiology, vol. 3, Elsevier Ltd, Academic Press, London, pp. 226–231.

Jarvis, B., 2014b. Microbiological reference materials. Batt, C.A., Tortorello, M.L. (Eds.), Encyclopaedia of Food Microbiology, vol. 2, Elsevier Ltd, Academic Press, London, pp. 614–620.

Jarvis, B., Corry, J.E.L., Hedges, A.J., 2007a. Estimates of measurement uncertainty from proficiency testing schemes, internal laboratory quality monitoring and during routine enforcement examination of foods. J. Appl. Microbiol. 103, 462–467.

Jarvis, B., Hedges, A.J., Corry, J.E.L., 2007b. Assessment of measurement uncertainty for quantitative methods of analysis: comparative assessment of the precision (uncertainty) of bacterial colony counts. Int. J. Food Microbiol. 116, 44–51.

Jarvis, B., Wilrich, C., Wilrich, P.Th., 2010. Reconsideration of the derivation of most probable numbers, their standard deviations, confidence bounds and rarity values. J. Appl. Microbiol. 109, 1660–1667.

Jarvis, B., Hedges, A.J., Corry, J.E.L., 2012. The contribution of sampling uncertainty to total measurement uncertainty in the enumeration of microorganisms in foods. Food Microbiol. 30, 362–371.

Jewell, K. (Ed.), 2004. Microbiological Measurement Uncertainty: A Practical Guide. Guideline No. 47. Campden & Chorleywood Food Research Association, Campden, UK.

Kilsby, D.C., Pugh, M.E., 1981. The relevance of the distribution of microorganisms within batches of food to the control of microbiological hazards from foods. J. Appl. Microbiol. 51, 345–354.

Macarthur, R., von Holst, C., 2012. A protocol for the validation of qualitative methods for detection. Anal. Methods 4, 2744–2754.

Mărgăritescu, I., Wilrich, P.T., 2013. Determination of the RLOD of a qualitative microbiological measurement method with respect to a reference measurement method. J. AOAC Int. 96, 1086–1091.

Niemelä, S.I., 2002. Uncertainty of Quantitative Determinations Derived by Cultivation of Microorganisms, second ed. Centre for Metrology and Accreditation, Advisory Commission for Metrology, Chemistry Section, Expert Group for Microbiology, Helsinki, Finland, Publication J3/2002.

Niemelä, S.I., 2003. Measurement uncertainty of microbiological viable counts. Accred. Qual. Assur. 8, 559–563.

Rousseeuw, P.J., Croux, C., 1993. Alternatives to the median absolute deviation. J. Am. Stat. Assoc. 88, 1273–1283.

Schneeweiß, H., Wilrich, P.T., 2014. A simple method of constructing a confidence interval for the mean probability of detection in collaborative studies of binary measurement methods. Accred. Qual. Assur. 19, 221–223.

Shapiro, S.S., Wilk, M.B., 1965. An analysis of variance test for Normality (complete samples). Biometrika 52, 591–611.

Smith, P., Kokic, P., 1997. Winsorisation for numeric outliers in sample surveys. Bull. Int. Stat. Inst. 57, 609–610.

Stephens, M.A., 1974. EDF statistics for goodness of fit and some comparisons. J. Am. Stat. Assoc. 69, 730–737.

Tabachnick, B.G., Fidell, L.S., 1996. Using Multivariate Statistics, third ed. Harper Collins, New York.

Uhlig, S., Krügener, S., Gowik, P., 2013. A new profile likelihood confidence interval for the mean probability of detection in collaborative studies of binary test methods. Accred. Qual. Assur. 18, 367–372.

Wehling, P., LaBudde, R.A., Brunelle, S.L., Nelson, M.T., 2011. Probability of detection (POD) as a statistical model for the validation of qualitative methods. J. AOAC Int. 94, 335–347.

Wilrich, P.Th., 2007. Robust estimates of the theoretical standard deviation to be used in interlaboratory precision experiments. Accred. Qual. Assur. 12, 231–240.

Wilrich, P.Th., 2010. The determination of precision of qualitative measurement methods by interlaboratory experiments. Accred. Qual. Assur. 15, 439–444.

Wilrich, P.Th., 2015. The precision of binary measurement methods. Beran, J. et al., (Ed.), Empirical Economic and Financial Research: Advanced Studies in Theoretical and Applied Econometrics, vol. 48, Springer International, Switzerland, pp. 223–235.

Wilrich, C., Wilrich, P.Th., 2009. Estimation of the POD function and the LOD of a qualitative microbiological measurement method. J. AOAC Int. 92, 1763–1772.

Youden, W.J., Steiner, E.H., 1975. Statistical Manual of the AOAC. AOAC International, Washington, DC.

CHAPTER

Statistical process control using microbiological data 12

Microbiological data are often filed without much consideration as to their further application. When one considers the costs of providing, staffing and running laboratories, of obtaining and examining samples, etc., it is at best short-sighted not to make effective use of data. Procedures for statistical process control (SPC) have been around for more than 80 years and have been the subject of many books and papers (see, eg, Shewhart, 1931; Anon, 1956; Juran, 1974; Duncan, 1986; Beauregard et al., 1992; Montgomery, 2012), although few refer to microbiological applications. The benefits of, and approaches to, SPC for trend analysis of microbiological data were included in the report (Anon, 2006) of a US FDA-funded programme undertaken by an AOAC Presidential Task Force programme on 'Best Practices in Microbiological Methods'.

WHAT IS STATISTICAL PROCESS CONTROL?

SPC is a practical approach to continuous control of quality (and safety) that uses various tools to monitor, analyse and evaluate the control of a process and is based on the philosophical concept that it is essential to use the best available data in order to optimise process efficiency. In this context, the term 'process' is any activity that is subject to variability and includes not just production, storage and distribution within an industry, such as food processing, but also other operations where it is necessary to ensure that day-to-day activities are in control. It is therefore relevant to laboratory monitoring and testing activities, where 'process' describes the production of analytical data and other forms of quantifiable activity.

The concept was introduced in the late 1920s by Dr William Shewhart whose work helped to improve efficiency in the US manufacturing industry by moving from traditional quality control to one based on quality assurance. SPC is based on the concept of using data obtained from process measurements and/or on samples of intermediate and finished products, rather than reliance on 100% final product inspection. In this context, quality assurance implies that appropriate checks are 'built in' to ensure that a production process is in control (a concept not dissimilar to that of the HACCP approach to food safety management). Assurance of production control comes from the regular monitoring of process parameters using tests to evaluate product samples, or other criteria for a process. The aim of SPC is to achieve a stable process, the operational performance of which is predictable, through monitoring, analysis and evaluation. It is therefore one of the tools of total quality management,

Statistical Aspects of the Microbiological Examination of Foods. http://dx.doi.org/10.1016/B978-0-12-803973-1.00012-7

a subject largely outside the scope of this book, but considered in the context of the definition and implementation of food safety objectives in Chapter 14.

At its simplest, SPC in manufacturing production is done by operative recording of simple physical (temperature and pressure), chemical (pH, acidity, etc.) or other parameters critical to effective operation of a process plant. Such values are plotted on appropriately designed record forms or charts. An inherent problem in any manufacturing operation is that both management and operatives have a tendency to 'twiddle' knobs probably with the best of intentions. If it is suspected that a process is not quite running properly, it is a natural human instinct to try to put matters right – but by so doing matters may get worse. Provided that control systems are operating properly, there should be no necessity for human intervention unless there is clear evidence of an 'out-of-control' situation developing when action needs to be taken to document the problems and to identify and correct the cause. SPC provides an objective approach to aid decision making.

Data obtained by analysis of critical product compositional parameters can also be charted. SPC provides a means to assess trends in the data over time in order to ensure that products conform to previously defined criteria. Evidence of significant trends away from the 'norm' requires remedial action, for example, to halt production, investigate the causes of the problem and return the process operation to a controlled state. In a similar way, SPC of internal laboratory quality performance monitoring data can be used to provide assurance that analyses are being undertaken properly and to identify when and where potential problems may arise.

TREND ANALYSIS

At its simplest, trend analysis involves plotting data values, for which a target value can be established, against values for time, production lot number or other identifiable parameter. Note that a target value is *not* a product specification; rather it is a value derived by analysis of product produced under conditions when the process is known to be 'in control'. Fig. 12.1 provides a simple schematic of trend lines that lack any form of defined limits. Although the direction of the trends is obvious, in the absence of control limits, there is no guidance on whether or not a process is 'going out of control'.

A philosophical aspect of SPC is that once control has been achieved, the approach can be converted to statistical process improvement whereby process and, subsequently, product improvement is based on the principles of total quality management (Beauregard et al., 1992). It is noteworthy that the European legislation on microbiological criteria for foodstuffs recommends that manufacturers should use trend analysis to examine results of routine testing (EU, 2005).

TOOLS FOR STATISTICAL PROCESS CONTROL

The original Shewhart (1931) concept of a process is depicted in Fig. 12.2 as a 'cause and effect' (fishtail) diagram comprising materials, methods, manpower, equipment and measurement systems that together make up the process environment. The extent

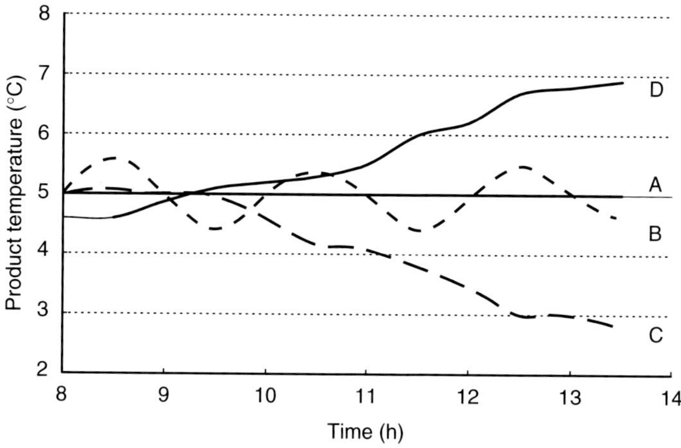

FIGURE 12.1 Hypothetical Illustration of Trends in Product Temperature

(A) Perfect control (never seen in practice); (B) regular temperature cycling within close limits; (C) continuous reduction in temperature; (D) continuous increase in temperature

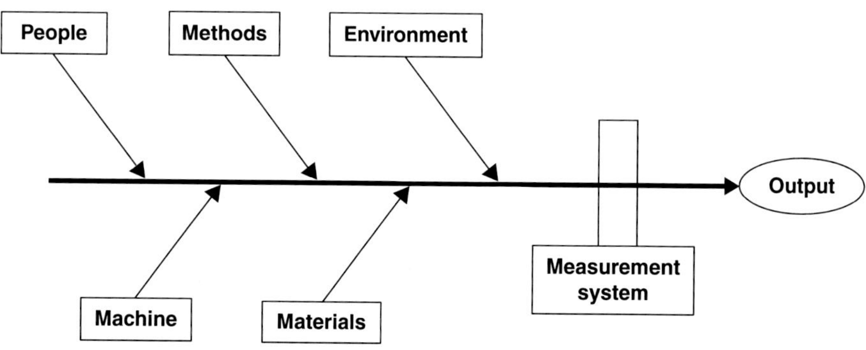

FIGURE 12.2 Schematic of Shewhart's Approach to Description of a Process

of variability (ie, lack of control) associated with each of these factors impacts on the subsequent stages of the process and the overall variability in a process can be determined from the individual factors, namely:

$$V_{\text{output}} = \sum_{i=1}^{K} V_K = V_{\text{materials}} + V_{\text{equipment}} + V_{\text{method}} + V_{\text{people}} + \cdots + V_K$$

where V_K is the variance associated with an individual factor K. Evidence of a high variance in some measurable quality of the finished product will be reflected in the sum of the variances of the manufacturing stages for which the causes of the variances may be either random or inherent.

Consequently, the basic statistical tools for SPC are the estimates of the mean (or median) value, the variance and the distribution characteristics of key parameters associated with a final, or intermediate, product and include the variances observed in the monitoring and recording of critical control point parameters, such as temperature and pressure (cf HACCP). From many perspectives, this is similar to the estimation of measurement uncertainty (Chapters 10 and 11). The use of a microbiological procedure to assess the quality or safety of a product can be used to verify that a production process is 'in control' but only if the microbiological test procedure is itself 'in control'.

The principle of SPC charts relies on the properties of the Normal distribution curve (Chapter 3), which has a symmetrical distribution around the mean value, with 95% of the population within the range of mean ± 2 SDs and 99.7% within the range of mean ± 3 SDs. However, as noted previously, the distribution of microbiological data is inherently asymmetrical. It is therefore necessary to use an appropriate transformation of the data, depending on its actual distribution characteristics.

Before any system of SPC can be used it is essential to define the process, the measurement system and the expected variability of the process characteristics to be measured. From a microbiological perspective, therefore, it is necessary to understand how a product is processed, the microbial associations of the ingredients and the product, the effects of processes on microbial populations and the procedures to be used to determine the microbial composition of the product. In terms of microbiological laboratory control, the product is the colony count or other output; the ingredients are the diluents, culture media and equipment (eg, macerators, pipettes) and the process is that of taking a test portion of a sample, diluting it, inoculating the media, incubation and final assessment of the output.

SPC PERFORMANCE STANDARDS

It is necessary to use 'performance standards' for implementation of SPC for microbiological data. Such 'performance standards' are not prescriptive procedures that should be used for evaluating processes, but provide guidance on the methodology to be used:

1. Charting of data is necessary to gain full benefit from SPC. Charts of output data provide a visual aid to detecting and identifying sources of unexpected variation and lead to their elimination.
2. Results are plotted in a control chart; when the process is known to be 'in control', the data should be Normally distributed. Where this is not the case, the data must be 'Normalised' by an appropriate mathematical transformation.
3. During the initial setting up of the charts it is essential that the process be operated in a stable and controlled manner, so that the data values used to establish the operational parameters of the SPC procedure are Normally distributed. The 'rule of thumb' requires at least 20 sets of individual results to compute the mean, standard deviation and other summary statistics in order

to estimate the distribution of the results and construct control limits and target values.

4. Rules for evaluating process control require the assessment of both Type I and Type II errors: a Type I error (probability = α) defines the process to be 'out of control' when it is not and a Type II error (probability = β) defines a process to be 'in control' when it is not. The probabilities α and β are based on the Normal distribution curve. In addition, the average run length, which is the expected number of samples seen before an 'out-of-control' signal occurs, is also an important parameter.
5. When a process is believed to be 'in control', the limits for assessing individual results are set at some distance from the mean or process target value. In general, the default distance is 3 SDs since the Type I error should be kept low (ie, $\alpha = 1\%$).
6. In using the SPC approach it is necessary to establish rules to assess whether a change in the process average has occurred. Such rules include the use of moving averages and the number of consecutive values moving in a single direction.
7. Process control limits used in SPC are *not* product criteria limits, set by customer or regulation; rather they are process-related limits. Product specifications should not be shown on control charts.

A fuller explanation of performance standards for SPC can be found in the BPMM report (Anon, 2006). For SPC of laboratory data it is desirable that anonymous reference samples should be tested during day-to-day work; such samples should have a known microbiological profile. For instance, in checking quantal methods for ability to detect salmonellae at a given level of contamination unknown reference samples with or without addition of specific inocula should be tested routinely.

SETTING CONTROL LIMITS

Assume, for a defined non-sterile food product, processed using a specific manufacturing system, that the microbial load is considered to be acceptable at the end of processing if the average aerobic colony count at 30°C is, say, not greater than 10^4 cfu/g ($4.0 \log_{10}$ cfu/g). Assume also that the transformed counts conform reasonably to a Normal distribution and that the standard uncertainty of the measurement method used to establish the levels of organisms from the product is $\pm 0.25 \log_{10}$ cfu/g. In most circumstances we are interested only in high counts, so a lower control limit is not necessary (the exception is when it is critical to have a minimal level of specific organisms, as in foods inoculated with probiotic bacteria). For a one-sided process control limit, with an acceptable mean colony count level of 10^4 cfu/g ($4.0 \log_{10}$ cfu/g) and SD = 0.25, for $\alpha = 0.01$, the upper 99% confidence interval will be $4.0 + 2.326 \times 0.25 = 4.58 \log_{10}$ cfu/g. It would then be expected that the probability of a value exceeding $4.6 \log_{10}$ cfu/g would occur on average only on 1 occasion in 100 tests. It is useful also to set a process control warning limit; for $\alpha = 0.05$ the warning limit would be 4.0 + 1.645 × 0.25 = 4.41 $\log_{10}$ cfu/g. These upper control limits (UCLs), together with the mean value of 4.0 $\log_{10}$ cfu/g, are used to construct the limit lines drawn on the one-sided SPC control chart for average values.

Data from tests on a large number of samples, taken during processing of a single batch of products, should conform (after transformation if necessary) to a Normal distribution (Fig. 12.3A) and should lie below the UCLs. Data conforming to a Normal distribution with an increased mean value (Fig. 12.3B) will include results that

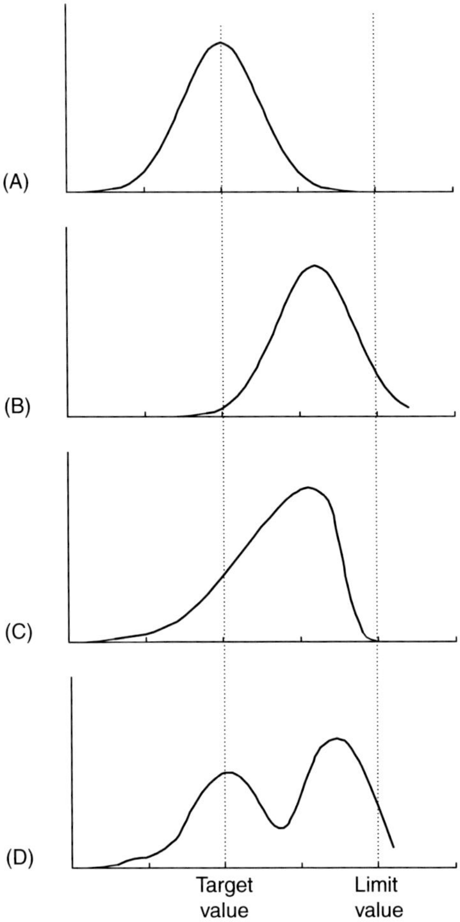

FIGURE 12.3 In Control or in Specification?

(A) In control and within limits; (B) in control but not totally within limits; (C) Within limits but not in control due to skewness; (D) Bi-modal distribution - some, but not all, data conform to the limits

exceed the upper limit, thus showing that some aspect of the process was not properly controlled (eg, sub-standard quality ingredients). However, a data distribution that is skewed (Fig. 12.3C) shows the process to have been out of control even though all the data values lie within the limits. Other variations could include data with larger variances, such that the distribution curve is spread more widely; data with smaller variances leading to a reduced spread of results; negatively skewed data where a majority of the values are below the control target; and multimodal distributions. A bimodal distribution (Fig. 12.3D) suggests either that the process used two different sources of a major ingredient or that a major change in process conditions occurred during some period of manufacture.

SHEWHART'S CONTROL CHARTS FOR VARIABLES DATA

Control charts present data over a period of time and compare these to limits devised using statistically based inferential techniques. The charts demonstrate when a process is in a state of control and quickly draw attention to even relatively small, but consistent, changes away from the 'steady-state' condition, leading to a requirement to stop or adjust a process before defects occur. Most control charts are based on the central limit theorem of Normal distribution using sample average and variance estimates, although in the case of microbiological data, levels of colony-forming units need to be 'Normalised' by an appropriate mathematical transformation.

There are many styles of control chart for variables data; the four major types are:

1. the $\bar{x}$ chart, based on the average result;
2. the R chart, based on the range of the results;
3. the s chart, based on the sample standard distribution; and
4. the x chart, based on individual results.

The R and s charts monitor variations in the process, whereas the x and $\bar{x}$ charts monitor the location of results relative to pre-determined limits. Often charts are combined into, for example, an $\bar{x}$ and an R chart, or an $\bar{x}$ and an s chart; in addition, special versions of control charts may be used for particular purposes such as batch operations.

$\bar{X}$ AND R CHARTS

These use the average and range of values determined on a set of samples to monitor the process location and process variation. Use of a typical $\bar{x}$ and R chart is illustrated in Example 12.1. The primary data for setting up a chart are derived by analysis, using the procedure to be used in process verification, of at least 5 sample units from each of at least 20 sets of samples from the specific process.

After 'Normalisation' of the data, the average ($\bar{x}$) and range (R) of values is derived for each set of samples and the data are examined to confirm that the pattern of distribution conforms reasonably to a Normal distribution. Assuming that the distribution is acceptable, the process average (target value = $\bar{\bar{x}}$) and the average process range ($\bar{R}$) are calculated from the individual sample set averages and ranges, using the following equations:

$$\text{Mean target value} = \bar{\bar{x}} = \frac{\sum \bar{x}}{n}; \qquad \text{average range target} = \bar{R} = \frac{\sum R}{n}$$

where $\bar{x}$ is the sample average, R is the sample range and n is the number of sample sets.

Control limits need to be established for both sample average and sample range data sets; if both upper and lower control limits are required, these are located symmetrically about the target (centre) line but, as noted earlier, most microbiological tests of quality/safety require only UCLs. Usually these will be set at $\text{UCL}_x = \bar{\bar{x}} + A_2\bar{R}$ and $\text{UCL}_R = D_4\bar{R}$ for the 99% UCLs, where A_2 and D_4 are factors that depend on the number of sample units tested (Table 12.2). The values for the target line and UCLs are drawn on the charts and are used to test the original sets of sample units for conformance (Example 12.1). If all data sets conform, then the chart can be used for routine testing, but if there is evidence of non-conformance, the cause must be investigated and corrected, the process resampled and new targets and UCLs established. The values for the control lines should be reviewed regularly; evidence that the limits are wider than previously determined is a cause for concern and the reasons must be investigated. If process improvements occur over time, the limits can be tightened accordingly. Procedures for assessing whether or not a process is going out of control are described in the following.

$\bar{x}$ AND s CHARTS

For sample unit sizes greater than $n = 12$, the standard deviation (s) should be used to monitor variation, rather than the range (R), because it is more sensitive to changes in the process. The target lines for the $\bar{x}$ and s charts are $\bar{\bar{x}}$ and $\bar{s}$, respectively. The calculation for $\bar{\bar{x}}$ was shown previously; $\bar{s}$ is determined by averaging the sample standard deviations of the individual sample sets:

$$\text{Target (centre) line for } \bar{s} = \frac{\sum_{i=1}^{k} s_i}{k} = \frac{s_1 + s_2 + \cdots + s_k}{k}$$

where k is the number of sample sets and s_i is the averaged standard deviation for sample sets 1, 2, 3,…, k. Note, however, that this is a biased average, since the true overall standard deviation is determined from

$$\bar{s} = \sqrt{\frac{\sum_{i=1}^{k} s_i^2}{k}} = \sqrt{\frac{s_1^2 + s_2^2 + \cdots + s_k^2}{k}}$$

The UCLs are calculated from

$$\mathrm{UCL}_{(\bar{x})} = \bar{\bar{x}} + A_3\bar{s} \quad \text{and} \quad \mathrm{UCL}_{(s)} = B_4\bar{s}$$

where A_3 and B_4 are standard factor values (Table 12.2). The charts are prepared and interpreted in a similar manner to $\bar{x}$ and R charts.

INTERPRETATION OF $\bar{\bar{x}}$, *s* AND *R* CHARTS

If a process is 'in control', the data from critical parameter tests should conform to the rules for Normal distribution, that is, symmetry about the centre distribution line and randomness; 95% of data should lie within ±2*s* and 99.7% should lie within ±3*s* of the centre line. A process complying with these rules is generally stable. If this is not the case, the process may be out of control and the cause(s) must be found and rectified.

The key question is: 'How do you recognise out-of-control situations from the chart?' At least 11 'out-of-control' patterns have been identified (Anon, 1956; Wheeler and Chambers, 1984; Beauregard et al., 1992). As a general rule, six patterns cover the majority of situations likely to be encountered. For a single-sided test, such as is used for microbiological data, these patterns can be summarised as follows:

1. a single point above the UCL that is set at +3*s* above the central line [the probability for any point to occur by chance above the UCL is less than 0.14% (about 1 in 700)];
2. seven consecutive points above the centre line [probability = 0.78%; about 1 in 125];
3. two consecutive points close to the UCL [the probability of 2 consecutive values occurring close to a limit in a Normal distribution is about 0.05% (1 in 2000)];
4. ten out of 11 consecutive points above the centre line [probability 0.54% (about 1 in 200); if reversed (ie, below the line), this pattern may also indicate a process change that warrants investigation];
5. a trend of seven consecutive points in an upward (or downwards) direction, indicating a lack of randomness;
6. *regular* cycling around the centre line.

Beauregard et al. (1992) suggest some possible engineering causes for these patterns but other causes may be equally plausible when examining microbiological data. When using microbiological colony counts or other quantitative measures, it is essential to ensure that the test is itself in control. Changes to the culture media, incubation conditions and other factors discussed previously (Chapter 11) must be taken into account when examining SPC and other forms of microbiological trend analysis to ensure that it is the process rather than the method of analysis that is 'out of control'! Examples 12.1 and 12.2 illustrate some 'out-of-control' patterns.

EXAMPLE 12.1 DATA HANDLING FOR PREPARATION OF SHEWHART'S CONTROL CHARTS FOR MEAN, STANDARD DEVIATION AND RANGE VALUES ($\bar{x}$, *s* AND *R* CHARTS)

Aerobic colony count data obtained by duplicate analysis of each of 5 random samples taken by Port Health inspectors from each of 82 consignments of frozen cooked prawns imported into the United Kingdom over a period of some months were used to determine measurement uncertainty (Jarvis et al., 2007). The objective of the analyses was to assess the 'quality' of the products and considerable variation was seen both within and between samples. Can we use SPC to assess the data retrospectively and set future acceptance criteria?

The first 24 data sets were used to set up the SPC system. The counts were $\log_{10}$-transformed and tested for conformance with Normality, skewness and kurtosis – although not perfect, the fit was 'adequate for purpose'. The mean, standard deviation and range of the $\log_{10}$ cfu/g for each sample set are given in Table 12.1. The parameter values were determined as the overall mean

Table 12.1 The Mean, Standard Deviation and Range of Aerobic Colony Counts for 5 Samples of Each of 24 Sets of Imported Frozen Cooked Prawns

	Aerobic Colony Count ($\log_{10}$ cfu/g)		
Occasion	**Mean ($\bar{x}$)**	**Standard Deviation (*s*)**	**Range (*R*)**
1	5.05	0.343	0.72
2	5.06	0.359	0.81
3	4.81	0.197	0.47
4	4.76	0.140	0.34
5	4.84	0.163	0.43
6	4.72	0.107	0.29
7	4.95	0.192	0.42
8	5.11	0.149	0.40
9	5.03	0.123	0.30
10	5.06	0.235	0.59
11	5.11	0.148	0.31
12	5.13	0.198	0.48
13	4.74	0.199	0.44
14	4.80	0.142	0.35
15	4.91	0.247	0.59
16	5.13	0.144	0.35
17	5.05	0.225	0.54
18	4.97	0.093	0.25
19	4.95	0.066	0.17
20	5.05	0.140	0.35
21	5.09	0.132	0.34
22	5.04	0.145	0.37
23	5.13	0.182	0.47
24	5.01	0.255	0.54
Overall mean	4.98	0.180	0.43

colony count $\bar{\bar{x}} = 4.98$ with an 'average' standard deviation $(\bar{s}) = 0.18$ and average range $(\bar{R}) = 0.43$ (all as $\log_{10}$ cfu/g). Note that $\bar{s}$ is the average of the 24 standard deviations; it is *not* an estimate of the combined standard deviations.

Setting control limits

'Mean value' ($\bar{x}$) chart

Target values were set as the overall mean count ($\bar{\bar{x}}$) and control limits (CLs) were established for the chart at 1*s*, 2*s* and 3*s* above and below the target line; the control lines at ±2*s* are the warning limits and those at ±3*s* are the absolute limits.

The target (centre line) was set at the overall mean value (4.98 $\log_{10}$ cfu/g). The upper and lower absolute (±3*s*) CLs (UCL and LCL, respectively) were determined as follows:

$$UCL_{\bar{x}} = \bar{\bar{x}} + A_2\bar{R} \quad \text{and} \quad LCL_{\bar{x}} = \bar{\bar{x}} - A_2\bar{R}$$

where $\bar{\bar{x}}$ is the overall mean value, A_2 is a constant (Table 12.2, column 2) and $\bar{R}$ is the mean range.

For these data with 24 sets each of 5 replicate test samples, $\bar{\bar{x}} = 4.98$, $\bar{R} = 0.43$ and $A_2 = 0.577$; so

$$UCL_{\bar{x}} = 4.98 + (0.577 \times 0.43) = 4.98 + 0.26 = 5.23$$

and

$$LCL_{\bar{x}} = 4.98 - (0.577 \times 0.43) = 4.98 - 0.26 = 4.74$$

CLs for ±1*s* and ±2*s* were determined similarly but the factors are $A_2 \times 1/3$ (ie, 0.577/3 = 0.192) and $A_2 \times 2/3$ (ie, 0.385), respectively.

An alternative calculation uses the average standard deviation ($\bar{s} = 0.180$) for which the constant $A_3 = 1.427$ (from column 5 of Table 12.2):

$$UCL_{\bar{x}} = \bar{\bar{x}} + A_3\bar{s} \quad \text{and} \quad LCL_{\bar{x}} = \bar{\bar{x}} - A_3\bar{s}$$

$$UCL_{\bar{x}} = 4.98 + (1.427 \times 0.180) = 5.24$$

Table 12.2 Some Constants for Control Chart Formulae

	$\bar{x}$ and *R* Charts			$\bar{x}$ and *s* Charts		
No. of Replicates in Test Group	**$\bar{x}$-Chart Limits**	***R*-Chart Limits[a]**		**$\bar{x}$-Chart Limits**	***s*-Chart Limits[a]**	
	A_2	D_3	D_4	A_3	B_3	B_4
2	1.880	—	3.267	2.659	—	3.267
3	1.023	—	2.574	1.954	—	2.568
4	0.729	—	2.282	1.628	—	2.266
5	0.577	—	2.114	1.427	—	2.089
6	0.483	—	2.004	1.287	0.030	1.970
7	0.419	0.076	1.924	1.182	0.118	1.882
8	0.373	0.136	1.864	1.099	0.185	1.815
10	0.308	0.223	1.777	0.975	0.284	1.716
15	0.223	0.347	1.653	0.789	0.428	1.572
20	0.180	0.425	1.585	0.680	0.510	1.490

[a]*If a value is not shown, use the next higher value.*

and

$$LCL_{\bar{x}} = 4.98 - (1.427 \times 0.180) = 4.72$$

Note that these limits are essentially the same as those derived from the range values.

Limit values for the ±1*s* and ±2*s* are calculated similarly, but using $A_3/3$ and $2A_3/3$, respectively, as constants for limits based on $\bar{s}$.

The CLs are drawn on the graph and the data plotted (Fig. 12.4). All data values lie within the range +2 SD to −2 SD. However, since in most microbiological situations lower counts are generally perceived to be beneficial, the LCL is rarely used (there are some exceptions, eg, determination of probiotic bacteria in foods).

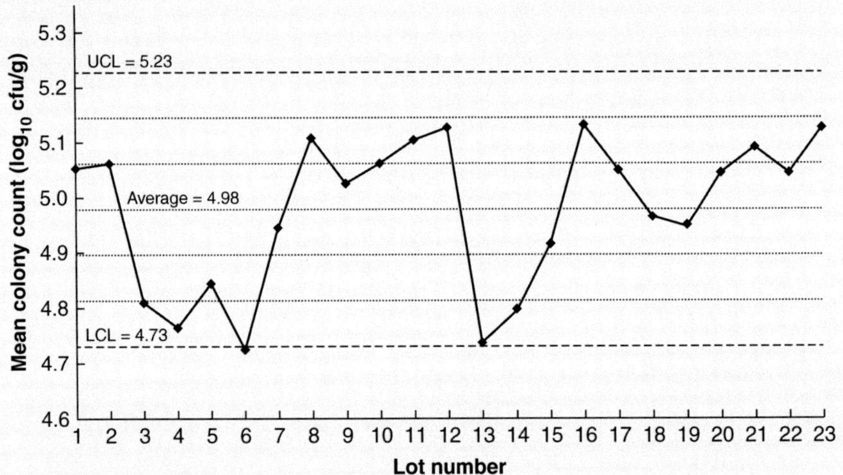

FIGURE 12.4 Statistical Control Chart for Mean ($\bar{x}$) Colony Count (as $\log_{10}$ cfu/g) on the First 24 Sets of Replicate Examinations of Imported Frozen Cooked Prawns Sampled at Port of Entry

The heavy dotted lines (--------) show the upper (UCL) and lower (LCL) 99% control limits. The lighter dotted lines (........) are the upper and lower limits for ±1 SD and ±2 SD around the centre line

Range (R) chart

The mean value for the range ($\bar{R}$) chart = 0.43; the UCL_R and LCL_R values were derived as:

$$UCL_R = D_4\bar{R} \quad \text{and} \quad LCL_R = D_3\bar{R}$$

where D_4 and D_3 are constants (2.114 and 0.076, respectively; from Table 12.2, columns 4 and 3). For these data:

$$UCL_R = 2.114 \times 0.43 = 0.909 \approx 0.91$$

and

$$LCL_R = 0.076 \times 0.43 = 0.033 \approx 0.03$$

Note that as no value is shown in Table 12.2 for D_3 with $n = 5$, the first higher value (0.076) must be used.

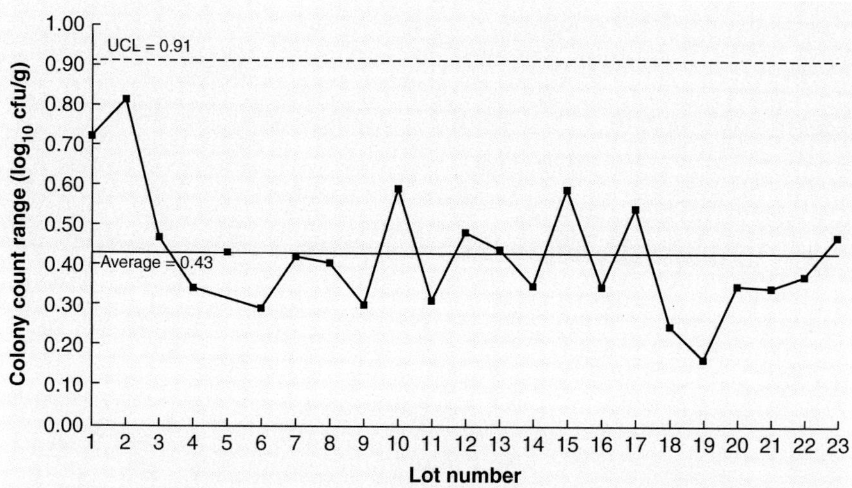

FIGURE 12.5 Statistical Control Chart for Range (R) of log_{10} cfu/g on the First 24 Sets of Replicate Examinations of Imported Frozen Cooked Prawns Sampled at Port of Entry

As before, the UCL value is entered on the graph as the CL above the target value of $\overline{R} = 0.43$ and the range data values plotted *but* the LCL is omitted since it is of little practical value for microbiological counts (Fig. 12.5). Note that the first two range values approach but do not exceed the upper CL.

Standard deviation (s) chart

The 'average' standard deviation is $(\overline{s}) = 0.180$. The upper and lower CLs are determined as follows: $UCL_s = B_4\overline{s}$ and $LCL_s = B_3\overline{s}$, where B_3 and B_4 are constants from columns 6 and 7, respectively, of Table 12.2. For these data, $B_4 = 2.089$ and $B_3 = 0.030$, so:

$$UCL_s = 2.089 \times 0.180 = 0.376 \approx 0.38$$

and

$$LCL_s = 0.030 \times 0.180 = 0.0054 \approx 0.01$$

As before, the LCL values are ignored; the average and UCL lines are drawn onto the chart and the standard deviation values plotted (Fig. 12.6). As with the range chart, the s chart shows that the standard deviations of each of the first two samples were high. Note that the range and s charts are very similar.

Interpretation of the standard reference charts

In a normal production situation, the data used to derive CLs would be obtained on samples taken when the process was believed to be 'in control'. In order to confirm that both the process and the measurements were 'in control' it is essential to chart these data before using the control charts for routine purposes.

In this example, the $\overline{x}$ chart (Fig. 12.4) shows that all 24 mean values are within the $+2s$ upper warning limit but that two values lie outside the $-2s$ LCL and one value is on the $-2s$ line. In a two-sided test this would indicate that the process was not absolutely 'in control' for these particular lots and the entire sampling process should be repeated. But in this instance, we are interested only in high bacterial counts so these low values can be ignored; the data values above the centre line are 'in control'. In Figs 12.5 and 12.6 most of the data values for the range and standard deviation are randomly distributed around the mid-line with no sequential runs of data on either side but the first two samples lie very close to the UCL values. The graphs were therefore re-drawn after re-calculation of reference values and CLs based on samples 3–24. In a normal production situation new data sets would have been generated before the charts were used.

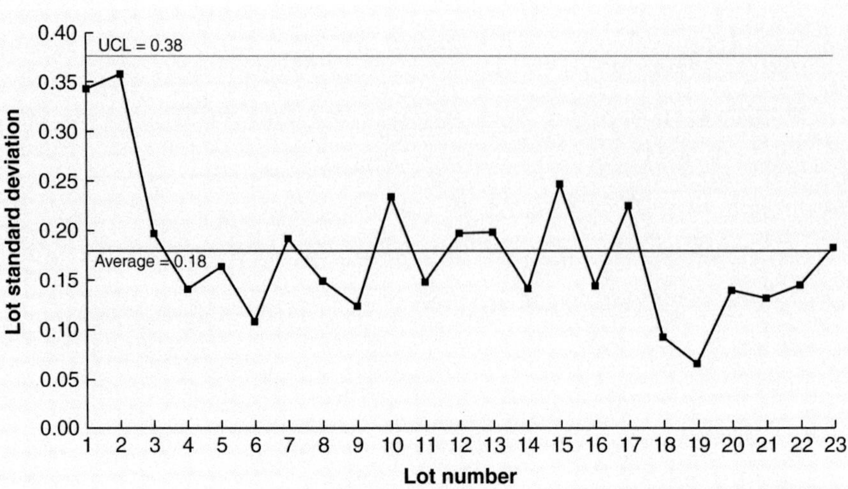

FIGURE 12.6 Statistical Control Chart for Standard Deviation (s) of $\log_{10}$ cfu/g on the First 24 Sets of Replicate Examinations of Imported Frozen Cooked Prawns Sampled at Port of Entry

Removal of the first two data values changes the parameters slightly: $\bar{\bar{x}}$ was reduced from 4.98 to 4.97, $\bar{s}$ from 0.180 to 0.160 and $\bar{R}$ from 0.43 to 0.40. Hence, the UCL of the range (UCL_R) changed from 0.91 to 0.85 and that of the standard deviation (UCL_s) from 0.38 to 0.33. The revised mean value of 4.97 ≈ 5.0 $\log_{10}$ cfu/g with a revised average range value of 0.40 $\log_{10}$ cfu/g are used as target values from which to calculate the UCLs in Example 12.2.

EXAMPLE 12.2 USE OF THE $\bar{x}$ AND R STATISTICAL PROCESS CONTROL CHARTS

Having prepared 'standard' SPC charts, how do we now use them to examine data from routine production and what should we look for to indicate that a process is not 'in control'? For this example we extend the database used in Example 12.1 by adding data for the remaining 58 samples. We plot the mean and range of the log-transformed counts for each sample (Table 12.3) on these charts using the centre line and UCL values derived for the revised mean log counts and revised UCL_R generated in Example 12.1. What do the results tell us?

The $\bar{x}$ chart (Fig. 12.7) provides a 'mountain range' pattern with numerous peaks and troughs. Note first that the mean colony counts on samples 27–28, 32, 48–53, 64–69, 75 and 80–82 all exceed the UCL, in some instances by a considerable amount, and are therefore 'out of control'. In this case, where retrospective sample data were being examined, the high counts indicate that the microbial load on some lots exceeds the statistically based UCL, although the tolerance for high counts given by the UCL is less than might be given in non-statistically based limits (see Chapter 14). If this were a production control chart, it would have been necessary to investigate the causes as soon as an 'out-of-control' situation was noted. For instance, were the colony counts recorded correctly or was there some other technical reason why high counts might have been obtained in these instances? Was one or more of the ingredients different? Had a large quantity of 're-worked' material been blended into a 'lot'? And so on.

Table 12.3 Aerobic Colony Count Data on 82 Sets of Samples of Imported Frozen Cooked Prawns Sampled at the Port of Entry

	Replicate Colony Counts (log_{10} cfu/g)						
Sample	**1**	**2**	**3**	**4**	**5**	**Mean**	**Range**
1[a]	4.73	5.37	4.65	5.17	5.35	5.05	0.72
2[a]	4.81	5.23	4.82	4.84	5.62	5.06	0.81
3	4.93	4.47	4.90	4.94	4.80	4.81	0.47
4	4.83	4.61	4.79	4.95	4.64	4.76	0.34
5	4.77	4.80	4.92	4.65	5.08	4.84	0.43
6	4.72	4.84	4.75	4.55	4.76	4.72	0.29
7	5.16	5.12	4.93	4.74	4.78	4.95	0.42
8	5.07	5.36	5.07	5.09	4.95	5.11	0.40
9	5.02	4.91	4.92	5.07	5.21	5.03	0.30
10	4.93	5.34	4.75	5.06	5.23	5.06	0.59
11	4.93	4.96	5.16	5.23	5.24	5.11	0.31
12	5.22	5.20	4.78	5.26	5.18	5.13	0.48
13	4.48	4.87	4.57	4.92	4.85	4.74	0.44
14	4.71	4.95	4.89	4.60	4.84	4.80	0.35
15	5.20	4.78	4.61	5.13	4.85	4.91	0.59
16	5.38	5.09	5.04	5.04	5.11	5.13	0.35
17	5.26	4.72	5.00	5.26	5.01	5.05	0.54
18	5.06	4.82	4.97	4.96	5.01	4.97	0.25
19	4.98	4.90	5.04	4.87	4.95	4.95	0.17
20	5.02	4.94	4.96	5.02	5.29	5.05	0.35
21	5.00	5.29	5.13	4.94	5.10	5.09	0.34
22	5.12	5.08	5.17	5.06	4.80	5.04	0.37
23	5.29	5.21	5.13	5.19	4.82	5.13	0.47
24	5.06	4.78	4.72	5.25	5.26	5.01	0.54
25	4.90	4.90	5.28	5.40	5.20	5.14	0.50
26	5.66	5.00	4.80	5.01	5.20	5.13	0.86
27	5.15	5.08	5.78	5.15	5.60	5.35	0.70
28	4.78	6.17	5.20	5.80	5.97	5.58	1.39
29	5.48	5.26	5.00	5.20	5.26	5.24	0.48
30	4.70	4.66	4.58	4.80	5.02	4.75	0.44
31	4.86	4.99	4.65	4.95	5.02	4.89	0.37
32	5.45	5.70	6.03	5.97	6.28	5.89	0.83
33	5.00	5.60	5.20	4.78	5.48	5.21	0.82
34	5.08	4.90	5.41	4.90	5.58	5.18	0.68
35	4.80	4.30	4.60	5.08	4.90	4.74	0.78
36	4.68	4.60	4.63	4.86	4.97	4.75	0.37

(Continued)

Table 12.3 Aerobic Colony Count Data on 82 Sets of Samples of Imported Frozen Cooked Prawns Sampled at the Port of Entry (*cont.*)

	Replicate Colony Counts (log_{10} cfu/g)						
Sample	**1**	**2**	**3**	**4**	**5**	**Mean**	**Range**
37	4.78	4.97	4.62	4.80	4.85	4.80	0.35
38	5.01	4.80	4.85	4.85	4.90	4.88	0.21
39	4.28	4.50	4.74	4.70	4.80	4.60	0.52
40	4.73	4.82	4.91	5.10	5.05	4.92	0.37
41	4.85	4.80	5.10	4.95	5.35	5.01	0.55
42	4.60	4.60	4.50	4.30	4.62	4.53	0.32
43	4.45	4.60	4.69	4.45	5.10	4.66	0.65
44	4.55	4.46	4.50	4.32	4.62	4.49	0.30
45	5.02	4.88	4.72	4.24	4.54	4.68	0.78
46	4.74	4.81	4.76	4.96	4.81	4.82	0.22
47	4.80	4.90	5.30	5.10	4.80	4.98	0.50
48	5.33	4.99	5.10	5.25	5.33	5.20	0.34
49	5.00	5.35	5.10	5.08	5.50	5.21	0.50
50	5.32	5.18	5.45	5.30	5.40	5.33	0.27
51	5.63	5.51	5.35	5.10	5.40	5.40	0.53
52	5.29	5.33	5.31	5.88	5.40	5.44	0.59
53	5.15	5.45	4.78	5.30	5.26	5.19	0.67
54	4.60	4.49	4.90	4.89	5.10	4.80	0.61
55	4.71	4.84	5.01	5.10	4.82	4.90	0.39
56	4.80	4.68	5.15	5.20	4.90	4.95	0.52
57	5.10	5.35	5.10	4.80	4.85	5.04	0.55
58	5.03	4.70	4.94	4.88	4.98	4.91	0.33
59	4.90	4.78	4.88	4.95	5.30	4.96	0.52
60	4.45	4.80	4.85	5.10	5.04	4.85	0.65
61	4.98	4.71	5.15	5.17	5.28	5.06	0.57
62	5.57	5.10	4.87	5.26	4.76	5.11	0.81
63	5.49	5.36	5.16	4.89	5.00	5.18	0.60
64	5.03	5.95	5.12	5.09	5.01	5.24	0.94
65	5.10	5.87	4.98	5.80	4.99	5.35	0.89
66	5.28	5.10	5.36	5.90	5.20	5.37	0.80
67	5.09	5.42	5.39	5.37	5.49	5.35	0.40
68	5.31	5.63	5.84	5.27	5.65	5.54	0.57
69	5.53	5.32	5.20	5.04	5.17	5.25	0.49
70	4.93	5.18	5.04	5.13	5.12	5.08	0.25
71	5.18	4.50	5.08	4.30	4.30	4.67	0.88
72	4.90	5.04	4.95	5.08	5.26	5.05	0.36

Table 12.3 Aerobic Colony Count Data on 82 Sets of Samples of Imported Frozen Cooked Prawns Sampled at the Port of Entry (*cont.*)

	Replicate Colony Counts ($\log_{10}$ cfu/g)						
Sample	**1**	**2**	**3**	**4**	**5**	**Mean**	**Range**
73	4.60	4.30	4.30	5.18	5.08	4.69	0.88
74	5.27	4.81	4.79	4.70	5.09	4.93	0.57
75	5.08	5.60	5.70	5.20	5.51	5.42	0.62
76	4.76	5.31	4.65	4.88	5.20	4.96	0.66
77	5.12	5.09	4.80	5.05	5.10	5.03	0.32
78	5.28	5.49	5.38	5.18	4.70	5.21	0.79
79	5.25	4.90	5.08	5.25	5.40	5.18	0.50
80	5.28	5.44	4.98	5.20	5.40	5.26	0.46
81	5.20	5.24	5.42	5.47	5.38	5.34	0.27
82	5.85	6.09	5.44	5.36	5.63	5.67	0.73

[a]*These two data sets were not used to determine the control limits used in the second group of SPC analyses because of extensive variance, but were graphed on the charts.*

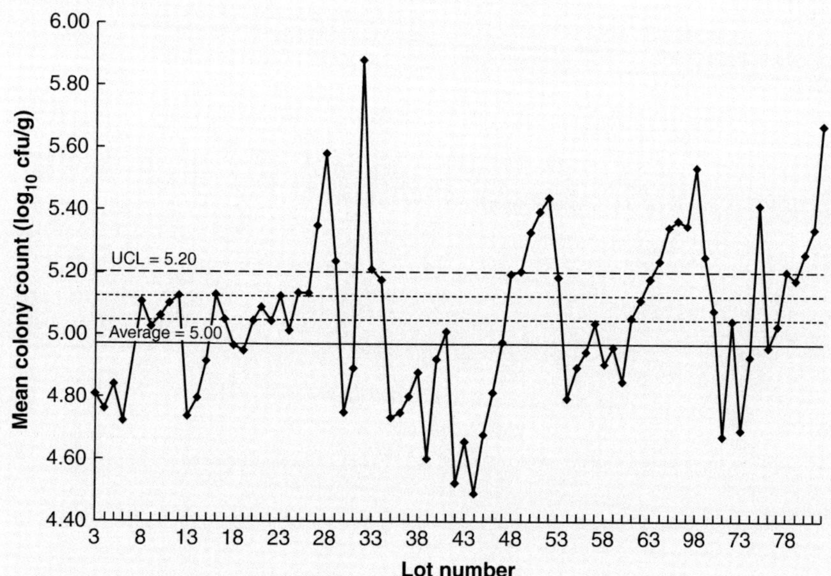

FIGURE 12.7 Statistical Control Chart for Mean ($\bar{x}$) Colony Counts (as $\log_{10}$ cfu/g) on 80 Sets of Replicate Examinations of Imported Frozen Cooked Prawns Sampled at Port of Entry

The control lines were derived from lot samples 3–24 and omit data values for lot samples 1 and 2

Note also that the values for lots 33–34, 48–49 and 78–79 all lie within the bounds of +2 SD to +3 SD (UCL) – another prime indicator of an 'out-of-control' situation that for a production chart would require investigation. In addition, several of these lots are linked to others where the colony counts exceed the UCL.

Then note that results for several consecutive lots of product lie above the average control line, for example, lots 20–29, 47–53 and 61–70; in each case there is a sequential run of 7 or more points above the line. However, it should be noted also that lots 20–24 were part of the 'in control' group used to set up the chart. If we include also some values below the average line (eg, 42–53), we have an even longer sequential run of increasing mean colony count values. For a production control chart such runs would clearly indicate a loss of control that could be of major importance.

But since these charts provide a retrospective analysis of data, we can only surmise that some 'lots' of the frozen cooked prawns were of significantly lower quality than others. Whether this might have been due to climatic conditions, preparation in different production plants or other reasons is not known.

The *R* chart (Fig. 12.8) is slightly less dramatic but still provides valuable information. It shows successive runs of range values above the average line for lots 24–28 and 58–66. In addition, the ranges on samples from lots 28 and 64 exceed the UCL for ranges, indicating that one or more of the colony counts was considerably larger than the other data values in the set.

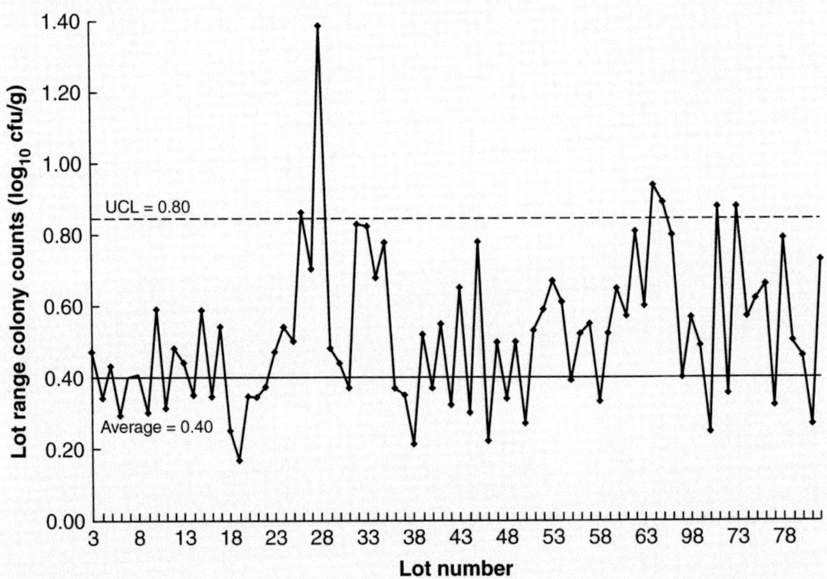

FIGURE 12.8 Statistical Control Chart for Range (*R*) of Colony Counts (as $\log_{10}$ cfu/g) on 80 Sets of Replicate Examinations of Imported Frozen Cooked Prawns Sampled at Port of Entry

The control lines were derived from lot samples 3–24 and omit data values for lot samples 1 and 2

How can we best interpret these data? First, any effects are more obvious when plotted as a chart than in a tabular format, but tabulation is part of the system not only for construction of the charts but also to enable retrospective evaluation of the source data. For instance, Table 12.3 shows that for sample 28 the recorded colony counts were 4.78, 6.17, 5.20, 5.80 and 5.97 – the analyst should have been suspicious of these wildly disparate results – were the results calculated correctly or was this a heterogeneous lot of product? During other investigations of imported frozen prawns we observed that a consignment often consisted of two or more smaller lots from different sources and the microbiological results varied between the 'lots'. Similarly, for sample 64 we have counts of

5.03, 5.95, 5.12, 5.09 and 5.01 $\log_{10}$ cfu/g. The second count is significantly higher than the others – should it have been 4.95 $\log_{10}$ cfu/g, not 5.95 $\log_{10}$ cfu/g? Had a simple calculation or transcription error resulted in a set of data that indicates an out-of-control situation? Or was it again a different 'lot' within the consignment? The retrospective analysis provided valuable insights into the lack of homogeneity of the frozen product and pointed to a need for more intensive evaluation of consignment information and for examination of a larger number of samples for mixed 'lot' consignments.

These are just some examples of how SPC charts can aid both microbiological laboratory control and process control. Rules for describing out-of-control situations within SPC charts are given in the text of this chapter.

CUSUM CHARTS

*Cu*mulative *sum* (CUSUM) charts provide a particularly useful means of process control. They are more sensitive than $\bar{x}$ and R charts for detection of small changes (0.5–1.5 SD) and signal changes in the process mean much faster than do conventional charts. CUSUM charts are constructed differently. The sum of the deviation of a 'Normalised' result from the target statistic is plotted (see Example 12.3) and they do not have control limits as such. Instead, they use decision criteria to determine when a process shift has occurred. For standard CUSUM charts a 'mask' is used to cover the most recent data point; if the mask covers any of the previously plotted points, it indicates that a shift in process conditions occurred at that point. Unfortunately, construction of the mask is complex and it is difficult to use (Montgomery, 2012). Beauregard et al. (1992) recommend that it is not used and that a CUSUM signal chart should be used instead.

THE CUSUM SIGNAL CHART

This consists of two parallel graphs of modified CUSUM values. The upper graph is a plot of cumulative 'upper signal' values in relation to an 'upper signal' alarm limit (USAL). The lower graph records the 'lower signal' values against a 'lower signal' alarm limit (LSAL). The cumulative signal values are calculated by adjusting the nominal mean by a signal factor (SF):

$$\text{Cumulative USAL} = \sum \tilde{x}_v - \text{SF}; \qquad \text{cumulative LSAL} = \sum \tilde{x}_v + \text{SF}$$

where $\tilde{x}_v$ is the nominalised mean ($\bar{x}$ – a target value). Limits may be applied to both the upper and lower signal values. The upper signal value can never be negative so that if the calculation indicates a negative value, then it is put equal to zero and is not plotted. Similarly, the lower signal value can never be positive so a positive value is put as zero and is not plotted. Example 12.4 illustrates the derivation of a CUSUM signal plot.

Interpretation of CUSUM signal charts is relatively straightforward. Any value that exceeds the USAL (or for some purposes the LSAL) is deemed to be 'out of control' since the signal value limits are set to enable the detection of a predetermined change, usually 10 times the SF. Smaller signals may be indicative of less extreme changes.

EXAMPLE 12.3 USE OF A CUSUM CHART TO MONITOR A CRITICAL CONTROL POINT

An ATP bioluminescence test was used to provide a rapid and simple method to monitor the hygienic status of a process plant critical control point (CCP). The data, as relative light units (RLU), measure the amount of microbial and/or other ATP after cleaning of a mandrel on a dairy filling plant (a CCP) in order to assess the hygienic status of the process plant before starting production. Status was assessed by comparison against a previously determined target level and a maximum acceptable RLU level on the basis of a pass/fail decision. If the test indicates that the plant had not been properly cleaned, then cleaning and testing should have been repeated until satisfactory results were obtained, before starting process operations. Can we use a CUSUM chart to examine ongoing performance in cleaning at this CCP?

Trend analysis of RLU data provides a simple means to monitor the efficiency of cleaning operations over a period of time. Hayes et al. (1997) used the CUSUM approach to analyse data obtained from a commercial dairy company. They argued that since the data set consists of only single measurements ($n = 1$), other forms of SPC were generally not suitable and the raw data do not conform to a Normal distribution. Whilst alternative procedures could have been used, their data provide a useful example for CUSUM analysis.

The raw data and several alternative transformations are shown in Table 12.4. The square root transformation (as used by Hayes et al., 1997) does not improve the Normality of the distribution and there is still considerable skewness and kurtosis; both forms of logarithmic transformation effectively normalise the data and remove both skewness and kurtosis (Table 12.5). Fig. 12.9 shows the distributions of the raw and the ln-transformed data: the 'box and whisker' plots show outlier values in the raw data and the 'Normality plots' show the differences between the raw and ln-transformed data; Table 12.5 shows the importance of choosing the correct transformation procedure to normalise the data.

An 'individuals' plot of the ATP measurements (as RLUs) for each day of testing (Fig. 12.10) shows a day-to-day lack of consistency amongst the readings with values in excess of the arbitrary UCL value of 110 RLUs, set prior to the routine use of the system, on days 46, 47, 58, 69, 70, 71, 74 and 87. The standard plot shows the individual values above the limit but fails to indicate any significant adverse trends.

The CUSUM plot is set up using a spreadsheet (Table 12.6) with a pre-determined target value (50 RLU ≡ 3.912 ln RLU) and a maximum acceptable value (110 RLU ≡ 4.70 ln RLU) as the UCL:

Column A lists the data reference values (ie, day numbers) and column B the actual data values (the RLUs measured on each day).

Column C shows the transformed data values, that is, the ln-RLU values.

Column D shows the ln-transformed 'process target' value (3.912 ln-RLU).

Column E shows the difference between the ln actual value (column C) and the ln target value (column D) on each day.

The cumulative sum (CUSUM; column F) is determined by sequentially adding the difference values (column E) 1 day at a time to the previous CUSUM value. For instance, the value shown in cell F4 equals the sum of F3 + E4 [ie, $-0.777 + (-0.0830) = -0.860$]; similarly the value in cell F5 = $-0.860 + (-0.248) = -1.108$, etc. The CUSUM values (column F) are then plotted against the day reference value (column A).

The CUSUM plot (Fig. 12.11) has been superimposed on a plot of the ln-transformed data, for which the target and UCL lines are also shown. The CUSUM plot shows several serious adverse trends including those starting on days 26, 46, 65 and 86, which show the process was going out of control, that is, development of a potential problem, before it actually occurred. This is where the CUSUM plot scores over other forms of trend analysis – it is predictive and provides early warning of changes before they actually happen. However, to use the CUSUM system effectively it is necessary to produce a CUSUM Mask (see Fig. 1 in Hayes et al., 1997), which is time consuming and needs to be used properly. An alternate procedure is shown in Example 12.4.

Table 12.4 ATP 'Hygiene Test' Measurements on a Critical Control Point (CCP) in a Dairy Plant, Before and After Transformation Using the Square Root Function, the Natural Logarithm (ln) and the Logarithm to Base 10 ($\log_{10}$)

		Transformed RLU					Transformed RLU		
Day	RLU (x)	$\sqrt{x}$	ln (x)	$\log_{10}$ (x)	Day	RLU (x)	$\sqrt{x}$	ln (x)	$\log_{10}$ (x)
1	23	4.80	3.14	1.36	47	147	12.12	4.99	2.17
2	46	6.78	3.83	1.66	48	13	3.61	2.56	1.11
3	39	6.24	3.66	1.59	49	12	3.46	2.48	1.08
4	62	7.87	4.13	1.79	50	15	3.87	2.71	1.18
5	33	5.74	3.50	1.52	51	14	3.74	2.64	1.15
6	21	4.58	3.04	1.32	52	76	8.72	4.33	1.88
7	20	4.47	3.00	1.30	53	28	5.29	3.33	1.45
8	56	7.48	4.03	1.75	54	29	5.39	3.37	1.46
9	69	8.31	4.23	1.84	55	44	6.63	3.78	1.64
10	26	5.10	3.26	1.41	56	15	3.87	2.71	1.18
11	27	5.20	3.30	1.43	57	16	4.00	2.77	1.20
12	24	4.90	3.18	1.38	58	123	11.09	4.81	2.09
13	22	4.69	3.09	1.34	59	5	2.24	1.61	0.70
14	69	8.31	4.23	1.84	60	41	6.40	3.71	1.61
15	42	6.48	3.74	1.62	61	56	7.48	4.03	1.75
16	97	9.85	4.57	1.99	62	9	3.00	2.20	0.95
17	56	7.48	4.03	1.75	63	82	9.06	4.41	1.91
18	33	5.74	3.50	1.52	64	22	4.69	3.09	1.34
19	33	5.74	3.50	1.52	65	22	4.69	3.09	1.34
20	69	8.31	4.23	1.84	66	63	7.94	4.14	1.80
21	24	4.90	3.18	1.38	67	85	9.22	4.44	1.93
22	34	5.83	3.53	1.53	68	34	5.83	3.53	1.53
23	29	5.39	3.37	1.46	69	158	12.57	5.06	2.20
24	29	5.39	3.37	1.46	70	164	12.81	5.10	2.21
25	19	4.36	2.94	1.28	71	155	12.45	5.04	2.19
26	57	7.55	4.04	1.76	72	79	8.89	4.37	1.90
27	81	9.00	4.39	1.91	73	94	9.70	4.54	1.97
28	99	9.95	4.60	2.00	74	319	17.86	5.77	2.50
29	43	6.56	3.76	1.63	75	38	6.16	3.64	1.58
30	80	8.94	4.38	1.90	76	19	4.36	2.94	1.28
31	68	8.25	4.22	1.83	77	64	8.00	4.16	1.81
32	112	10.58	4.72	2.05	78	50	7.07	3.91	1.70
33	25	5.00	3.22	1.40	79	16	4.00	2.77	1.20
34	52	7.21	3.95	1.72	80	25	5.00	3.22	1.40
35	24	4.90	3.18	1.38	81	28	5.29	3.33	1.45
36	35	5.92	3.56	1.54	82	35	5.92	3.56	1.54
37	15	3.87	2.71	1.18	83	52	7.21	3.95	1.72
38	47	6.86	3.85	1.67	84	28	5.29	3.33	1.45
39	104	10.20	4.64	2.02	85	35	5.92	3.56	1.54
40	53	7.28	3.97	1.72	86	52	7.21	3.95	1.72
41	81	9.00	4.39	1.91	87	139	11.79	4.93	2.14
42	34	5.83	3.53	1.53	88	89	9.43	4.49	1.95
43	17	4.12	2.83	1.23	89	62	7.87	4.13	1.79
44	62	7.87	4.13	1.79	90	55	7.42	4.01	1.74
45	23	4.80	3.14	1.36	91	12	3.46	2.48	1.08
46	123	11.09	4.81	2.09	92	67	8.19	4.20	1.83

Based on the data of Hayes et al. (1997).

Table 12.5 Descriptive Statistics of Data From Table 12.4

		Transformed RLU		
Parameter	**RLU (x)**	$\sqrt{x}$	**ln(x)**	$\log_{10}(x)$
Number of tests (*n*)	92	92	92	92
Mean	54.6	6.90	3.726	1.618
Median	41.5	6.44	3.726	1.618
Variance	2109.8	7.044	0.558	0.105
Standard deviation	45.93	2.654	0.747	0.324
Normality test[a]	4.69 ($P < 0.0001$)	1.33 ($P = 0.0019$)	0.205 ($P = 0.8723$)	0.205 ($P = 0.8723$)
Skewness	2.67 ($P < 0.0001$)	1.13 ($P < 0.0001$)	−0.004 ($P = 0.9864$)	−0.004 ($P = 0.9864$)
Kurtosis	11.45 ($P < 0.0001$)	2.18 ($P = 0.0055$)	−0.025 ($P = 0.8904$)	−0.025 ($P = 0.8904$)

Note that the tests for Normality, skewness and kurtosis are identical for the two alternative logarithmic transformations for which conformance with Normality could not be rejected; note also that neither the original data nor the square root transformed data conform to a Normal distribution and show marked skewness and kurtosis.

[a]Anderson–Darling test.

Based on the data of Hayes et al. (1997).

Table 12.6 Spreadsheet Layout for Data Used to Derive a Cumulative Sum (CUSUM) Chart for ATP Levels Determined Against an ATP Target Level (50 RLU)

	A	B	C	D	E	F
1				Target	Difference	
2	Day	RLU	ln RLU	(ln RLU)	(C − D)	CUSUM
3	1	23	3.135	3.912	−0.777	−0.777
4	2	46	3.829	3.912	−0.083	−0.860
5	3	39	3.664	3.912	−0.248	−1.108
6	4	62	4.127	3.912	0.215	−0.893
7	5	33	3.497	3.912	−0.416	−1.309
8	6	21	3.045	3.912	−0.868	−2.176
9	7	20	2.996	3.912	−0.916	−3.093
10	8	56	4.025	3.912	0.113	−2.979
11	9	69	4.234	3.912	0.322	−2.657
12	10	26	3.258	3.912	−0.654	−3.311
	…	…	…	…	…	…
	…	…	…	…	…	…
91	90	55	4.007	3.912	0.095	−12.090
92	91	12	2.485	3.912	−1.427	−13.517
93	92	67	4.205	3.912	0.293	−13.224

Based on the data of Hayes et al. (1997).

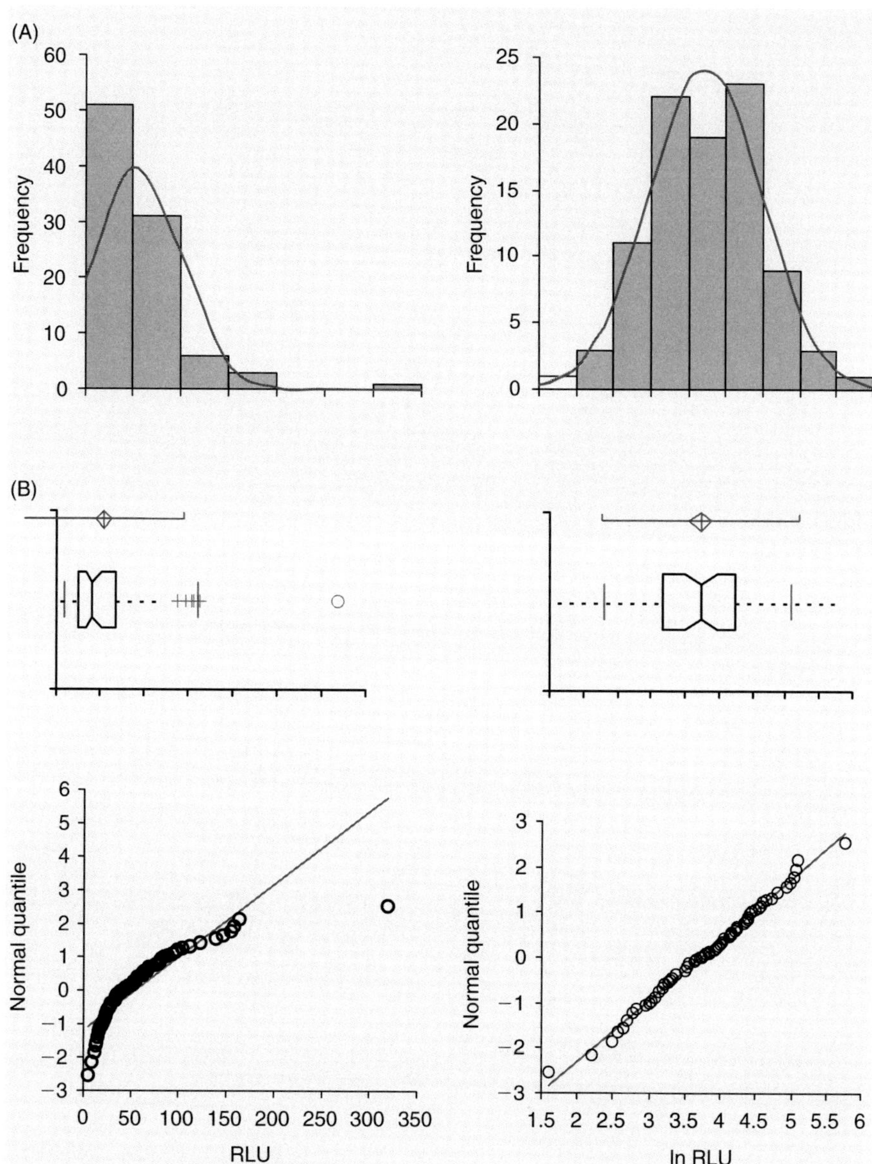

FIGURE 12.9 Data Distributions, 'Box and Whisker' Plots and 'Normality' Plots for Untransformed and ln-Transformed ATP Estimates (as Relative Light Units, RLU) Determined on the Mandrel of a Dairy Bottling Machine

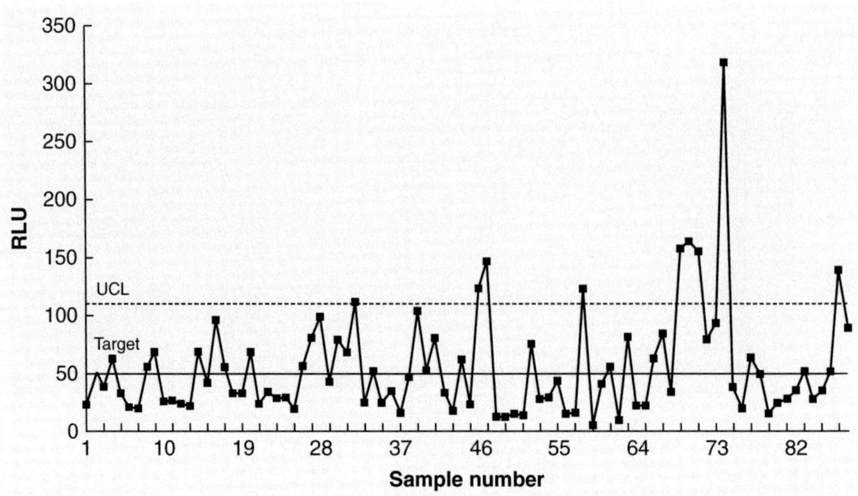

FIGURE 12.10 Individuals Plot of the ATP Data From Hygiene Tests on the Mandrel of a Dairy Bottling Machine

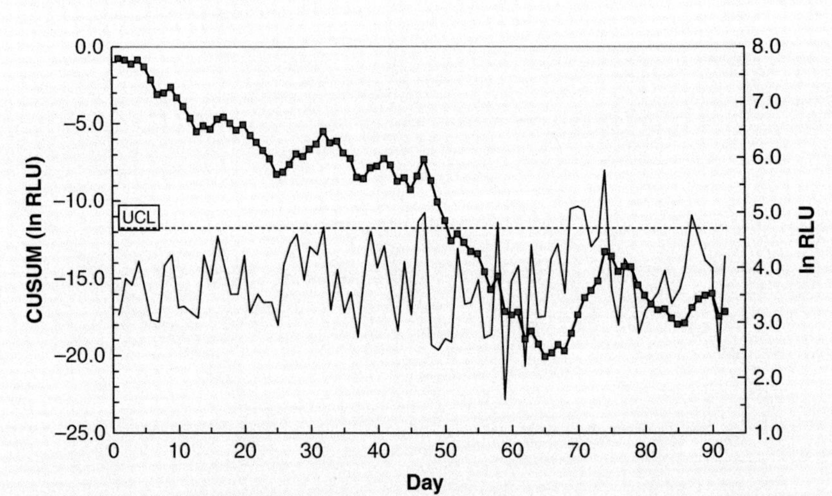

FIGURE 12.11 CUSUM Plot of the ATP Data (as ln-RLU) From Hygiene Tests on the Mandrel of a Dairy Bottling Machine Overlaid on an Individuals Plot of the ln-RLU for Which the Upper Acceptable Limit Line is Shown

EXAMPLE 12.4 CUSUM SIGNAL PLOT

Can we modify the CUSUM chart to provide more definitive information on 'out-of-control' situations?

The CUSUM signal chart consists of two parallel graphs of modified cumulative sum values. The upper graph plots a 'cumulative upper signal (CUS) value' that is monitored against an upper signal alarm limit (USAL). The lower graph monitors a 'cumulative lower signal (CLS) value' against a lower signal alarm limit (LSAL). A pre-determined target value is subtracted from each individual day value to derive a 'nominalised' mean determined as follows: $\bar{x}_v = \bar{x} -$ target value. The upper and lower signal values are calculated by adjusting the 'nominalised' mean value by a signal factor (SF), which is a function of the extent of change (level of shift) to be detected and the SD of the data. It is calculated as follows: SF = (change to be detected) × (SD/2). If a change of 1 standard deviation is to be signalled by the CUSUM chart, then the SF = 1 × s/2. The cumulative signal values are calculated as follows:

$$\text{Cumulative upper signal value} = \Sigma(\bar{x}_v - \text{SF})$$
$$\text{Cumulative lower signal value} = \Sigma(\bar{x}_v + \text{SF})$$

The USAL and LSAL, against which any change is monitored, are usually set at +10SF and −10SF, respectively. To use the system, the data are 'nominalised', adjusted by the SF and then summed cumulatively, although in this version summation is done only for each set of definitive values. Table 12.7 shows how the data for analysis are set out in a spreadsheet; for the purpose of this example, the data are the individual ln-transformed values used in Example 12.3.

Note in columns E and G, respectively, that cumulative US values ≤0 and cumulative LS values ≥0 are shown as [0]. Zero values are not plotted for either value. In deriving the table, the SD used (s = 0.75) is that of the ln-transformed group (see Table 12.5) and the 'nominal value' used to 'nominalise' the data is the approximate mean value (3.7) of the entire group.

The CUS values (column E) and the CLS values (column G) are plotted against sample number, to give a series of individual points. Only consecutive points are joined with lines. It is simple to plot the data using normal Excel graphics, although SPC add-in programmes for Excel are available. Fig. 12.12 shows a continuous rise in CUS values starting from day 69 to day 74, with days 74–78 all exceeding the USAL that is set at +10SF (ie, 10 × s/2 = 3.75). This indicates that for some days increasing ATP levels were detected at the test site but insufficient notice was taken before unacceptable ATP levels had built up!

Table 12.7 Data Layout for Deriving CUSUM Signal Values

	A	B	C	D	E	F	G
1	Day	ln RLU	Nominalised mean (nominal = 3.7)	Nominalised − SF	CUS value[a]	Nominalised + SF	CLS value[b]
2	v	$\bar{x}$	$\bar{x}_v = \bar{x} - 3.7$	$\bar{x}_v - \frac{s}{2}$	$\sum\left[\bar{x}_v - \frac{s}{2}\right]$	$\bar{x}_v + \frac{s}{2}$	$\sum\left[\bar{x}_v + \frac{s}{2}\right]$
3	1	3.135	−0.56	−0.94	[0]	−0.19	−0.19
4	2	3.829	0.13	−0.25	[0]	0.50	[0]
5	3	3.664	−0.04	−0.41	[0]	0.34	[0]
6	4	4.127	0.43	0.05	0.05	0.80	[0]
7	5	3.497	−0.20	−0.58	[0]	0.17	[0]

(Continued)

Table 12.7 Data Layout for Deriving CUSUM Signal Values (*cont.*)

	A	B	C	D	E	F	G
8	6	3.045	−0.66	−1.03	[0]	−0.28	−0.28
9	7	2.996	−0.70	−1.08	[0]	−0.33	−0.61
10	8	4.025	0.33	−0.05	[0]	0.70	[0]
11	9	4.234	0.53	0.16	0.16	0.91	[0]
12	10	3.258	−0.44	−0.82	[0]	−0.07	−0.07
13	11	3.296	−0.40	−0.78	[0]	−0.03	−0.10
14	12	3.178	−0.52	−0.90	[0]	−0.15	−0.24
15	13	3.091	−0.61	−0.98	[0]	−0.23	−0.48
16	14	4.234	0.53	0.16	0.16	0.91	[0]
17	15	3.738	0.04	−0.34	[0]	0.41	[0]
18	16	4.575	0.87	0.50	0.50	1.25	[0]
19	17	4.025	0.33	−0.05	0.45	0.70	[0]
20	18	3.497	−0.20	−0.58	[0]	0.17	[0]

Only the first 20 values are shown. The nominal value is taken as 3.7 (mean ln value for all 92 data points) and s = 0.75 (from Table 12.5); the signal factor is assumed to be 1s so SF = $s/2 \simeq 0.38$. *Note: The value of ln-RLU is shown as* $\bar{x}$, *since in most CUSUM signal charts this would be the mean of two or more values. In this version of the signal value chart the cumulative signals are determined only for definitive values <0 or >0. The full data set is shown in Fig. 12.13.*

[a]*CUS, cumulative upper signal value; values ≤0 are shown as [0] but not included in the chart.*

[b]*CLS, cumulative lower signal value; values ≥0 are shown as [0] but not included in the chart.*

Based on the data of Hayes et al. (1997).

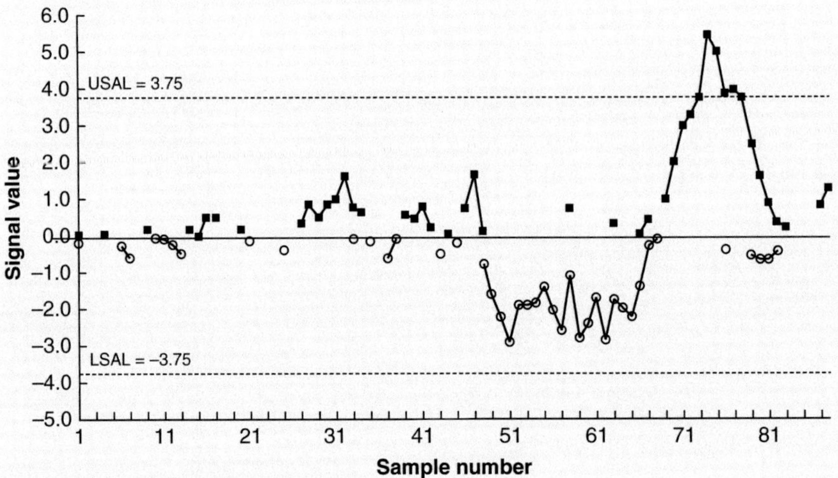

FIGURE 12.12 CUSUM Signal Plot of the ATP Data (as ln RLU) From Hygiene Tests on the Mandrel of a Dairy Bottling Machine

THE MOVING WINDOWS AVERAGE

The moving windows average (MWA; or rolling average) approach to SPC provides an alternative to the Shewhart and CUSUM techniques that rely on the use of defined criteria to assess whether or not a process is going out of control. The MWA is a technique used extensively in financial analyses where it is used to 'smooth out' extreme responses. It has been recommended by CODEX (Anon, 2013) as 'a practical and cost beneficial way of checking continuous microbiological performance of a process or a food safety control system. As in the traditional point-in-time approach commonly used in connection with microbiological criteria, the moving window determines the acceptability of the performance so that appropriate interventions can be made in case of unacceptable shifts in control'.

The procedure essentially compares microbiological results obtained by averaging (or 'smoothing') results over a period of time. For instance, colony counts determined daily are averaged over a defined time period of, say, 5 days (this is the window period). On day 6 results obtained for that day are added and those for day 1 are removed from the average; similarly, on day 7 the results for that day are added and those for day 2 are removed; and so on. The averaged results can be compared, for instance, against microbial limits set as a two-class or a three-class plan to identify whether performance is within agreed tolerances. Heggum et al. (2015) have reported on the use of the MW approach in monitoring staphylococci in cheese manufacture whilst Lee et al. (2015) have used the MW approach to verify compliance with a Performance Objective in order to reduce the incidence of *Campylobacter* in poultry. The procedure is illustrated in Example 12.5. The benefit of this technique is that it provides a simple means to analyse trends that may point to development of a potential 'out-of-control' situation that may not be obvious using a more traditional approach to SPC.

EXAMPLE 12.5 THE MOVING WINDOWS AVERAGE

Suppose that you use a colony count procedure for Enterobacteriaceae to verify performance at a critical control point in a HACCP scheme by testing a number of product samples taken during a working day (eg, such samples might be of a non-pasteurised product such as sprouted seeds). Suppose further that a 3-class sampling plan is used with $n = 5$, $c = 2$, $m = 4.5\ log_{10}$ cfu/g and $M = 5.5\ log_{10}$ cfu/g. Hypothetical results of the replicate sample analyses and of the daily average count are shown in Table 12.8 for a 20-day period.

On most days, colony counts on one or two samples exceed the acceptability limit (m) and on day 6 one count exceeded M. But on days 3 and 15–17 the acceptance control limit c was exceeded, as also were the daily average counts for these days. A simple graph (Fig. 12.13A) confirms visually that the daily average counts exceeded the limit (m) on these days.

We therefore use a graph of the moving windows (MW) average to examine the data with a window period of 5 days. For days 1–5, we take the average of the daily counts (see Table 12.8) and plot this on the graph at day 5; MW average is subsequently plotted for day 6 (average count for days 2–6), day 7 (average count for days 3–7), etc. The graph (Fig. 12.13B) of the averages for the MW provides a 'smoothed' trend that can be compared against the daily mean counts; note that the MW graph does not start until day 5.

Fig. 12.13 shows the daily average counts and a trend of increasing MW average counts starting on day 9 and rising to a maximum on day 17. This shows that although the average daily counts did not exceed the acceptable limit threshold number (c) for the period from day 10 to day 16, an adverse trend was occurring that could have been detected before unacceptable results were obtained. Although in this example no daily average or MW average count exceeded the absolute limit value (M) (although one individual count did so), high results on 1 or more days would have increased the slope of the MW graph. The CODEX guidelines (Anon, 2013), which advocate this technique, state that the MW average approach is not a trend analysis, but that is incorrect since the MW clearly provides evidence of trends. The technique is easy to use through a simple spreadsheet analysis.

An algorithm can be used to determine each new MW average. The MW average for the first window is determined as follows: $MW_k = (x_1 + x_2 + x_3 + \cdots + x_k)/k$, where k is the number of data periods in the window (in this example $k = 5$).

For the second MW average, $MW_{k+1} = MW_k + (x_{k+1} - x_1)/k$; similarly the third is $MW_{k+2} = MW_{k+1} + (x_{k+2} - x_2)/k$, and so on. As shown in Table 12.8, the first MW average is given by $MW_k = (3.874 + 3.874 + 4.560 + 4.140 + 3.638)/5 = 20.086/5 = 4.017$.

The second MW average is $MW_{k+1} = 4.017 + (4.108 - 3.874)/5 = 4.017 + 0.047 = 4.064$.

The third MW average is $MW_{k+2} = 4.064 + (3.778 - 3.874)/5 = 4.07 - 0.019 = 4.045$; and so on.

The procedure can be used for analysis of any data that conform to a logical time series.

Table 12.8 The Moving Windows Average

	Colony Count ($\log_{10}$ cfu/g)							Number of Counts	
Day	Sample 1	Sample 2	Sample 3	Sample 4	Sample 5	Daily Average	MW Average	$>c$	$>M$
1	3.8	**4.69**	3.37	3.08	4.43	3.874		1	0
2	4.12	3.93	3.48	3.86	3.98	3.874		0	0
3	4.21	**4.70**	4.39	**4.81**	**4.69**	4.560		2	0
4	3.87	3.80	**4.57**	3.77	**4.69**	4.140		2	0
5	3.98	3.52	3.64	3.93	3.12	3.638	4.017	0	0
6	5.54	4.02	3.73	3.10	4.15	4.108	4.064	1	1
7	3.72	3.75	3.4	3.28	**4.74**	3.778	4.045	1	0
8	3.41	3.41	3.15	3.94	3.49	3.480	3.829	0	0
9	3.23	4.00	3.42	4.29	**4.69**	3.926	3.786	1	0
10	3.75	3.19	3.24	4.4	**4.86**	3.888	3.836	1	0
11	**4.89**	4.2	4.45	3.02	**4.62**	4.236	3.862	2	0
12	**4.65**	4.12	4.42	3.41	3.66	4.052	3.916	1	0
13	4.33	4.42	3.51	**4.95**	3.78	4.198	4.060	1	0
14	3.76	3.78	4.23	3.91	3.99	3.934	4.062	0	0
15	4.67	5.34	5.30	5.05	4.03	4.878	4.260	4	0
16	5.31	5.36	4.15	4.99	4.94	4.950	4.402	4	0
17	4.45	4.89	4.81	4.14	4.66	4.590	4.510	3	0
18	3.38	3.02	3.69	**4.56**	**4.89**	3.908	4.452	2	0
19	3.76	3.95	**4.78**	3.94	4.44	4.174	4.500	1	0
20	4.07	3.36	3.57	**4.58**	**4.69**	4.054	4.335	2	0

Hypothetical data for total Enterobacteriaceae in five samples of a product taken at different times each day to verify compliance at a critical control point on each of 20 days. The sampling plan and criteria for compliance are $n = 5$, $c = 2$, $m = 4.5 \log_{10}$ cfu/g and $M = 5.5 \log_{10}$ cfu/g – highlighted data indicate non-compliant sample results. Derivation of the MW average is described in the text.

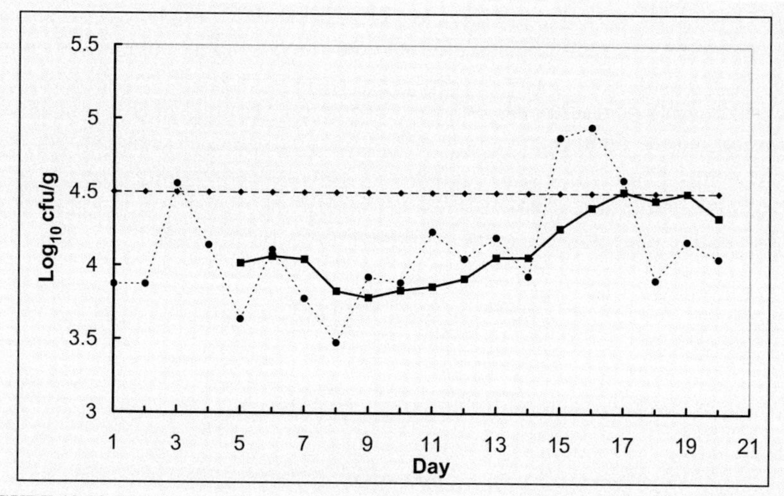

FIGURE 12.13 The Moving Window Approach

Enterobacteriaceae colony counts ($\log_{10}$ cfu/g) from Table 12.8 are shown as: A. The daily average (●----●); and B. The moving average count over a 5-day period (■—■); note that the moving average increases progressively from day 9 indicating a deteriorating trend that reaches the critical limit value at day 17

CONTROL CHARTS FOR ATTRIBUTE DATA

Attribute control charts can be used for the results of quantal tests but are probably worthwhile only if many sample units are tested, albeit in SPC testing of microbiological methods, results obtained on anonymous reference samples should always be assessed. As with SPC for variables data, there are four main types of charts for attribute data:

1. *p* charts, for the proportion of defective items in a binomial series;
2. *np* charts, for the number of defective items in a binomial series;
3. *c* charts, for the number of defective items in a fixed-size sample from Poisson data; and
4. *u* charts, for the rate of production of defective items in Poisson data.

The *p* chart shows the variations in the rate at which defective items occur. Sample j consists of n_j observations, each of which is either defective (eg, a positive result for a target organism) or not defective. The sample proportion of defectives within each lot, $\hat{p}_j$, is calculated as $\Sigma p_j/n_j$. The chart is used to plot these $\hat{p}_j$ against the sample test number or date. The target line of the graph is given by $\overline{p}$, which is determined as the acceptable defect rate, for example, 0.1 (ie, 1 positive test out of 10). Control lines for sample j are plotted at $\overline{p}+2\hat{s}_j$ and $\overline{p}+3\hat{s}_j$, where $\hat{s}_j$ is the estimate of standard deviation given by $\sqrt{\overline{p}(1-\overline{p})/n_j}$. For 1 positive from 10 tests,

$\hat{s}_j = \sqrt{(0.1 \times 0.9)/10} = \sqrt{0.009} = 0.095$; so $2\hat{s}_j = 0.19 \approx 0.2$. Anon (2006) provides special measures for examining and plotting data when different numbers of samples are tested.

For microbiological tests, the *np* chart for non-conforming units is probably the more useful but the number of sample units tested must remain constant. For an *np* chart the centre line is the average number of non-conforming units ($n\bar{p}$), where $\bar{p}$ is the average proportion of positive results, $(np)_i$ is the number of non-conforming samples in sample set $i = 1 \ldots k$ and k is the number of sample sets:

$$n\bar{p} = \frac{\sum_{i=1}^{k}(np)_i}{k} = \frac{(np)_1 + (np)_2 + \cdots + (np)_k}{k}$$

The upper and lower control limits are calculated from

$$\mathrm{UCL}_{np} = n\bar{p} + 3\sqrt{\frac{n\bar{p}(1-n\bar{p})}{n}}$$
$$\mathrm{LCL}_{np} = n\bar{p} - 3\sqrt{\frac{n\bar{p}(1-n\bar{p})}{n}}$$

where n is the sample size. Applications of the procedure are shown in Examples 12.6 and 12.7.

EXAMPLE 12.6 SPC OF ATTRIBUTE DATA

How can I set up a SPC for data on the prevalence of organisms from quantal tests?

Special versions of SPC are available for analysis of attribute data. Attributes can include the occurrence of defective products (Example 12.7), the prevalence of pathogenic or indicator organisms in food samples and other measurable attributes.

Assume that tests were done for *Campylobacter* on swabs of 30 randomly drawn poultry carcases per flock in a processing plant; assume further that several flocks of poultry had been slaughtered and processed sequentially and that the first 8 flocks had previously been designated as being free from campylobacters ('camp −ve') flocks whilst the *Campylobacter* history of the other flocks was unknown. The results of the examinations are summarised in Table 12.9.

We examine the data using an *np* control chart, where p is the proportion of positive samples and n is the total number of replicate tests per flock ($n = 30$), to compare the number of positive samples for each set of tests (Fig. 12.14). The chart shows the low prevalence of *Campylobacter*-positive results from the initial eight flocks and a high, but variable, prevalence of campylobacters in the remaining flocks. The average number of all positive tests is $n\bar{p} = 10.3$ out of 30 tests (n) for which the upper (UCL) and lower (LCL) control limits can be determined as UCL = 18.1 and the LCL = 2.5, using the equations $\mathrm{UCL}_{np} = n\bar{p} + 3\sqrt{n\bar{p}[1-n\bar{p}]/n}$ and $\mathrm{LCL}_{np} = n\bar{p} - 3\sqrt{n\bar{p}[1-n\bar{p}]/n}$. These values are shown in Fig. 12.14.

However, we need to set control limits using data from the uncontaminated flocks (1–8) for which the value of $n\bar{p} = \sum np/k = 8/8 = 1$, where k is the number of trials. Since this value of $n\bar{p}$ is <5, we cannot use the standard equations given earlier to derive the values for the UCL and LCL which will be asymmetrical about the mean value ($n\bar{p}$). We must derive these estimates from the binomial distribution.

The probability with which we can determine the occurrence of defective samples is based on the following equation: $P_x = \{n!/[(n-i)!i!]\}p^i(1-p)^{n-i}$, where i is the number of positive results, n is the number of replicate tests = 30 and p is the average proportion of positive results = $1/30 = 0.03\dot{3}\dot{3}$. The cumulative binomial probability is the sum of the individual probabilities for $i = 0 \ldots x$. The probability for zero positives (ie, $i = 0$) is given by

$$P_{i=0} = \frac{30!}{(30-0)!0!}0.03333^0(1-0.03333)^{30-0} = 0.9667^{30} = 0.3617$$

Similarly, the probability for one positive ($x = 1$) is given by

$$P_{i=1} = \frac{30!}{(30-1)!1!}0.03333^1(1-0.03333)^{30-1} = 30 \times 0.0333 \times 0.9667^{29} = 0.3741$$

Hence, the cumulative probability of finding up to, and including, one positive is given by

$$P_{i \le 1} = P_{i=0} + P_{i=1} = 0.3617 + 0.3741 = 0.7358$$

The individual probabilities of occurrence for values of i from 2 to 6 are calculated similarly and the cumulative probabilities are determined:

No. of Positive Tests (*i*)	**Probability of Occurrence**	**Cumulative Probability**
0	0.3617	0.3617
1	0.3741	0.7358
2	0.1871	0.9229
3	0.0602	0.9831
4	0.0140	0.9971
5	0.0025	0.9996
6	0.00036	>0.9999

For a single-sided limit, the UCL must be set at an α value of 0.0027 such that $1 - \alpha = 0.9973$. We must therefore use the value of $x = 5$ for the UCL, since this is the first value for which the cumulative probability (0.9996) exceeds the α limit probability of 0.9973. By definition, the LCL is zero. We can now set the control limits based on the initial 8 data sets: the average number of positive tests = 1 and the UCL = 5. However, if the chart were to be based on two-sided limits, the UCL would be based on $\alpha = 0.9956$ in which case we would use the value for $x = 4$.

The control chart (Fig. 12.15) shows that all of the succeeding data sets exceed this UCL and therefore demonstrate that the prevalence of campylobacters in the 'camp +ve' flocks is 'out of control' by comparison with the 'camp −ve' flocks.

The control charts used in this example were prepared using the 'SPC for Excel' add-in programme but the charts could have been derived manually using the procedures described earlier.

Table 12.9 Occurrence of Positive *Campylobacter* Samples in 30 Swabs Taken From Processed Chickens in Sequential Flocks Passing Through a Processing Plant

	Tests for *Campylobacter* (n = 30)		
Flock Number	**No. of Positive, p**	**No. of Negative, $1 - p$**	**Proportion of Positive, p/n**
1	1	29	0.03
2	0	30	0.00
3	2	28	0.07
4	1	29	0.03
5	1	29	0.03
6	2	28	0.07
7	0	30	0.00
8	1	29	0.03
9	13	17	0.43
10	20	10	0.67
11	12	18	0.40
12	12	18	0.40
13	15	15	0.50
14	16	14	0.53
15	14	16	0.47
16	23	7	0.77
17	18	12	0.60
18	11	19	0.37
19	19	11	0.63
20	25	5	0.83
Total 1–8	8	232	0.03
Total 9–20	198	162	0.55

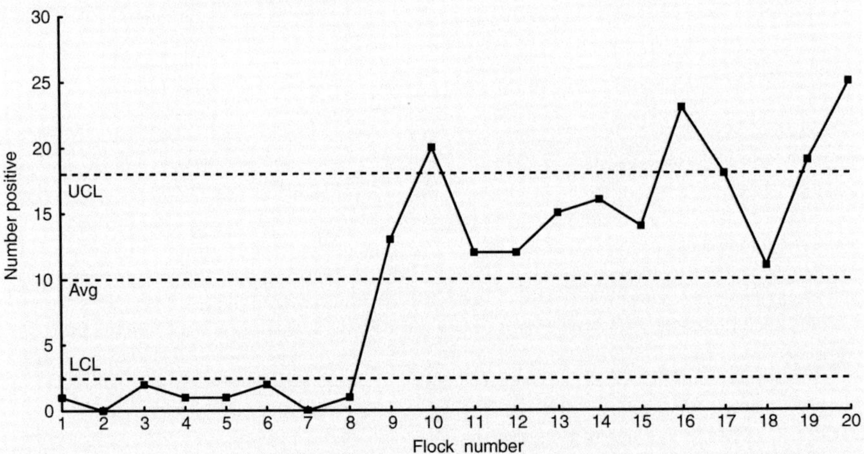

FIGURE 12.14 Attributes $n\bar{p}$ Chart for Prevalence of *Campylobacter* in Poultry Processed Sequentially Through a Commercial Plant

The first eight flocks were believed to be *Campylobacter*-free flocks but the status of the remainder was not known prior to testing. Note the overall average number of positive tests was 10.3 out of 30. The average (Avg), upper control limit (UCL) and lower control limit (LCL) are shown by dotted lines. Avg = 10.3; UCL = 18.1; LCL = 2.5; n = 30; for subgroups 1–20

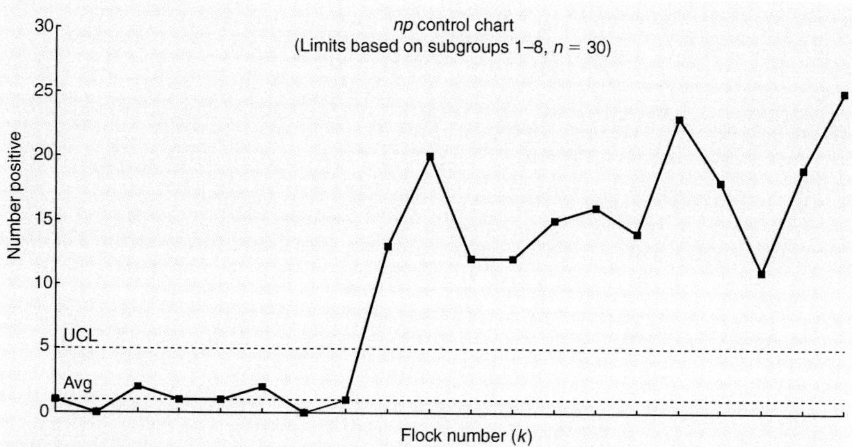

FIGURE 12.15

Attribute $n\bar{p}$-Chart for numbers of positive tests for *Campylobacter* in a poultry processing plant. The average and UCL (dotted lines) were derived for the first 8 flocks of 'camp –ve' birds giving an average number of positive tests of 1 and an upper control limit of 5. Note that the prevalence of contamination in all of the unknown flocks was greater than the UCL.

EXAMPLE 12.7 USE OF ATTRIBUTE DATA CHARTS IN PROBLEM INVESTIGATION

*During a spell of hot weather, a manufacturer of cider (*sic *hard cider) received a high number of complaints regarding spoilage in several lots of one particular brand of a flash-pasteurised product packaged in 1-L PET bottles. Spoilage was evidenced by cloudiness and gas formation. Production was halted and the product was recalled from the marketplace. The frequency of overt spoilage in bottles of product from consecutive lots and the presence of spoilage yeasts in apparently unspoiled product were determined to identify the cause(s) of the problem. The results are summarised in Table 12.10.*

It is clear from the results that the spoilage problem appeared to start suddenly – probably as a consequence of increased ambient temperatures. The plot of the spoilage data using an *np* attribute chart without control limits (Fig. 12.16) shows a sudden steep rise in incidence for lots 4037 onwards, followed by a gradual fall-off in lots 4042–4047. However, even in unspoiled product there was evidence of significant levels of contamination by spoilage yeasts in the majority of the unspoiled products in all lots from 4030 onwards. Examination of cool-stored reference stock from the earlier lots also showed evidence of yeast contamination at an average level of slightly greater than 1 organism/1 L bottle (Fig. 12.17).

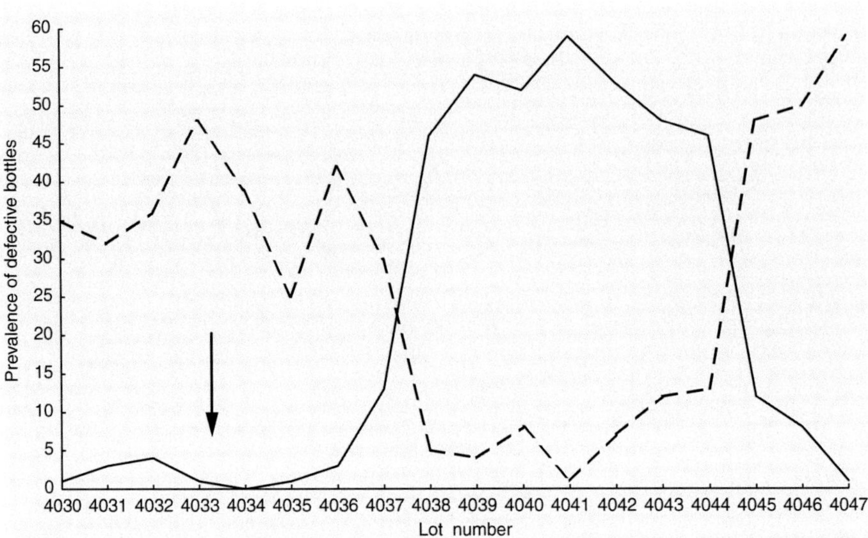

FIGURE 12.16

Multiple $n\bar{p}$-Charts showing the prevalence of spoiled bottled cider (—) and of unspoiled products with evidence of yeast contamination (-----). The arrow indicates the onset of the heat wave

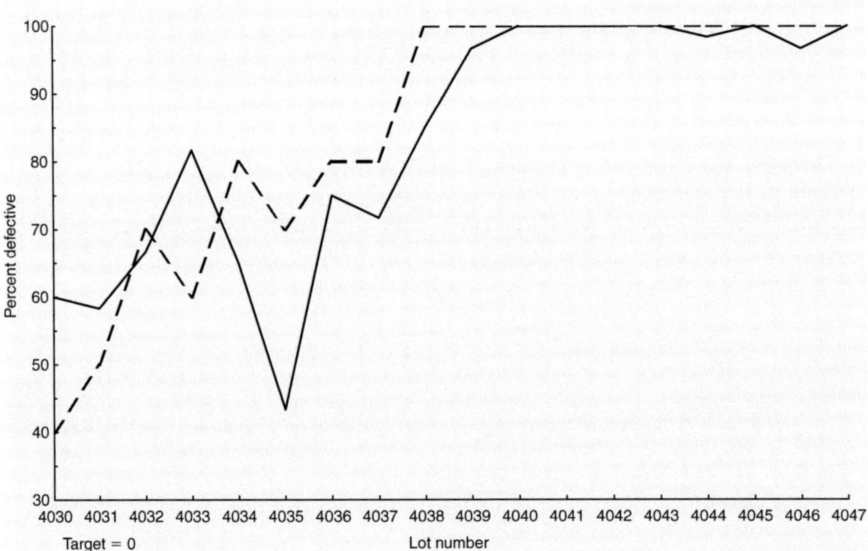

FIGURE 12.17

Comparison of prevalence of actual product spoilage (—) and the prevalence of low-level yeast contamination in cool-stored reference samples of the cider (-----)

Table 12.10 Incidence of Overt Spoilage and Yeast Contamination in a Bottled Beverage

Lot Number	Number of Products		Spoiled or Contaminated (%)	Incidence of Low-Level Contamination[a] (%)
	Spoiled	Contaminated		
4030	1	35	60	40
4031	3	32	58	50
4032	4	36	67	70
4033	1	48	82	50
4034	0	39	65	80
4035	1	25	43	60
4036	3	42	75	90
4037	13	10	38	60
4038	46	5	85	100
4039	54	4	97	100
4040	52	8	100	100
4041	59	1	100	100
4042	53	7	100	100
4043	48	12	100	100
4044	46	13	98	100
4045	12	48	100	100
4046	8	50	97	100
4047	1	59	100	100
4048	0	60	100	100
4049	0	60	100	100
4050	0	57	95	100

[a] *10 cool-stored reference products tested per lot.*

A cause and effect study was undertaken prior to examination of the process plant; Fig. 12.18 shows some potential causal factors, to illustrate the way in which a 'cause and effect' diagram can be used as an aid to problem solving. After much investigation it was concluded that the actual cause was a major breakdown in pasteurisation efficiency, although the process records (not shown) indicated that the pasteurisation plant had operated within defined time and temperature limits. A strip-down examination of the pasteurisation plant identified a defective plate seal in the pre-heat section that had permitted a low-level contamination of pasteurised product by unpasteurised material. The presence of low levels of viable yeast cells in reference samples of earlier lots of product indicated that previous lots had also been contaminated but that the organisms had not grown sufficiently to cause overt spoilage at the ambient temperatures then occurring.

Why had these yeasts not been detected in the regular checks made on products? Simply, because the initial level of contamination was only slightly more than one viable yeast cell per bottle of product and the methodology used for routine testing relied on membrane filtration of 100 mL of product from each bottle tested. Hence, the likelihood of detecting such a low prevalence of yeasts was extremely low.

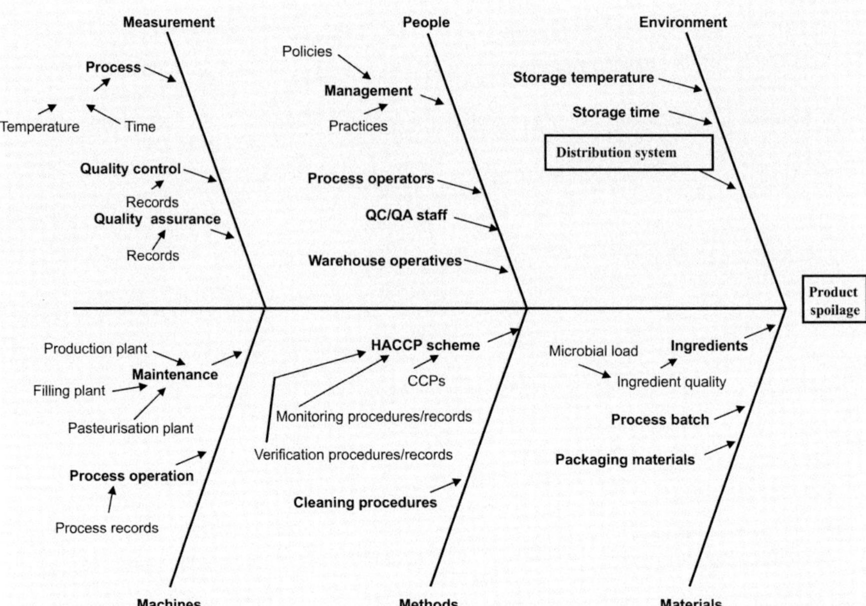

FIGURE 12.18 Example of a simplified "Cause and Effect" (Fishbone) diagram used in the investigation of product spoilage

In this example the causal effect of contamination was a pasteurisation plant defect, but a major contributory factor was the elevated ambient temperature. Another major factor, albeit not a causal factor, was the inadequacy of the microbiological test system that was not 'fit for purpose' since it could not detect the presence of such a low level of product contamination. This example serves also as a warning that it is not possible to 'test quality into a production process'!

RECENT DEVELOPMENTS IN STATISTICAL PROCESS CONTROL

SPC continues to evolve, especially as part of quality management systems for manufacturing processes. Some concepts are developments of existing process control systems whilst others are new approaches to total quality management that include the use of 'Lean' and 'Six Sigma' approaches to improve performance through a 'root and branch' analysis of causal effects associated with defective product. Increasingly such procedures are being used to improve health care systems, including the work of clinical laboratories (Schweikart and Denbe, 2009). Whilst not strictly statistical procedures *per se*, they aim to improve outcomes in, for example, laboratory handling of samples, turn-around times and automation of laboratory practices, and impact on quality of service.

Many of the techniques for SPC can be handled by computer software that is suitable for use on Windows™ or Mackintosh™ operating systems. The Windows software can be downloaded from suppliers as an 'add-in' tool for use in Excel™

(see, eg, Buttrey, 2009). My personal experience of such software is that, once the basic procedures have been learned, they are very simple to use and provide all the support that is necessary for satisfactory and efficient SPC operation.

Finally, a word of warning: if you use a computer programme to generate control charts, it is essential, once the control limits have been set, to ensure that the control limits for a chart are not updated automatically when new data are added!

REFERENCES

Anon, 1956. Statistical Quality Control Handbook, second ed. Delco Remy, Anderson, IN.

Anon, 2006. Final Report and Executive Summaries From the AOAC International Presidential Task Force on Best Practices in Microbiological Methodology: Part II F Statistical Process Control – 1. Appendices for Statistical Process Control. USFDA. <http://www.fda.gov/downloads/Food/FoodScienceResearch/UCM088754.pdf> (accessed 18.07.15).

Anon, 2013. Principles and Guidelines for the Establishment and Application of Microbiological Criteria Related to Foods. CAC/GL 21 (revised 2013). <http://www.codexalimentarius.org/standards/list-of-standards/en/?no_cache¼1> (accessed 19.10.15).

Beauregard, M.R., Mikulak, R.J., Olson, B.A., 1992. A Practical Guide to Statistical Quality Improvement. Van Nostrand Reinhold, New York.

Buttrey, S.E., 2009. An Excel add-in for statistical process control. J. Stat. Software 30 (13), 1–12, <http://www.jstatsoft.org/v30> (accessed 18.07.15).

Duncan, A.J., 1986. Quality Control and Industrial Statistics, fifth ed. Richard D. Irwin, Homewood, IL.

EU, 2005. Commission Regulation (EC) No. 2073/2005 of 15 November 2005 on Microbiological Criteria for Foodstuffs. Official J. Eur. Union L138, 1–26.

Hayes, G.D., Scallan, A.J., Wong, J.H.F., 1997. Applying statistical process control to monitor and evaluate the hazard analysis critical control point hygiene data. Food Control 8, 173–176.

Heggum, C., Vallejos, J.G., Nijie, O.B., Adegboye, A.O., 2015. Application of the moving window approach in the verification of the performance of food safety management systems. Food Control 58, 17–22.

Jarvis, B., Corry, J.E.L., Hedges, A.J., 2007. Estimates of measurement uncertainty from proficiency testing schemes, internal laboratory quality monitoring and during routine enforcement examination of foods. J. Appl. Microbiol. 103, 462–467.

Juran, J.M., 1974. Quality Control Handbook, third ed. McGraw Hill Book Co., New York.

Lee, J., et al., 2015. Example of a microbiological criterion (MC) for verifying the performance of a food safety control system: *Campylobacter* performance target at end of processing of broiler chickens. Food Control 58, 23–28.

Montgomery, D.C., 2012. Introduction to Statistical Quality Control, sixth ed. John Wiley & Sons, New York.

Schweikart, S.A., Denbe, A.E., 2009. The applicability of Lean and Six Sigma techniques to clinical and translational research. J. Invest. Med. 59, 748–755.

Shewhart, W.A., 1931. Economic Control of Quality of Manufactured Product. Van Nostrand Co. Inc., New York.

Wheeler, D.J., Chambers, D., 1984. Understanding Statistical Process Control, second ed. Statistical Process Control Inc., Knoxville, TN.

CHAPTER

Validation of microbiological methods for food 13

Sharon Brunelle

Brunelle Biotech Consulting, AOAC International and AOAC Research Institute, Woodinville, WA, United States

Microbiological methods, either *qualitative* (quantal; presence/absence) or *quantitative*, are composed of multiple steps, including sample preparation, sample analysis, data interpretation and confirmation (Fig. 13.1). The optimal integration of these steps defines the method and is critical to providing reliable results. Nowadays, microbiological methods range from fully manual methods to partially and fully automated methods.

Quantal methods for the detection of pathogens in food are generally aimed at detecting 1 colony-forming unit (cfu) per test portion. Test portions are typically 25 g of food matrix, but can be as large as 375 g as in the case of composited test portions (15 × 25 g test portions analysed as one). In order to achieve this level of detection (LOD), the target organisms in the test portion of food must be enriched to a level detectable by the analytical assay. The detectable level for polymerase chain reaction (PCR) methods is in the range of 10^3–10^4 cfu/mL and for immunoassays is 10^5–10^6 cfu/mL. Both raw foods and processed foods set challenges at a detection level of 1 cfu/test portion. Raw foods carry a higher bacterial load than processed foods, so the challenge is to detect 1 cfu of target organism amongst 10^4–10^6 cfu of background flora and, as a result, the enrichment media used often contain selective agents to suppress non-target organisms. For processed foods, the challenge lies in the state of the organism. After processing, any surviving microbial cells are likely to be injured, so the method must allow for recovery of injured cells. In this case, a pre-enrichment broth without selective agents can be employed to allow the organisms to repair and reproduce. Transfer to a selective broth or addition of selective agents can then be used to suppress non-target organisms once repair and recovery have taken place. Examples of qualitative methods include use of chromogenic agars, enzyme-linked immunosorbent assays (ELISAs), lateral flow immunoassays, enzyme-linked gene probe assays and PCR techniques.

Quantitative methods are aimed at estimating the level of pathogens, coliforms and *E. coli*, yeast and mould, or total aerobic bacterial load. A test portion, typically 50 g of food matrix, is diluted into a broth or buffer at a ratio of 1:9. The diluted homogenised test portion is analysed directly without enrichment. Quantitative

Statistical Aspects of the Microbiological Examination of Foods. http://dx.doi.org/10.1016/B978-0-12-803973-1.00013-9

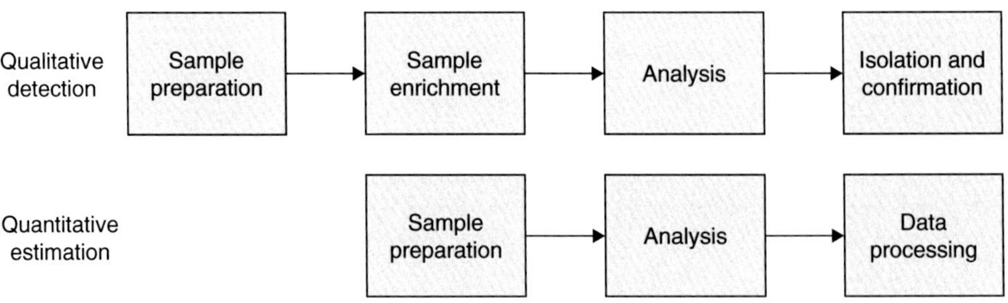

FIGURE 13.1 Generic Stages in Qualitative and Quantitative Microbiological Analyses

methods include direct methods (eg, enumeration by plate count) or indirect methods [eg, most probable number (MPN), impedance measurement, dissolved oxygen measurement and real-time PCR (RT-PCR)]. MPN procedures require replicate tubes at multiple dilutions with a qualitative readout of each tube, followed by the calculation of the 'most likely' estimate of the contamination level based on the number of tubes positive at each dilution level. Impedance methods measure the time required, under controlled growth conditions, to cross a threshold impedance value. The time required to cross the threshold value is related to the contamination level of the sample using a standard curve. Likewise, methods measuring dissolved oxygen rely on the principle that the higher the bacterial count, the faster the dissolved oxygen will decrease. Measurements are compared to a standard curve of bacterial load versus time to reach a threshold level of oxygen. RT-PCR measures the number of cycles required to cross a threshold value (C_T value). The C_T value relates to the contamination level of the sample in that the higher the contamination level, the fewer PCR cycles are required to reach the threshold. C_T values are converted to cfu/test portion using a standard curve usually built into the RT-PCR software.

THE STAGES OF METHOD DEVELOPMENT

Method development begins with a concept of the *intended use* of the method. This defines the type of test, the target analyte, the applicable foods or matrices and the user of the method. The type of test may be qualitative detection or quantitative estimation of contamination. The target analyte, meaning the microorganism that the method detects, quantifies or identifies, can be a genus, a species or a serovar. Matrices are generally foods or environmental samples. Foods are typically divided into categories, such as meat, poultry, seafood, fruits and vegetables, and these categories are subdivided into raw, heat processed, frozen, fermented, etc., as appropriate. Examples of environmental surfaces are those that can be found in a food manufacturing facility, such as stainless steel, rubber, sealed concrete, ceramic or glass, plastic, air filter material and cast iron. In microbiological food examination, the end user is typically a trained laboratory technician. However, there are instances where this may not be the case, such as detection of biological threat agents where the end user could be a trained 'first responder' in the field. As an example of an intended use statement, the method developer might state, 'this method is intended for the detection of *Salmonella* species in raw poultry in a microbiological laboratory'.

Method development can begin once the concept of the method is defined by the intended use. First, the 'assay critical' reagents are identified. If the method is an immunoassay, antibodies are screened against pure culture target and non-target organisms to identify highly selective candidates. For molecular-based assays, DNA sequences are identified and targeted with primers and probes. These primers and probes are screened against DNA derived from target and non-target organisms. Performance parameters, such as limit of detection, specificity (SP), selectivity, C_T values and the like, are measured under various assay conditions to optimise performance and to determine the best candidate critical reagents. Assays are then constructed around the 'best candidate' critical reagents, based on the initial pure-culture screening. The assays are challenged with pure cultures and a few selected foods representative of the breadth of the intended matrices. During this process, sample preparation methodology is also varied to ensure that the target analyte is presented to the critical reagent(s) in an optimal manner that facilitates detection. For immunoassays, this means ensuring that the antigenic target is accessible to the antibody.

In the final stages of method development, the optimised assay configuration is tested against matrix samples inoculated at various levels of target analyte. Final optimisation of assay conditions (buffers, critical reagent concentrations, reaction times, reaction temperatures, etc.) occurs at this stage. Statistically designed experiments should be carried out to verify optimal assay conditions, to examine the ruggedness of the assay and to establish 'guard bands' on the critical assay parameters. 'Guard bands' are defined in this context as the variations of assay conditions that are tolerated by the method without significantly affecting the method results.

Once method development is complete, the method is transferred to the manufacturing facility and process development occurs. The goal of process development is to devise a manufacturing scheme that produces assay components consistent with the final assay design in a reproducible manner. Some re-optimisation of the assay design may be required on scaled-up production in order to achieve the assay performance observed in the final method development stage. Additionally, new 'guard bands' may need to be established on full-scale manufactured assay components. The manufactured assay components and method instructions are now ready for validation.

WHAT IS VALIDATION?

Validation is the establishment of method performance in a single laboratory or multiple laboratories under controlled conditions. A method can be validated to demonstrate that it performs as claimed, that it performs at least as well as a validated standard method or that it performs to a set of established criteria. In general, current practices in food microbiology (AOAC, 2012a; Anon, 2015a) validate by comparison to a standard method, typically an official or regulatory method, such as AOAC (2015), FDA (2015), USDA (2015), Anon (2015b), APHA (1985) or others. Since appropriate reference methods are not always available, AOAC has established stakeholder panels, such as the Stakeholder Panel on Agent Detection Assays (SPADA, 2015), to generate standard method performance requirements (SMPRs), which are voluntary consensus standards establishing testing conditions and acceptance criteria for specific intended uses. This has worked well for detection of biothreat agents in the environment, for example, SMPR *2010.003* (AOAC, 2011), which provided the validation parameters for OMA *2012.06*, *Bacillus anthracis in Aerosol Collection Samples* (Hadfield et al., 2013). The establishment of SMPRs may soon be applied to food microbiology through AOAC's International Stakeholder Panel on Alternative Methods (ISPAM, 2015).

Validation of a microbiological method for food generally includes a test of inclusivity and exclusivity to establish the analytical selectivity of the method for the analyte. In other words, is the method *inclusive* in the scope of the target analyte and is it *exclusive* of cross-reactivity to closely related non-target analytes? Such pure culture studies establish the analytical scope of the method.

The method is then challenged with a range of artificially or naturally contaminated food matrices to establish (1) that the food matrices do not interfere with either the growth of the organism or its detection and (2) that background flora found naturally in foods do not suppress the enrichment or detection of the target organism. It may be discovered, for example, that a particular method well suited to detection of analyte in processed foods where background flora is generally very low does not perform well with unprocessed or raw foods that tend to have much higher levels of competing background flora. It is for these reasons that a method should be considered validated only for those foods or food types that have been tested successfully in the validation study.

Whilst validation establishes the method performance in one or multiple laboratories, 'verification' establishes proper implementation of a method in the user's laboratory. Using a specified verification protocol, the laboratory performs the method and ensures that results are comparable to the method performance established in the validation study.

HARMONISATION

Over the past decade or more, there has been interest in harmonising validation schemes between Europe and the United States with the goal of performing one validation study and using the data as the basis for both AOAC and European (AFNOR, MicroVal or other validation organisation) certifications. Despite enormous effort on both sides of the Atlantic, no agreement has been reached on the statistics for method comparison studies, but the study designs were aligned. This allows for data collection on a single set of test portions, followed by statistical analysis according to each validation guideline (AOAC, 2012a; Anon, 2015a). The one remaining hurdle to harmonised studies is the choice of reference method. EU law requires comparison to ISO reference methods, whilst US regulatory agencies require comparison to US reference methods if the alternative method is to be used for regulatory purposes. The first step towards overcoming this hurdle is the establishment of the

Harmonization of *Salmonella* Methods Working Group by ISPAM whose goal is to make recommendations on how to harmonise the US FDA BAM, US FSIS MLG, Health Canada and ISO *Salmonella* reference culture methods (ISPAM, 2015). In like manner, ISO Technical Committee 34/Subcommittee 9 put forth an initiative on Harmonized Incubation Temperatures With Selective Enrichment Broth with a goal of evaluating the impact of 35 ± 1°C versus 37 ± 1°C incubation on selective enrichment using predictive microbiology. Overall, the goal of harmonisation is to reduce the costs associated with validation of methods by aligning the technical aspects of method validation, but allowing the various validating organisations to maintain their independent processes and acceptance criteria.

VALIDATION OF QUALITATIVE METHODS

SELECTIVITY TESTING

For inclusivity testing of qualitative and quantitative methods, 50 or more strains are chosen for analysis based on the target scope of the method. For example, if the target is a species, then a range of strains within that species should be chosen to represent the antigenic or genetic diversity. In the case of a genus-level test for *Salmonella*, 100 distinct serotypes sampled from all species and sub-species should be chosen due to the much greater size and variation of the genus compared to other pathogens. For AOAC qualitative methods, the inclusivity organisms are cultured in the method-specific broth(s) and diluted to approximately $100 \times LOD_{50}$ (see Chapter 11) for testing. For ISO 16140-2 qualitative methods, inclusivity organisms are inoculated into method-specific broth at 10–100 times the LOD and enriched according to the most challenging enrichment protocol prescribed by the method and tested at the end of enrichment. For ISO 16140-2 and AOAC quantitative methods, organisms are tested at 100 times the level of quantification (LOQ) or $100 \times LOD_{50}$, respectively.

Exclusivity testing requires at least 30 organisms closely related to the target group. For both AOAC and ISO 16140-2, this testing is done at a high concentration to challenge the alternative method. The test results (positive or negative) of each inclusivity and exclusivity organism are reported and any unexpected results are noted. No statistical analyses are performed. Selectivity testing is not applicable to quantitative methods that measure general categories of organisms, such as total aerobic counts.

METHOD COMPARISON

Qualitative method validation studies can be divided into two types – paired sample design and unpaired sample design (Fig. 13.2). A *paired sample design* is one in which a single test portion is enriched and the enriched test portion is analysed by both the *alternative method* (the new method being validated) and the *reference method* (the established official or regulatory method). This design results from a common enrichment or pre-enrichment of the two methods. When the initial

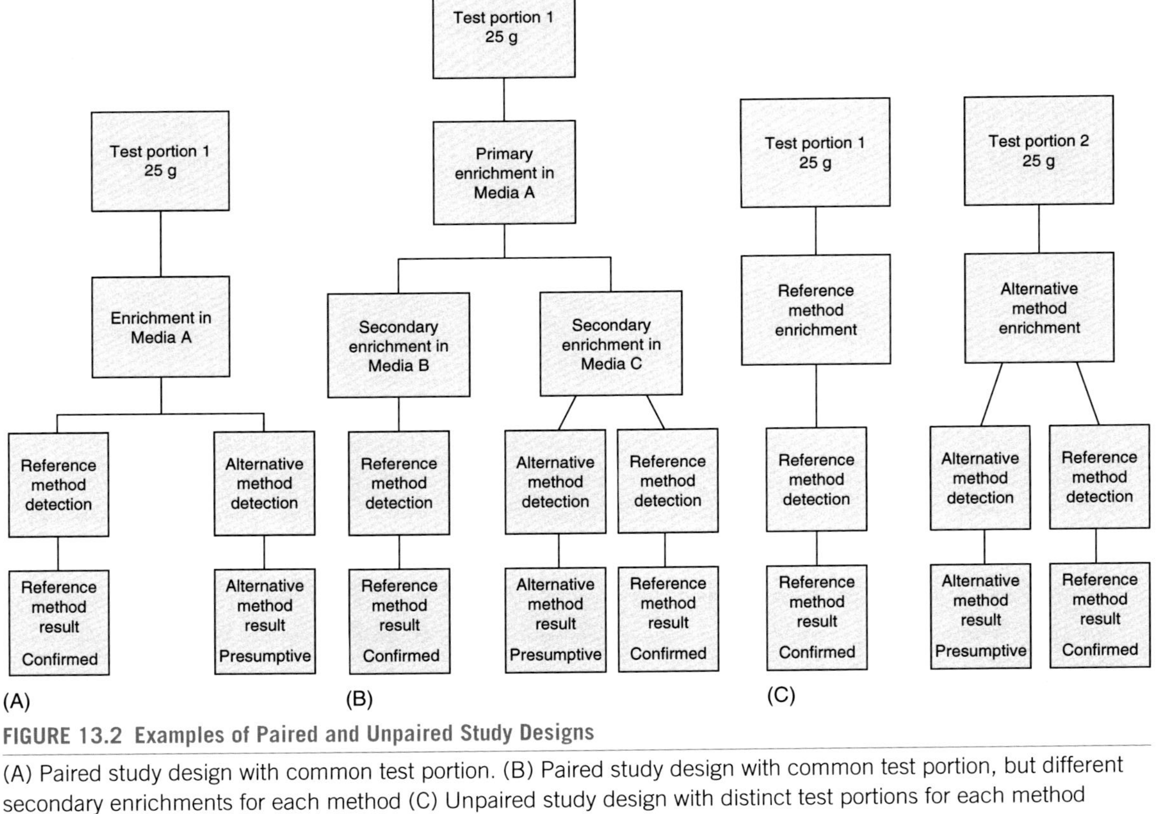

FIGURE 13.2 Examples of Paired and Unpaired Study Designs

(A) Paired study design with common test portion. (B) Paired study design with common test portion, but different secondary enrichments for each method (C) Unpaired study design with distinct test portions for each method

enrichment conditions differ between the two methods, an *unpaired sample design* is used. In the unpaired design, distinct test portions are enriched and analysed by the two methods. It must be noted that, regardless of presumptive results, the alternative method test portion enrichment cultures must be subjected to the confirmatory procedures outlined in the reference method in order to establish the true status (positive or negative) of the test portions.

Comparison of two qualitative methods is best accomplished near the limit of detection of one or both of the methods. It is only at low levels of contamination (approximately 0.2–2 cfu/test portion) that differences between the methods can be observed. Thus, the goal of the method comparison study is to achieve an artificial or natural contamination level that yields fractional positive results across all replicates (25–75% positive) for one of the methods. The data can then be statistically analysed for a significant difference between the methods. If fractional positive results are not obtained for at least one of the methods, then the study is repeated at a higher or lower contamination level as needed to achieve fractional positive results. To avoid having to repeat a study, some analysts will inoculate at more than one level to increase the chances of obtaining fractional positive results. The inoculation of food matrices is carried out using a single strain of target organism. Food isolates are preferred and a different strain is used for each food matrix tested. The inoculating strain is cultured in an appropriate enrichment broth and the concentration is determined by plate count. Enough food matrix is inoculated at one level to carry out testing of replicate test portions by the alternative and reference methods as well as the test portions required for MPN analysis (Blodgett, 2010; Jarvis et al., 2010). Typically, these amounts of inoculated matrix are approximately 900 g for paired samples and approximately 1400 g for unpaired samples when the test portion size is 25 g. Because batch inoculation is preferred, it is imperative that the inoculated batch is well mixed to ensure homogeneity.

Three main statistical procedures are used to analyse qualitative method comparison data. These are probability of detection (POD), relative level of detection (RLOD) and sensitivity. POD (Wehling et al., 2011) is used by AOAC, whilst RLOD (Mărgăritescu and Wilrich, 2013) is used by ISO. Both use the same single-laboratory study design, so harmonisation of data collection can easily be accomplished. For artificially contaminated matrix, 20 replicate test portions are analysed per method at a low contamination level (fractional positive results required), 5 replicates per method at a high contamination level (eg, 2–3 × low level) and 5 replicates per method of uninoculated matrix. If naturally contaminated matrix can be obtained, then 20 replicate test portions of each of 2 contaminated lots are analysed by each method. In this case, one of the lots must achieve fractional positive results for one of the methods.

ISO (Anon, 2015a) includes an additional sensitivity study that AOAC does not require. In the sensitivity study, a wide variety of food samples is included at various natural or artificial contamination levels. Sixty samples per food category, including 20 samples per each of 3 food types, are analysed by the alternative and reference methods.

Multi-laboratory study designs are similar between AOAC and ISO 16140-2 and harmonisation of data collection can be accomplished by following the more challenging design. Both guidelines require at least 10 valid data sets including a high, a low and an uncontaminated level, but AOAC requires 12 replicates per level per data set and ISO requires 8 replicates. The harmonised study would follow the 12-replicate study design in order to satisfy both requirements. The data from multi-laboratory studies are analysed by the same statistical approaches as single-laboratory data – POD, RLOD and sensitivity.

Probability of detection – AOAC

POD is defined as the probability of a positive response at a given concentration of analyte (Wilrich and Wilrich, 2009; Wehling et al., 2011). It is calculated simply as the number of positive responses divided by the total number of replicates tested at a given analyte concentration:

$$\mathrm{POD} = \frac{x}{N}$$

where x is the number of positive results and N is the number of replicates tested.

If one thinks of qualitative data in quantitative terms, with 0 representing a 'not detected' result and 1 representing a 'detected' result, then the POD value is analogous to the mean of replicate results. When POD is calculated for a range of concentrations, a response curve of POD by concentration can be generated, and, further, multiple response curves could be generated to compare methods, foods or laboratories for detection of a particular analyte. Ideally, one would like to define the full response curve for the analyte in each food being validated and compare those to response curves generated for the reference method. Whilst this would provide a wealth of information about each method and each matrix, it is not a practical approach as it would be cost-prohibitive to test enough replicates at enough concentrations for each food to define the response curves. Instead, the compromise is to focus the efforts on the fractional positive range (25–75% positive) and compare POD values at a single concentration of analyte within this range. In the single-laboratory study, 20 replicates at the low fractional level are tested by each method for comparison. In addition, 5 replicates at a higher level expected to yield all positives (POD = 1) and 5 replicates of uncontaminated control matrix expected to yield no positives (POD = 0) are also tested for each method. So, a crude response curve can be estimated with POD near 0, POD near 0.5 and POD near 1 with corresponding analyte concentrations determined by MPN analysis. For each POD value, the error is expressed as a binomial 95% confidence interval (CI) (Wehling et al., 2011; Wilrich, 2015). Excel® spreadsheets are available to ensure the proper calculations of the CIs (LaBudde, 2013) and MPN values (LaBudde, 2009).

For comparison of methods, the difference of two POD values (dPOD) is calculated, much like the bias parameter for quantitative methods. Like the POD values, dPOD will have an associated 95% CI, but the method of calculation of that CI will depend on whether the validation study design included paired or unpaired samples (AOAC, 2012a). In the paired sample design, a stronger correlation of results is expected since the same test portion enrichment is being analysed by both methods. Unpaired study designs may have more variability in method results since distinct test portion populations are analysed by each method. The strength of the correlation of results from unpaired studies will depend not only on the methods of analysis but also to a large degree on the homogeneity of the analyte in the food matrix. Therefore, the CI for dPOD for paired study designs is more stringent (narrower) than for unpaired study designs. Recently, however, the procedure for determining dPOD CIs for paired data has been shown to produce aberrant results and is the subject of a revision of the guideline. In the meantime, the calculations for unpaired data can be used to provide indicative CI limits. These dPOD CIs are used to determine whether the two methods are significantly different at the 95% level. If the dPOD CI contains '0', that is, if the lower confidence limit (CL) is at or below zero and the upper CL is at or above zero, then the two methods are not significantly different. If both limits of the 95% CI are positive, or both negative, then the CI does not include zero and the methods are statistically different.

Whilst POD is the mean result of replicate analyses within a laboratory, LPOD is the mean result across all laboratories in a multi-laboratory study. Likewise, dLPOD is the difference of two LPOD values, for example, the LPOD values for the reference and alternative methods and the 95% CI of the dLPOD determine whether the two methods are significantly different within the experimental design of the multi-laboratory study. Like dPOD, the dLPOD CI calculation will depend on whether the study uses a paired or unpaired sample design (AOAC, 2012a). In addition, from the multi-laboratory data, one can make estimates of repeatability (s_r), reproducibility (s_R) and the laboratory effect (s_L), again analogous to quantitative method statistics. A *t*-test is used to determine whether the laboratory effect is significant within the design of the study.

POD statistics are currently used for validating a qualitative method against either a reference method (Example 13.1; AOAC, 2012a) or SMPRs (Hadfield et al., 2013; AOAC, 2011; AOAC, 2012b).

EXAMPLE 13.1 AOAC QUALITATIVE METHOD VALIDATION

In an AOAC *Performance Tested Methods*℠ (PTM) study, the *mericon*® *Listeria monocytogenes* Pathogen Detection Assay with manual DNA preparation was compared to FSIS MLG 8.08 (USDA, 2015) for hot dogs and deli turkey; to FDA BAM Chapter 10 (FDA, 2015), for smoked salmon and mung bean sprouts; and to OMA 993.12 (AOAC, 2015) for pasteurised whole milk. Each matrix was inoculated with a different strain of *L. monocytogenes* at a low level to produce fractional positives and at a high level. Test portions were analysed by the *mericon* and reference methods in an unpaired study. All *mericon* test portions, regardless of presumptive result, were subjected to the confirmation procedures from the appropriate reference method. Table 13.1 presents the *mericon* presumptive and confirmed results. Fractional positives were achieved at the low level for all matrices. The dPODcp, the difference of POD values between the candidate presumptive and candidate confirmed results, was 0 at every inoculation level for every matrix, indicating perfect agreement between the presumptive and confirmed *mericon* results. Table 13.2 presents the comparison of the *mericon* method and the appropriate reference method for each matrix. Differences are observed at the low level of each matrix as indicated by the non-zero dPODc values. This is not unexpected for an unpaired study design. The 95% confidence intervals on the dPODc values demonstrate, however, that these small differences are not statistically significant since all confidence intervals include 0. Thus, the *mericon Listeria monocytogenes* Pathogen Detection Assay with manual DNA preparation was shown to be not statistically different from the three reference methods for the matrices studied and was awarded PTM certification.

Data by permission of QIAGEN GmbH.

Table 13.1 POD Results for the *mericon*® *Listeria monocytogenes* Pathogen Detection Assay With Manual DNA Preparation – Presumptive Versus Confirmed

Matrix	Inoculum Level	*N*	Presumptive			Confirmed				
			x	PODcp	95% CI	*x*	PODcc	95% CI	dPODcp	95% CI
Hot dogs	Uninoculated	5	0	0.00	0.00, 0.43	0	0.00	0.00, 0.43	0.00	−0.43, 0.43
	Low	20	8	0.40	0.22, 0.61	8	0.40	0.22, 0.61	0.00	−0.28, 0.28
	High	5	5	1.00	0.57, 1.00	5	1.00	0.57, 1.00	0.00	−0.43, 0.43
Deli Turkey	Uninoculated	5	0	0.00	0.00, 0.43	0	0.00	0.00, 0.43	0.00	−0.43, 0.43
	Low	20	5	0.25	0.11, 0.47	5	0.25	0.11, 0.47	0.00	−0.26, 0.26
	High	5	5	1.00	0.57, 1.00	5	1.00	0.57, 1.00	0.00	−0.43, 0.43
Smoked salmon	Uninoculated	5	0	0.00	0.00, 0.43	0	0.00	0.00, 0.43	0.00	−0.43, 0.43
	Low	20	10	0.50	0.30, 0.70	10	0.50	0.30, 0.70	0.00	−0.28, 0.28
	High	5	5	1.00	0.57, 1.00	5	1.00	0.57, 1.00	0.00	−0.43, 0.43
Mung bean sprouts	Uninoculated	5	0	0.00	0.00, 0.43	0	0.00	0.00, 0.43	0.00	−0.43, 0.43
	Low	20	7	0.35	0.18, 0.57	7	0.35	0.18, 0.57	0.00	−0.28, 0.28
	High	5	5	1.00	0.57, 1.00	5	1.00	0.57, 1.00	0.00	−0.43, 0.43
Pasteurised whole milk	Uninoculated	5	0	0.00	0.00, 0.43	0	0.00	0.00, 0.43	0.00	−0.43, 0.43
	Low	20	6	0.30	0.15, 0.52	6	0.30	0.15, 0.52	0.00	−0.27, 0.27
	High	5	5	1.00	0.57, 1.00	5	1.00	0.57, 1.00	0.00	−0.43, 0.43

N, *number of test portions;* x, *number of positive test portions; PODcp, candidate method presumptive positive outcomes divided by* N; *PODcc, candidate method confirmed positive outcomes divided by* N; *dPODcp, PODcp-PODcc; 95% CI, if the CI of a dPOD does not contain zero, then the difference is statistically significant at the 5% level.*

Excerpts from Tables 39 and 43 reproduced from QIAGEN mericon® Listeria monocytogenes and mericon® Listeria species Pathogen Detection Assays: AOAC Performance Tested Methods℠ 061401 and 061402, unpublished report., by permission of AOAC International, and QUIAGEN Gbmh

Table 13.2 POD Results for the *mericon*® *Listeria monocytogenes* Pathogen Detection Assay With Manual DNA Preparation – *mericon* Method Versus Reference Method

			***Mericon L. monocytogenes* Pathogen Detection Kit**			**Reference Method**				
Matrix	**Inoculum Level**	***N***	***x***	**PODc**	**95% CI**	***x***	**PODr**	**95% CI**	**dPODc**	**95% CI**
Hot dogs	Uninoculated	5	0	0.00	0.00, 0.43	0	0.00	0.00, 0.43	0.00	−0.43, 0.43
	Low	20	8	0.40	0.22, 0.61	6	0.30	0.15, 0.52	0.10	−0.18, 0.36
	High	5	5	1.00	0.57, 1.00	5	1.00	0.57, 1.00	0.00	−0.43, 0.43
Deli Turkey	Uninoculated	5	0	0.00	0.00, 0.43	0	0.00	0.00, 0.43	0.00	−0.43, 0.43
	Low	20	5	0.25	0.11, 0.47	6	0.30	0.15, 0.52	−0.05	−0.31, 0.22
	High	5	5	1.00	0.57, 1.00	5	1.00	0.57, 1.00	0.00	−0.43, 0.43
Smoked salmon	Uninoculated	5	0	0.00	0.00, 0.43	0	0.00	0.00, 0.43	0.00	−0.43, 0.43
	Low	20	10	0.50	0.30, 0.70	11	0.55	0.34, 0.74	−0.05	−0.33, 0.24
	High	5	5	1.00	0.57, 1.00	5	1.00	0.57, 1.00	0.00	−0.43, 0.43
Mung bean sprouts	Uninoculated	5	0	0.00	0.00, 0.43	0	0.00	0.00, 0.43	0.00	−0.43, 0.43
	Low	20	7	0.35	0.18, 0.57	6	0.30	0.15, 0.52	0.05	−0.23, 0.32
	High	5	5	1.00	0.57, 1.00	5	1.00	0.57, 1.00	0.00	−0.43, 0.43
Pasteurised whole milk	Uninoculated	5	0	0.00	0.00, 0.43	0	0.00	0.00, 0.43	0.00	−0.43, 0.43
	Low	20	6	0.30	0.15, 0.52	5	0.25	0.11, 0.47	0.05	−0.22, 0.31
	High	5	5	1.00	0.57, 1.00	5	1.00	0.57, 1.00	0.00	−0.43, 0.43

N, *number of test portions; x, number of positive test portions; PODc, candidate method confirmed positive outcomes divided by* N; *PODr, reference method confirmed positive outcomes divided by* N; *dPODc, PODc-PODr; 95% CI, if the CI of a dPOD does not contain zero, then the difference is statistically significant at the 5% level.*

Excerpts from Tables 39 and 43 reproduced from QIAGEN mericon® Listeria monocytogenes and mericon® Listeria species Pathogen Detection Assays: AOAC Performance Tested Methods℠ 061401 and 061402, unpublished report., by permission of AOAC International, and QUIAGEN Gbmh

Relative level of detection – ISO 16140-2

The RLOD study in ISO 16140-2 (Anon, 2015a) has the same study design as the AOAC method comparison study, that is, 20 replicate test portions at a low level yielding fractional positives (25–75% positive), 5 replicate test portions at a high level aiming for all positives and 5 replicates of uncontaminated food matrix expected to yield all negatives. The RLOD is defined as the ratio of LODs of the alternate method and the reference method:

$$\text{RLOD} = \frac{\text{LOD}_{\text{alt}}}{\text{LOD}_{\text{ref}}}$$

LOD is defined as the lowest level (concentration of analyte) yielding reliable detection. In the context of ISO 16140-2, LOD_{50} is used to indicate the level of analyte yielding 50% positive results. If analyte concentrations are known, for example, through MPN estimation, LOD_{50} can be estimated for each method and RLOD then determined. If analyte concentrations are not known, then RLOD can be estimated only directly. The model uses a complementary log–log (CLL) fit of the binomial data in the fractional positive range and an Excel spreadsheet is available to aid in the calculations of RLOD (Anon, 2015c). Once RLOD is determined, it is compared to an acceptability limit (AL), a criterion determined by a voluntary consensus process and which differs for paired and unpaired sample designs for the reasons described earlier. In a paired sample study, the RLOD must be ≤ 1.5, meaning that the alternative method LOD cannot be more than 1.5 times greater than the reference method LOD. For unpaired sample studies the AL is set at 2.5, meaning RLOD must be ≤ 2.5. An LOD value for the alternative method smaller than the LOD value for the reference method is always accepted, as this means that the alternative method is likely to detect lower levels of contamination than the reference method (Example 13.2).

In the multi-laboratory study, RLOD is calculated for each laboratory producing fractional positive results (one to seven positives per eight replicates tested) at the low contamination level for the reference method. Initially, a CLL model that includes laboratory effects is used and the differences between laboratories are subjected to an analysis of deviance test to determine statistical significance. If the differences are not significant, then an overall RLOD can be determined using a CLL model without laboratory effects and compared to an AL of 1.6. A 90% CI on the overall RLOD is also determined and, since RLOD is a ratio of terms, if the CI contains the value '1' (ie, the lower CL is at or below 1 and the upper CL is at or above 1), then the methods are not statistically different within the study design of the multi-laboratory study.

So, whilst the POD model compares methods using a *difference* between methods (analogous to a bias estimate), the RLOD model uses a *ratio* of methods (analogous to an accuracy estimate). It is noteworthy that both AOAC and ISO encourage the analyst to perform both sets of calculations in an effort to collect additional information about the methods being compared and to compare the two statistical approaches.

EXAMPLE 13.2 ISO 16140-2 QUALITATIVE METHOD VALIDATION

The same method as in Example 13.1 (*mericon*® *Listeria monocytogenes* Pathogen Detection Assay with manual DNA preparation) was validated for AFNOR certification according to ISO 16140-2:2015. Similar to the AOAC study design, the RLOD study compared the *mericon* method to ISO 11290-1 (Anon, 2015b) for two matrices, rillettes and raw milk, in an unpaired study design. Each matrix was inoculated with a different strain of *L. monocytogenes*. Table 13.3 shows that the fractional positive criterion was met for the low contamination level with both matrices. The RLOD values for each matrix and for the two matrices combined are presented with 95% CIs. For an unpaired study, the acceptability limit (AL) on RLOD is 2.5. This AL was met for both matrices individually as well as combined.

For the sensitivity study, 144 meat and dairy samples were examined. Of those, 28 were artificially contaminated with 1 of 12 different strains of *L. monocytogenes* at levels below 5 cfu/test portion. The remainder were naturally contaminated samples at unknown levels. Samples yielding presumptive positive results were confirmed by streaking on chromogenic agar. The *mericon* and reference method results are summarised in Table 13.4 for each category and for both categories

Table 13.3 RLOD Results for the *mericon*® *Listeria monocytogenes* Pathogen Detection Assay With Manual DNA Preparation

Matrix	Level	Contamination (cfu/Test Portion)	*N*	ISO 11290-1 Positive	*Mericon* Positive	RLOD	95% CI
Rillettes	0	0	5	0	0	1.000	(0.434, 2.304)
	Low	0.7	20	10	10		
	High	2	5	5	5		
Raw milk	0	0	5	0	0	1.229	(0.586, 2.576)
	Low	0.6	20	14	11		
	High	1.7	5	3	4		
Combined						1.082	(0.632, 1.854)

Reproduced with permission from EN ISO 16140 validation study of the mericon® Manual Listeria monocytogenes Detection in meat and dairy products, unpublished report, by permission of QUIAGEN Gbmh

Table 13.4 Summary of Sensitivity Study Results for the *mericon*® *Listeria monocytogenes* Pathogen Detection Assay With Manual DNA Preparation

	Meat Category		Dairy Category		All Categories	
	Reference Method Positive (R+)	Reference Method Negative (R−)	Reference Method Positive (R+)	Reference Method Negative (R−)	Reference Method Positive (R+)	Reference Method Negative (R−)
Alternative method positive (A+)	PA = 21	PD = 8	PA = 22	PD = 5	PA = 43	PD = 13
Alternative method negative (A−)	ND = 8	NA = 35	ND = 5	NA = 40	ND = 13	NA = 75

PA, positive agreement = A+/R+; PD, positive deviation = A+/R−; ND, negative deviation = A−/R+; NA, negative agreement = A−/R−.

Reproduced with permission from EN ISO 16140 validation study of the mericon® Manual Listeria monocytogenes Detection in meat and dairy products, unpublished report, by permission of QUIAGEN Gbmh

Table 13.5 Relative Accuracy (AC), Relative Sensitivity (SE), False-Positive (FP) Ratio and Acceptability Limit (AL) for the *mericon*® *Listeria monocytogenes* Pathogen Detection Assay With Manual DNA Preparation

Category	Type	PA	NA	PD	ND	SE_{alt} (%)	SE_{ref} (%)	AC (%)	FP (%)	ND − PD	AL
Meat products	Raw meat	10	7	4	2	87.5	75.0	73.9	0.0	−2	
	Raw and cooked delicatessen	4	16	3	2	77.8	66.7	80.0	0.0	−1	
	Ready to eat, ready to re-heat	7	12	1	4	66.7	91.7	79.2	0.0	3	
	Total	21	35	8	8	78.4	78.4	77.8	0.0	0	+3
Dairy products	Raw and fermented milk	9	16	2	2	84.6	84.6	86.2	0.0	0	
	Cheese (raw and pasteurised)	9	13	0	1	90.0	100.0	95.7	0.0	1	
	Desserts and ice cream	4	11	3	2	77.8	66.7	75.0	0.0	−1	
	Total	22	40	5	5	84.4	84.4	86.1	0.0	0	+3
All categories		43	75	13	13	81.2	81.2	81.9	0.0	0	+4

Reproduced with permission from EN ISO 16140 validation study of the mericon® Manual Listeria monocytogenes Detection in meat and dairy products, unpublished report, by permission of QUIAGEN Gbmh

combined. Overall, 118 samples showed agreement between the method results (PA + NA). Twenty-six samples yielded deviations, with 13 being positive deviations (PD) and 13 being negative deviations (ND). Deviations are not unexpected in an unpaired study design, but the distribution between PD and ND should be approximately equal if the methods have similar performance. Table 13.5 presents the performance indicators including relative sensitivity (SE), relative accuracy (AC), false-positive (FP) ratio and the difference ND − PD. The relative sensitivities, when calculated for each category or for the combined categories, are identical between the alternative and reference methods. The relative accuracy of the alternative method was 81.9% overall with a 0% FP ratio. Examining the deviations, ND − PD was 0 for each category as a whole and for all categories combined. The AL for 1 category in an unpaired study is 3, and for 2 categories is 4 (Anon, 2015a).

With the AL criteria met in the RLOD and sensitivity studies and supported by the SE, AC and FP performance indicators, the *mericon* method was determined to be not significantly different from the ISO reference method for meat and dairy products and awarded AFNOR certification.

Data by permission of QIAGEN GmbH.

Sensitivity - ISO 16140–2

ISO 16140-2 includes a second method comparison study with a very different study design. In this case, rather than examining each food matrix with strict replication at distinct concentrations as for the studies given earlier, the sensitivity study includes a broad range of foods including as many naturally contaminated samples as possible and data are analysed by food category, not by food item. The level of contamination of each sample is not necessarily known and samples within a food category will have differing levels of contamination, either natural or artificial. A minimum of 60 food samples (20 samples from each of 3 food types) within each food category are tested by the alternate and reference methods. Results from the two methods are then tabulated and scored as positive agreement (PA, both methods agree the sample was positive), negative agreement (NA, both methods agree the sample was negative), positive deviation (PD, the alternative method gave a positive result where the reference method was negative) or negative deviation (ND, the alternative method gave a negative result where the reference method was positive). False-positive (FP, positive presumptive result but negative confirmation) and false-negative (FN, negative presumptive result but positive confirmation) results are noted for the alternative method. Sensitivity of the alternate method (SE_{alt}) is calculated as the number of positive samples by the alternative method (PA + PD) divided by the total number of positive samples by either method (PA + PD + ND) expressed as a percentage. Likewise, sensitivity of the reference method (SE_{ref}) is calculated as the number of positive samples by the reference method (PA + ND) divided by the total number of positive samples by either method (PA + PD + ND) expressed as a percentage. Relative accuracy (AC) is calculated as the total number of method agreements (PA + NA) divided by the total number of samples tested (*N*), expressed as a percentage. The FP ratio of the alternative method is calculated as the number of FP samples by the alternative method (regardless of reference method result) divided by the number of samples with NA, expressed as a percentage.

Taking a closer look at the rate of deviations between the methods, for paired sample studies both the sum of deviations (ND + PD) and the difference of deviations (ND − PD) are calculated for each food category and for all food categories combined. For unpaired sample studies, only the difference between deviations (ND − PD) is considered. Anon (2015a) tabulates AL values, based on consensus expert opinion, by number of food categories for determination of method acceptability for individual food categories or overall (Example 13.2).

In the multi-laboratory study (with an RLOD-type study design), SP of the alternative method (after the confirmatory step) and of the reference method are calculated from the uninoculated test portions across all laboratories as $[1-(P_0/N_-)]$, where P_0 is the number of positive results at the uninoculated level and N_- is the number of uninoculated test portions tested. The data across all laboratories from levels where fractional positives were obtained (low level and possibly high level) are then tabulated as earlier for determination of PA, NA, PD, ND and FP. SE_{alt}, SE_{ref}, AC and FP ratio are determined as earlier. For paired studies, the sum and difference of deviations are compared to AL based on the number of laboratories participating. For unpaired studies, rather than by consensus expert opinion, the AL is determined at each level individually where fractional positives are obtained and is based on the proportion positive by the reference method (p^+_{ref}) and alternative method (p^+_{alt}) using the following equation:

$$AL = (ND - PD)_{max} = \sqrt{3N[(p^+_{ref} + p^+_{alt}) - 2(p^+_{ref} \times p^+_{alt})]}$$

where *N* is the number of samples tested at that level.

VALIDATION OF QUANTITATIVE METHODS

Using either naturally contaminated or artificially contaminated food, three levels of contamination at least 1 $\log_{10}$ unit apart and covering the range of the alternative method are analysed by both the alternative and reference methods. If artificially contaminated food is tested, then one lot of uninoculated food matrix must also be included. The lowest contamination level should be close to, but not at, the limit of detection of the method. In the single-laboratory study, five replicate test portions at each level are examined by each method. Collaborative studies require a minimum of 8 valid data sets for each food type, so it is recommended that 10–14 laboratories participate in the study. Each laboratory must analyse two test portions per contamination level per food matrix.

EXAMINATION OF DATA FOR OUTLIERS

The data are first subjected to visual inspection for obvious aberrant data. If aberrant data are observed, the laboratory is contacted to determine whether there is cause for removal of data points or the whole data set. The data can also be examined for statistical outliers using the Grubbs, Cochran or Dixon tests (see Chapter 11), but removal of statistical outliers is not acceptable without assignable cause.

GRAPHICAL REPRESENTATION OF DATA

Quantitative data are first Normalised by logarithmic transformation. In the AOAC case, the following equation is used: $Y = \log_{10}(x + 0.1f)$, where x is the colony count (cfu/unit) and f is the lowest reportable number of cfu/unit for the alternative method. If, for example, the smallest reportable result of a direct plating method is 1 cfu/g, then $f = 1$. The addition of the relatively small factor '$0.1f$' allows for inclusion of 'zero' results in the statistical analyses. The data for each food matrix are then graphed with the alternative method results on the y-axis and the reference method results on the x-axis. Linear regression is performed to determine the slope and linear correlation coefficient of the line. This is most easily done using a program such as Microsoft Excel. Fig. 13.3 shows an example of pooled raw meat data from a total aerobic count method (Kodaka, 2004) graphed against the reference standard plate count method (Official Method 966.23 in AOAC, 2015). Visual inspection will reveal any obvious outliers.

PERFORMANCE PARAMETERS

For quantitative methods, the performance parameters are repeatability, reproducibility and difference of means in both AOAC and ISO studies. This allows comparison of the method variability at different concentrations. In addition, ISO 16140-2 (Anon, 2015a) includes an accuracy profile study for analysis of bias and the beta expectation tolerance interval (β-ETI). For each contamination level of each food

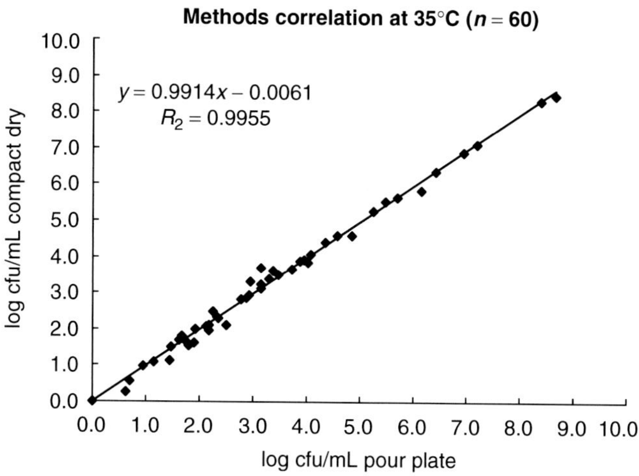

FIGURE 13.3 Graph of Pooled Raw Meat Data From the Validation of the Nissui Compact Dry TC

Reproduced from Kodaka (2004), by permission of AOAC International

in a single laboratory, the mean of the $\log_{10}$-transformed values is calculated and the standard deviation, S_r, determined. From collaborative study data, the mean of the $\log_{10}$-transformed data across all laboratories and the standard deviation, S_R, are calculated for each contamination level of each food.

Difference of means—AOAC

The mean $\log_{10}$ values for the alternative and reference methods are compared for each lot or contamination level of each matrix in the AOAC study (AOAC, 2012a). The difference in mean values is calculated as the mean of the alternative method minus the mean of the reference method and the 95% CI of the difference is determined. A CI outside the range (−0.5 to 0.5), meaning the true mean results of the two methods differ by more than 0.5 $\log_{10}$ units, has been proposed as an acceptance criterion based on expert opinion (Nelson et al., 2013). An Excel spreadsheet calculator is available to aid in the calculations (Example 13.3; LaBudde, 2010).

Accuracy profile- ISO 16140–2

The accuracy profile study follows a very similar study design to the AOAC difference of means using artificially contaminated samples (Anon, 2015a). Six samples per food type are prepared: two at a low level, two at an intermediate level and two at a high level of contamination. The samples can be the same matrix or two different matrices within the same food type. For each sample, five replicate test portions are tested by the alternative and reference methods. All data are Normalised by $\log_{10}$

transformation. The accuracy profile starts with the same premise as the AOAC difference of means in that the alternative method results should not differ from the reference method results by more than 0.5 $\log_{10}$ units. Therefore, an AL is set at ± 0.5 $\log_{10}$ units around the reference method value. In this case, however, it is the median of replicates, not the mean value, that is used for comparison, and bias (or lack of trueness) is calculated as the median result for the reference method minus the median for the alternative method for each sample: $B_i = Y_i - X_i$, where B_i is the bias of the ith sample, Y_i is the reference method median result for the ith sample and X_i is the alternative method median result for the ith sample. A beta expectation tolerance interval (β-ETI) is then calculated for the alternative method (Gutman, 1970;

EXAMPLE 13.3 AOAC QUANTITATIVE METHOD VALIDATION

In this example, the Compact Dry EC method (Nissui Pharmaceuticals Ltd) for enumeration of *E. coli* and coliforms was compared to two reference methods, ISO 16649-2 for *E. coli* and ISO 4832 for coliforms (Mizuochi et al., 2016). Four matrices were examined in this single-laboratory study: pre-packaged cooked sliced chicken, pre-washed bagged shredded iceberg lettuce, frozen cod fillets and instant non-fat dry milk (NFDM). Each matrix was artificially inoculated as needed with *E. coli* and/or a non–*E. coli* coliform and analysed at five contamination levels (including the unaltered matrix, level 1). Five replicate 10-g test portions from each contamination level were homogenised in 90-mL Maximum Recovery Diluent (MRD), followed by subsequent serial dilutions in MRD. In this paired study, each dilution was plated in duplicate on Compact Dry EC and according to ISO 16649-2 and ISO 4832 and each method was carried out according to the method instructions. The results for *E. coli* are presented in Table 13.6 and for total coliforms in Table 13.7. Using the Grubb's test, three outliers were detected in the Compact Dry EC data and one outlier was detected in the ISO 16649-2 data. No justifiable causes were found for these outliers, so they were not removed from the statistical analyses.

The acceptance criterion applied was that the confidence intervals on the difference between mean values must be within (-0.5, 0.5), in order to demonstrate with at least a 95% confidence that the true values for each method do not differ by more than 0.5 $\log_{10}$ (Nelson et al., 2013). For *E. coli*, there were four instances where the confidence interval was outside the acceptance criterion. These were level 2 of cooked chicken, levels 2 and 5 of lettuce, and level 4 of NFDM. For level 2 of cooked chicken and lettuce, there were only a few colonies recovered on the plates for the candidate and/or reference method, so these levels are not representative of the true performance of the methods. At level 5 of lettuce the Compact Dry EC recovered more colonies than the reference method, and at level 4 of NFDM the reference method recovered more colonies than the Compact Dry. Thus, no systematic bias was observed. For coliforms, there were six instances of confidence limits exceeding the criterion, level 2 of cooked chicken, levels 2 and 4 of lettuce, level 1 of cod, and levels 1 and 2 of NFDM. For level 2 of chicken, and level 1 of cod and NFDM, only a few colonies were recovered by one or both methods, so again these levels are not indicative of the performance of the methods. In each of the remaining 3 cases, the colony counts varied by more than 10-fold within the 5-replicate data sets, as can be seen by the unusually high standard deviations for both methods. This indicates that the inoculated matrix in those instances were not homogeneous prior to removing the test portions for analysis. As with the *E. coli* results, there was no observed systematic bias. For *E. coli* and total coliforms, the squared correlation coefficients were all >0.92. Based on these results, pre-packaged cooked sliced chicken, pre-washed bagged shredded iceberg lettuce, frozen cod fillets and instant NFDM were added to the PTM certification of the Compact Dry EC method.

Table 13.6 AOAC Single Laboratory Quantitative Matrix Study: Compact Dry EC Versus ISO 16649-2 – *E. coli*

		Compact Dry EC		ISO 16649-2		Difference of	95% CI		
Matrix	**Contamination Level**	**Mean**[a]	s_r[b]	**Mean**	s_r	**Means**[c]	**LCL**[d]	**UCL**[e]	r^2[f]
Cooked chicken	1	0.000	0.000	0.000	0.000	0.000	0.000	0.000	0.95
	2	0.577	0.617	0.341	0.554	0.236	−0.336	0.809	
	3	2.470	0.126	2.312	0.184	0.158	−0.015	0.300	
	4	3.511	0.111	3.340	0.283	0.172	−0.025	0.318	
	5	4.473	0.147	4.310	0.207	0.163	0.020	0.305	
Lettuce	1	0.000	0.000	0.000	0.000	0.000	0.000	0.000	0.94
	2	0.754	0.663	0.521	0.549	0.233	−0.374	0.841	
	3	2.676	0.303	2.558	0.167	0.118	−0.067	0.304	
	4	3.737	0.404	3.559	0.300	0.178	0.034	0.322	
	5	4.635	0.429	4.273	0.275	0.362	−0.124	0.600	
Frozen cod	1	0.000	0.000	0.000	0.000	0.000	0.000	0.000	0.99
	2	1.833	0.260	1.821	0.332	0.012	−0.271	0.294	
	3	2.832	0.135	2.827	0.101	0.005	−0.085	0.095	
	4	3.796	0.215	3.859	0.143	−0.063	−0.133	0.006	
	5	4.845	0.189	4.749	0.222	0.095	0.005	0.185	
Instant NFDM[g]	1	0.000	0.000	0.000	0.000	0.000	0.000	0.000	0.99
	2	2.837	0.307	2.838	0.218	−0.001	−0.146	0.144	
	3	3.520	0.139	3.551	0.182	−0.030	−0.111	0.050	
	4	4.201	0.028	4.564	0.217	−0.363	−0.519	−0.207	
	5	5.561	0.267	5.521	0.252	0.040	−0.069	0.149	

[a]*Mean of five replicate portions, plated in duplicate, after logarithmic transformation:* $log_{10}[cfu/g + (0.1)f]$.
[b]*Repeatability standard deviation.*
[c]*Difference of means between the candidate and reference methods, expressed as candidate minus reference.*
[d]*95% lower CL for difference of means.*
[e]*95% upper CL for difference of means.*
[f]*Square of correlation coefficient.*
[g]*Non-fat dry milk.*
Reproduced with permission from Mizuochi, S., Nelson, M., Baylis, C., Green, B., Jewell, K., Monadjemi, F., Chen, Y., Salfinger, Y., Fernandez, M.C., 2016. Matrix extension study: validation of the compact dry EC method for enumeration of Escherichia coli *and non-*E. coli *coliform bacteria in selected foods. J. AOAC Int., fast track publication, DOI: http://dx.doi.org/10.5740/jaoacint.15-0268 by permission of IAOAC International and HyServe GmbH.*

Table 13.7 AOAC Single Laboratory Quantitative Matrix Study: Compact Dry EC Versus ISO 4832 – Coliforms

Matrix	Contamination Level	Compact Dry EC		ISO 4832		Difference of Means[c]	95% CI		
		Mean[a]	s_r[b]	Mean	s_r		LCL[d]	UCL[e]	r^2[f]
Cooked chicken	1	0.000	0.000	0.000	0.000	0.000	0.000	0.000	0.98
	2	0.577	0.617	0.774	0.552	−0.197	−0.525	0.131	
	3	2.473	0.123	2.448	0.134	0.026	−0.017	0.068	
	4	3.514	0.112	3.428	0.248	0.085	−0.038	0.209	
	5	4.473	0.147	4.353	0.133	0.120	0.013	0.253	
Lettuce	1	0.000	0.000	0.000	0.000	0.000	0.000	0.000	0.94
	2	1.700	0.829	2.216	0.531	−0.756	−1.057	0.025	
	3	3.813	0.197	4.026	0.153	−0.212	−0.287	−0.138	
	4	4.635	0.429	4.241	0.202	0.394	0.058	0.730	
	5	5.053	0.111	4.955	0.126	0.098	−0.019	0.177	
Frozen cod	1	0.945	0.516	1.748	0.296	−0.802	−1.078	−0.527	0.96
	2	2.042	0.233	2.169	0.180	−0.127	−0.262	0.008	
	3	3.030	0.130	3.098	0.116	−0.068	−0.176	0.039	
	4	4.010	0.193	4.090	0.135	−0.080	−0.150	−0.010	
	5	5.091	0.146	5.097	0.179	−0.006	−0.056	0.045	
Instant NFDM[g]	1	0.653	0.568	0.000	0.000	0.653	−0.246	1.060	0.92
	2	1.797	1.058	2.041	0.735	−0.244	−0.690	0.202	
	3	2.661	0.119	2.719	0.236	−0.058	−0.190	0.074	
	4	3.960	0.072	3.972	0.079	−0.012	−0.057	0.033	
	5	4.648	0.225	4.951	0.177	−0.303	−0.389	−0.218	

[a] *Mean of five replicate portions, plated in duplicate, after logarithmic transformation:* $log_{10}[cfu/g + (0.1)f]$.
[b] *Repeatability standard deviation.*
[c] *Difference of means between the candidate and reference methods, expressed as candidate minus reference.*
[d] *95% lower CL for difference of means.*
[e] *95% upper CL for difference of means.*
[f] *Square of correlation coefficient.*
[g] *Non-fat dry milk.*

Reproduced with permission from Mizuochi, S., Nelson, M., Baylis, C., Green, B., Jewell, K., Monadjemi, F., Chen, Y., Salfinger, Y., Fernandez, M.C., 2016. Matrix extension study: validation of the compact dry EC method for enumeration of Escherichia coli *and non-E. coli coliform bacteria in selected foods. J. AOAC Int., fast track publication, DOI: http://dx.doi.org/10.5740/jaoacint.15-0268 by permission of IAOAC International and HyServe GmbH.*

Feinberg et al., 2009). With β set at 80%, the β-ETI is the interval where 80% of future results are expected to fall. To calculate β-ETI for each sample, the combined standard deviation of the alternative method across all samples must be determined. First, calculate the standard deviation of the alternative method for each sample i as follows:

$$S_{\text{alt},i} = \sqrt{\frac{1}{n-1}\sum_{j=1}^{n}(y_{ij} - \overline{y}_i)^2}$$

where n is the total number of replicates j of sample i, y_{ij} is the alternative method result for replicate j of sample i and $\overline{y}_i$ is the mean of alternate method results for sample i. Then the combined standard deviation for the alternate method is calculated as follows:

$$S_{\text{alt}} = \frac{1}{q}\sum_{i=1}^{q} S^2_{\text{alt},j}$$

The β-ETI limits for each sample are then calculated as follows:

$$B_i \pm T \times S_{\text{alt}}\sqrt{1+\frac{1}{n}}$$

where the value of T is obtained for a probability of $0.5(1 - \beta) = 0.10$ [ie, $0.5(1 - 0.80)$] from the Student's t distribution with $q(n - 1)$ degrees of freedom. A graph is then prepared of bias (B_i) versus reference method median for each sample i. The upper and lower limits of the β-ETI are plotted for each sample and the bias points, upper limits of the β-ETI and lower limits of the β-ETI are connected by straight lines. Once graphed, it is easy to see whether the tolerance intervals are contained within the AL. If this is the case, the methods are considered to be equivalent for the food categories tested. If the upper or lower limits of the β-ETI for any of the samples exceed the AL and the $S_{\text{ref}} > 0.125$, then new ALs are determined based on the reference method standard deviation as follows:

$$\text{AL}_S = 4 \times S_{\text{ref}}$$

If the derived tolerance limits are contained within the limits ($+\text{AL}_S$, $-\text{AL}_S$), then the methods are considered equivalent for the food categories tested. If the tolerance limits fall outside ($+\text{AL}_S$, $-\text{AL}_S$), then a root cause analysis is recommended to determine corrective action (Example 13.4).

In the multi-laboratory study, β-ETI is determined in a slightly different way to take into consideration the reproducibility conditions under which the study is conducted as compared to the repeatability conditions of the single-laboratory study, but the overall concept is the same. Accuracy profile calculation spreadsheets are available for free download for the method comparison study and the inter-laboratory study (Anon, 2015c).

EXAMPLE 13.4 ISO 16140-2 QUANTITATIVE METHOD VALIDATION

In a MicroVal single-laboratory study, the Compact Dry ETC method for determination of *enterococci* in a variety of foods was compared to the Nordic Committee on Food Analysis (NMKL) method 68 for *Enterococcus* (NMKL, 2011). Five food categories were validated: dairy products, fresh produce, raw meat, ready-to-eat (RTE) foods and multi-component foods. In the accuracy profile study, 10 foods consisting of 2 foods in each category were selected. These included chilled custard, chilled pasteurised whipping cream, pre-washed bagged fresh parsley, pre-washed bagged iceberg lettuce, raw steak, frozen ground beef patties, tuna paté, frozen cooked and peeled prawns, chilled cheddar and bacon deli pasta salad, and pre-packaged egg, cress and mayonnaise sandwich on wheat germ bread. The foods in each category were inoculated with a different species of *Enterococcus* at each of 3 levels: low (50–500 cfu/g), medium (100–5000 cfu/g) and high (1000–50,000 cfu/g). Five 10-g replicate test portions from each level were analysed by the Compact Dry ETC and NMKL-68 methods in a paired study. After $\log_{10}$ transformation, the

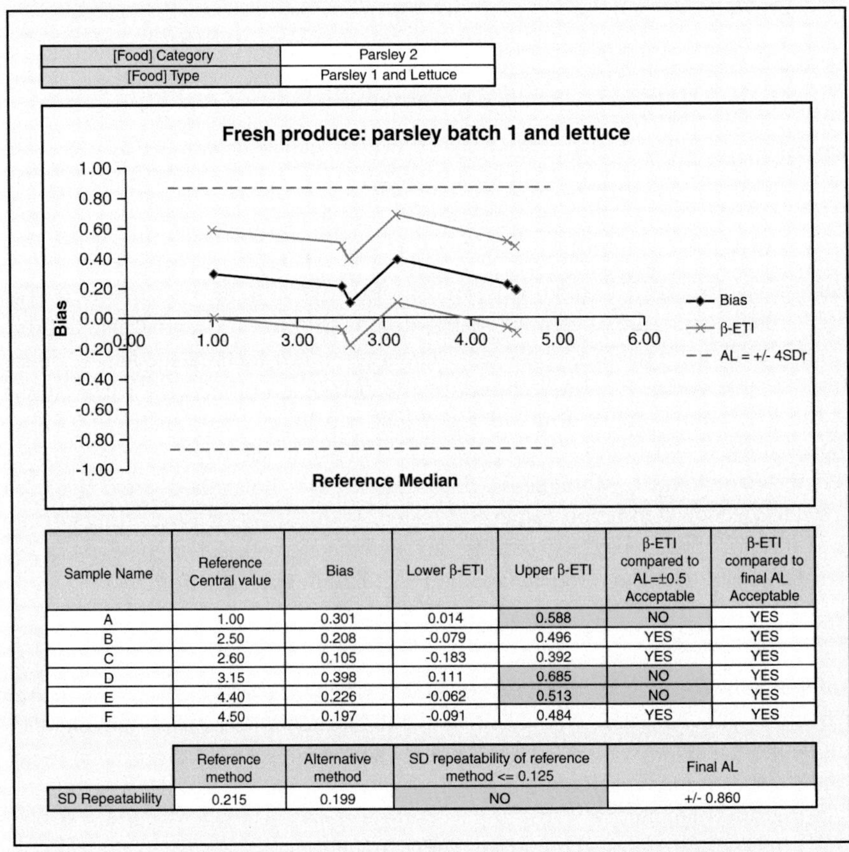

[Food] Category	Parsley 2
[Food] Type	Parsley 1 and Lettuce

Sample Name	Reference Central value	Bias	Lower β-ETI	Upper β-ETI	β-ETI compared to AL=±0.5 Acceptable	β-ETI compared to final AL Acceptable
A	1.00	0.301	0.014	0.588	NO	YES
B	2.50	0.208	-0.079	0.496	YES	YES
C	2.60	0.105	-0.183	0.392	YES	YES
D	3.15	0.398	0.111	0.685	NO	YES
E	4.40	0.226	-0.062	0.513	NO	YES
F	4.50	0.197	-0.091	0.484	YES	YES

	Reference method	Alternative method	SD repeatability of reference method <= 0.125	Final AL
SD Repeatability	0.215	0.199	NO	+/- 0.860

(A)

FIGURE 13.4

Two examples of accuracy profile graphs for Nissui Compact Dry ETC validation: (A) fresh produce; (B) fish products. For further detail see Example 13.4

EN ISO 16140-2 validation study of Compact Dry ETC for enumeration of Enterococci in foods, unpublished report. Figures 1f, 2, 6a-e, and Table 4. By permission of HyServe GmbH

median values for each method at each contamination level for each matrix were calculated. The use of the median rather than the mean eliminates the need for removal of outliers, so no outlier analysis was needed. The difference of medians, or bias, was determined at each level. Using the accuracy profile calculation spreadsheet, the beta expected tolerance interval (β-ETI) limits were calculated and the data were graphed as shown in Fig. 13.4. For the raw meat, RTE foods and multi-component foods categories, the tolerance intervals all fell within the (-0.5, 0.5) AL, so no further analysis was needed. The remaining two categories yielded tolerance intervals that exceeded the AL of (-0.5, 0.5). In these cases, S_{ref} was >0.125, allowing for the calculation of a new AL_S as $4 \times S_{ref}$. For dairy, this was $4 \times 0.311 = 1.244$ and no tolerance intervals exceeded the AL_S. For leafy greens, $AL_S = 4 \times 0.215 = 0.860$ and no tolerance intervals exceeded this new limit. Thus, the Compact Dry ETC passed the accuracy profile requirements for all five categories.

In the relative trueness study, 78 artificially inoculated samples were analysed. An additional 30 samples were screened for natural contamination, but were found to be negative and therefore were not utilised. Scatter plots of the Compact Dry ETC versus NMKL-68 $\log_{10}$-transformed data were constructed for each category and for all categories combined. The combined plot is shown in Fig. 13.5. Two obvious outliers can be seen, but as no assignable cause could be attributed, these outliers were not removed. Using the relative trueness spreadsheet (Anon, 2015c), the method means and difference of results D for each sample were determined. In addition, the mean difference $\bar{D}$, standard deviation

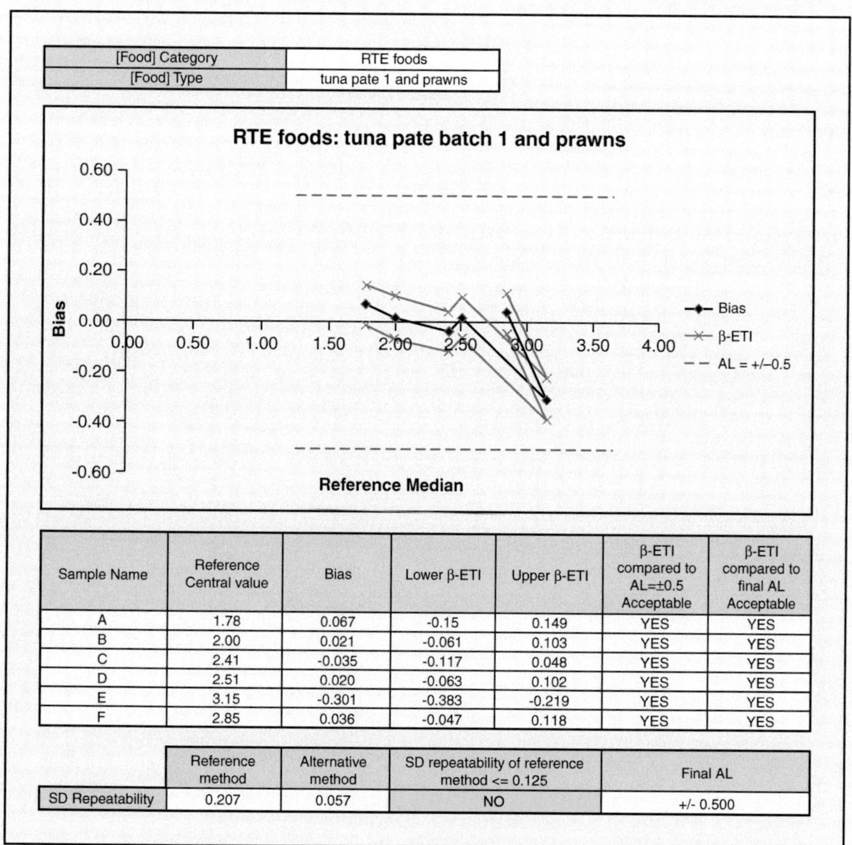

[Food] Category	RTE foods
[Food] Type	tuna pate 1 and prawns

Sample Name	Reference Central value	Bias	Lower β-ETI	Upper β-ETI	β-ETI compared to AL=±0.5 Acceptable	β-ETI compared to final AL Acceptable
A	1.78	0.067	-0.15	0.149	YES	YES
B	2.00	0.021	-0.061	0.103	YES	YES
C	2.41	-0.035	-0.117	0.048	YES	YES
D	2.51	0.020	-0.063	0.102	YES	YES
E	3.15	-0.301	-0.383	-0.219	YES	YES
F	2.85	0.036	-0.047	0.118	YES	YES

	Reference method	Alternative method	SD repeatability of reference method <= 0.125	Final AL
SD Repeatability	0.207	0.057	NO	+/- 0.500

(B)

FIGURE 13.4 (*Cont.*)

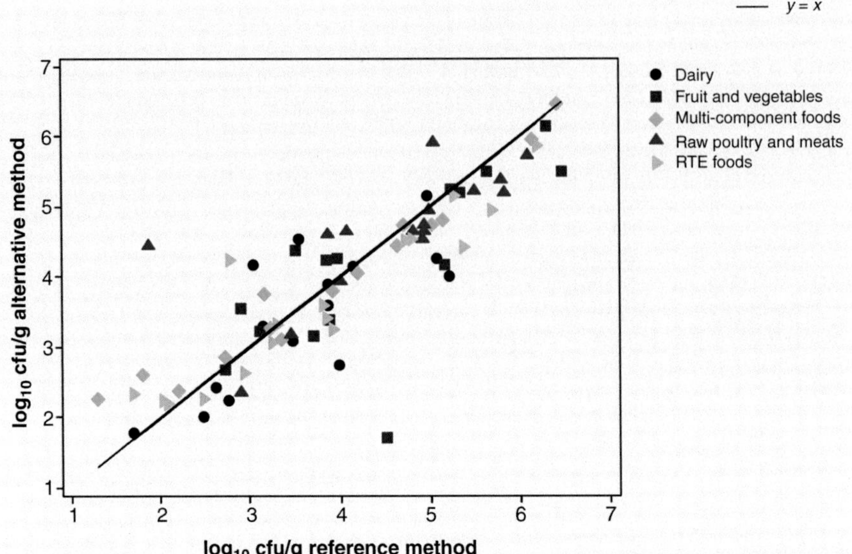

FIGURE 13.5 Scatter Plot of Nissui Compact Dry ETC Versus NMKL-68 Results for All Categories Combined

EN ISO 16140-2 validation study of Compact Dry ETC for enumeration of Enterococci in foods, unpublished report. Figures 1f, 2, 6a-e, and Table 4. By permission of HyServe GmbH

Table 13.8 Mean Difference ($\bar{D}$), Standard Deviation of Differences (S_D) and Confidence Limits on $\bar{D}$ for Compact Dry ETC Trueness Study

Category	N	$\bar{D}$	S_D	95% LCL[a]	95% UCL[b]
Dairy	15	−0.184	0.669	−1.665	1.297
Fruit and vegetables	15	−0.276	0.876	−2.215	1.664
Multi-component foods	15	0.076	0.395	−0.798	0.951
Raw poultry and meats	16	0.064	0.801	−1.696	1.824
RTE foods	17	−0.142	0.553	−1.349	1.065
All categories	78	−0.092	0.676	−1.446	1.263

[a]*Lower 95% CL.*
[b]*Upper 95% CL.*
EN ISO 16140-2 validation study of Compact Dry ETC for enumeration of Enterococci in foods, unpublished report. Figures 1f, 2, 6a-e, and Table 4. By permission of HyServe GmbH

of differences S_D and 95% confidence interval on $\bar{D}$ were calculated by category and for all categories combined (see Table 13.8). The individual sample differences were plotted against the mean values in a Bland–Altman plot (Fig. 13.6). Three data points out of 78 fell outside the confidence limits, including the 2 outliers previously noted. Each was from a different food category. As this does not exceed the 1 in 20 allowed, and no trend in food category was observed, the data were considered satisfactory and the Compact Dry ETC passed the trueness study requirements for all five categories.

Data by permission of HyServe GmbH

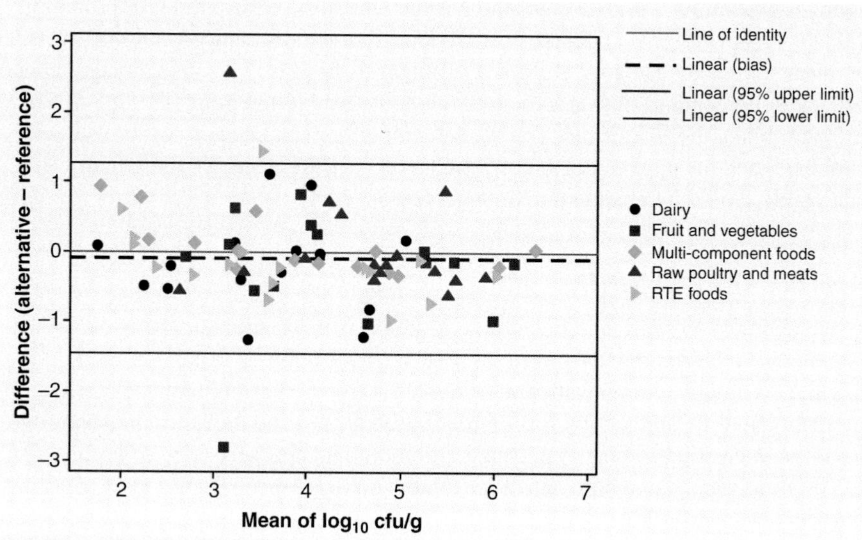

FIGURE 13.6 Bland–Altman Plot for Bias Assessment of All Food Categories for the Compact Dry ETC and NMKL-68 Method Comparison

EN ISO 16140-2 validation study of Compact Dry ETC for enumeration of Enterococci in foods, unpublished report. Figures 1f, 2, 6a-e, and Table 4. By permission of HyServe GmbH

Relative trueness - ISO 16140–2

This study is conducted using a wide variety of food items, food types and food categories, depending on the scope of the alternative method (Anon, 2015a). For each food category, three food types are selected and for each food type, five food samples (not necessarily the same food item) are tested. Naturally contaminated food samples are used as much as possible, but artificial contamination can be performed if needed. The contamination levels should represent the natural variation in each food type or category for the analyte.

The Bland and Altman (1986) method is used to analyse the data. For each category, and for all categories combined, plot for each sample the $\log_{10}$-transformed count for the alternative method on the *y*-axis against the $\log_{10}$-transformed count for the reference method on the *x*-axis. Draw the line of identity, representing the line if the alternative method and reference method results were identical. The relative trueness, or agreement, between methods will be visually apparent. Next, tabulate the paired data and for each sample, determine the mean result [(alternative result + reference result)/2] as well as the difference D (alternative result minus reference result). For each category and for all categories combined, calculate the mean difference $\bar{D}$ and the standard deviation of differences S_D. The CLs on the mean difference are then calculated as follows:

$$\bar{D} = T \times S_D \sqrt{1 + \frac{1}{n}}$$

where n is the number of data pairs and T is the Student's t distribution for probability at $0.5(1 - \beta)$ with $(n - 1)$ degrees of freedom. Plot the difference values (D) versus the mean values by category and for all categories to again visualise the agreement (or lack of agreement) between the two methods. Add to the graph the line of identity (zero bias), the line of bias (mean difference for each category or all categories combined) and the upper and lower 95% CLs on the difference. No more than 1 in 20 data points should lie outside the limits (Example 13.4).

FUTURE DIRECTIONS

As we continue to share and debate ideas amongst international experts and stakeholders, we will continue to move closer to harmonisation of validation methodologies and eventually harmonisation of reference methods. This will reduce barriers for international trade and aid in cross-border outbreak investigations. In the meantime, it will be interesting to begin analysing common data sets according to both AOAC and ISO statistics to compare outcomes relative to the respective statistical measures and acceptance criteria.

REFERENCES

Anon, 2015a. Microbiology of the Food Chain – Method Validation – Part 2: Protocol for the Validation of Alternative (Proprietary) Methods Against a Reference Method. ISO/FDIS 16140-2:2015. International Organisation for Standardization, Geneva (final committee draft).

Anon, 2015b. ICS 07.100.30 Food Microbiology. International Organisation for Standardization, Geneva. <http://www.iso.org/iso/iso_catalogue/catalogue_ics/catalogue_ics_browse.htm?ICS1=07&ICS2=100&ICS3=30&development=on> (accessed 30.11.15).

Anon, 2015c. Excel Spreadsheets for Statistical Analyses in ISO 16140-2:2015. <http://standards.iso.org/iso/16140> (accessed 26.11.15).

AOAC, 2011. AOAC SMPR 2010.003, Standard Method Performance Requirements for Polymerase Chain Reaction (PCR) Methods for Detection of *Bacillus anthracis* in Aerosol Collection Filters and/or Liquids. J. AOAC Int. 94, 1347–1351, <http://www.aoac.org/imis15_prod/AOAC_Docs/SMPRs/SMPR_2010.003.pdf> (accessed 30.11.15).

AOAC, 2012a. AOAC International Methods Committee Guidelines for Validation of Microbiological Methods for Food and Environmental Surfaces, Official Methods of Analysis, Appendix J. <http://www.aoac.org/imis15_prod/AOAC_Docs/StandardsDevelopment/eoma_appendix_j.pdf> (accessed 26.11.15).

AOAC, 2012b. AOAC International Methods Committee Guidelines for Validation of Biological Threat Agent Methods and/or Procedures, Official Methods of Analysis, Appendix I. <http://www.aoac.org/imis15_prod/AOAC_Docs/StandardsDevelopment/eoma_appendix_i.pdf> (accessed 26.11.15).

AOAC, 2015. Official Methods of Analysis. AOAC International, Rockville, MD. <http://eoma.aoac.org> (accessed 26.11.15).

APHA, 1985. Standard Methods for the Examination of Dairy Products. American Public Health Association, Washington, DC.

Bland, J.M., Altman, D.G., 1986. Statistical methods for assessing agreement between two methods of clinical measurement. Lancet i, 307–310.

Blodgett, R., 2010. Most Probable Number Determination From Serial Dilutions, Bacteriological Analytical Manual, Appendix 2. <http://www.fda.gov/Food/FoodScienceResearch/LaboratoryMethods/ucm109656.htm> (accessed 26.11.15).

FDA, 2015. Bacteriological Analytical Manual. <http://www.fda.gov/Food/FoodScienceResearch/LaboratoryMethods/ucm114664.htm> (accessed 26.11.15).

Feinberg, M., Sohier, D., David, J.F., 2009. Validation of an alternative method for counting Enterobacteriaceae in foods based on accuracy profile. J. AOAC Int. 92, 527–537.

Gutman, I., 1970. Statistical Tolerance Regions: Classical and Bayesian. Hafner Publishing, Darian, CT.

Hadfield, T., Ryan, V., Spaulding, U.K., Clemens, K.M., Ota, I.M., Brunelle, S.L., 2013. RAZOR™ EX Anthrax Air Detection System for detection of *Bacillus anthracis* spores from aerosol collection samples: collaborative study. J. AOAC Int. 96, 392–398.

ISPAM, 2015. AOAC International Stakeholder Panel on Alternative Methods. <http://www.aoac.org/iMIS15_Prod/AOAC/SD/ISPAM/AOAC_Member/SH/ISPAMCF/ISPAM_M.aspx?hkey=bb6632fb-9f4e-4c36-b065-78b13ef9b2fd> (accessed 26.11.15).

Jarvis, B., Wilrich, C., Wilrich, P.-T., 2010. Reconsideration of the derivation of most probable numbers, their standard deviations, confidence bounds and rarity values. J. Appl. Microbiol. 109, 1660–1667.

Kodaka, H., 2004. Nissui Pharmaceutical and Neogen kits granted PTM status. Inside Lab. Manage. 8 (4), 19–22.

LaBudde, R., 2009. MPN Calculator. Least Cost Formulations Ltd., Virginia Beach, VA. <http://www.lcfltd.com/customer/LCFMPNCalculator.exe> (accessed 26.11.15).

LaBudde, R., 2010. Paired Method Analysis for Micro Testing, Version 1.0. Least Cost Formulations Ltd., Virginia Beach, VA. <http://lcfltd.com/AOAC/paired-method-analysis-for-micro.xlsx> (accessed 30.11.15).

LaBudde, R., 2013. POD Calculator, Version 2.3. Least Cost Formulations Ltd., Virginia Beach, VA. <http://lcfltd.com/AOAC/aoac-binary-v2-3.xls> (accessed 26.11.15).

Mărgăritescu, I., Wilrich, P.Th., 2013. Determination of the RLOD of a qualitative microbiological measurement method with respect to a reference measurement method. J. AOAC Int. 96, 1086–1091.

Mizuochi, S., Nelson, M., Baylis, C., Green, B., Jewell, K., Monadjemi, F., Chen, Y., Salfinger, Y., Fernandez, M.C., 2016. Matrix extension study: validation of the compact dry EC method for enumeration of *Escherichia coli* and non-*E. coli* coliform bacteria in selected foods. J. AOAC Int., fast track publication, DOI: http://dx.doi.org/10.5740/jaoacint.15-0268.

Nelson, M.T., LaBudde, R.A., Tomaisino, S.F., Pines, R.M., 2013. Comparison of 3M™ Petrifilm™ aerobic count plates to standard plating methodology for use with AOAC antimicrobial efficacy methods 955.14, 955.15, 964.02, and 966.04 as an alternative enumeration procedure: collaborative study. J AOAC Int. 96, 717–722.

NMKL, 2011. NMKL Method 68, 5th Edition: *Enterococcus*. Determination in Foods and Feeds. <http://www.nmkl.org/index.php?option=com_content&view=article&id=37&Itemid=366&lang=en> (accessed 30.11.15).

SPADA, 2015. AOAC International Stakeholder Panel on Agent Detection Assays, Rockville, MD. <http://www.aoac.org/iMIS15_Prod/AOAC_Member/SH/SPADACF/SPADAM.aspx?WebsiteKey=2e25ab5a-1f6d-4d78-a498-19b9763d11b4&hkey=ba02632d-abd0-4a54-be1e-60873eb13a1f&CCO=1> (accessed 30.11.15).

USDA, 2015. Microbiology Laboratory Guidebook. <http://www.fsis.usda.gov/wps/portal/fsis/topics/science/laboratories-and-procedures/guidebooks-and-methods/microbiology-laboratory-guidebook/microbiology-laboratory-guidebook> (accessed 26.11.15).

Wehling, P., LaBudde, R.A., Brunelle, S.L., Nelson, M.T., 2011. Probability of detection (POD) as a statistical model for the validation of qualitative methods. J. AOAC Int. 94, 335–347.

Wilrich, P.-Th., 2015. The precision of binary measurement methods. Beran, J., Feng, Y., Hebbel, H. (Eds.), Empirical Economic and Financial Research: Advanced Studies in Theoretical and Applied Econometrics, vol. 48, Springer International, Switzerland, pp. 223–235.

Wilrich, C., Wilrich, P.-Th., 2009. Estimation of the POD function and the LOD of a qualitative microbiological measurement method. J. AOAC Int. 92, 1763–1772.

CHAPTER

Risk assessment and microbiological criteria for foods

14

The objective of ensuring safe food for the world's constantly growing population has been a major pre-occupation of national and international organisations [eg, WHO/FAO Codex Committee on Food Hygiene (CCFH), ILSI, ICMSF] and professional and trade bodies over many years. Yet, in deprived areas of the world, there remains a basic need to ensure even a reliable food supply. In all countries, especially in developed consumer-oriented countries, there is a perceived need to ensure that foods do not present an unacceptable risk to the health and well-being of the consumer. This can lead to a clash in priorities: foods exported from third-world countries support the national economy, but the foods are required to comply with the rules imposed by international trade and import regulations of developed countries. Meanwhile, the indigenous population of the producer country often consume foods that could never meet those same criteria. In third-world countries approximately 4.5 billion people are exposed to high levels of aflatoxins in foods; for instance, in Malawi, 73% of samples of groundnut are contaminated with aflatoxins at levels much greater than the EU import limit. Attempts to improve food quality and safety are important for all consumers – but if you are starving, the importance of quality and safety may be less important than having enough food for your family. In developed countries, people who constantly demand ever-increasing quality and safety, ignore this ethical paradigm; we also waste an enormous amount of food – according to Anon (2010) approximately 89 million tonnes of food was wasted across Europe in 2006, with consequential environmental effects!

Following the publication by Accum (1820) of his *Treatise on Adulteration of Food and Culinary Poisons* and subsequent work in the middle 19th century by the Lancet Analytical Sanitary Commission and other bodies (see Amos, 1960) the need to improve food safety in the United Kingdom led to the introduction of food legislation concerned with diverse areas such as food composition, chemical contaminants, food additives and, later, microbiological contamination. For the past few decades legislation has been concerned primarily with seeking to control aspects of food production that are necessary to assure the safety of foods at all stages from 'farm to fork'. In the area of food microbiology, such legislative control has been targeted at improving the safety and quality of foods processed by the food and catering industries since it is rightly perceived that large-scale production impacts on many more consumers than does traditional domestic production. Yet it is often mishandling of

Statistical Aspects of the Microbiological Examination of Foods. http://dx.doi.org/10.1016/B978-0-12-803973-1.00014-0

foods by small producers, by caterers and by the consumer that presents the greatest risk to consumer well-being.

Throughout the world, food legislation imposes a duty of care and responsibility for the safety and quality of foods on business organisations involved in their procurement, processing, distribution and retail sale. For instance, in Europe, the basic premise of food law is enshrined in the General Regulation on Food Safety (Anon, 2002a), with subsidiary legislation on key issues, including microbiological aspects of safety. An important facet is the requirement for risk assessment by governments to ensure a legislative framework within which all food producers and suppliers should operate.

RISK ASSESSMENT AND FOOD SAFETY OBJECTIVES

Modern approaches to food safety include the identification of actual, or potential, hazards from microbial contamination, assessing the risk that such contamination may cause disease in the consumer and then seeking to employ processes that will control and minimise such risks. 'Hazard' is defined as something that has the *potential* to cause harm, for instance, the contamination of food by pathogenic bacteria. 'Risk' is defined as the *likelihood* of harm in a specific situation, for instance, consumption of food contaminated with specific pathogens and/or their toxins. *Risk assessment* of foods is therefore concerned with assessing the potential risk that consumption of a food may harm consumers. As is well demonstrated by ICMSF (2002, 2011), risk assessment requires an understanding of both microbial contamination *per se* and the food process operations and domestic food handling practices that may reduce or increase the risk from a defined hazard for a particular group of vulnerable consumers (infants, children, the aged, the immuno-compromised, etc.). Guidelines on hazard and risk assessment in foods are published by WHO/FAO (Anon, 2003b, 2014, 2016).

Rather than seeking to control food safety on an *ad hoc* basis, specific objectives should be set to ensure, so far as is practicable, that food does not threaten the health of consumers. The phytosanitary measures (SPS) of the World Trade Organisation (WTO) require member states to ensure that their sanitary and phytosanitary requirements are based on scientific principles whilst not unnecessarily restricting international trade. This means that member countries must establish appropriate measures on the basis of the actual risks likely to occur; originally, this requirement was primarily for those risks arising from chemical contaminants. The concept of quantitative microbiological risk assessment was introduced in the 1990s following development of predictive models for growth and survival of microbial pathogens in foods based on fundamental studies of microbial growth and survival (Haas, 1983, 2002; Roberts and Jarvis, 1983; Baranyi and Roberts, 1994; Whiting, 1995). This approach is based on the application and extension of the concept of microbiological compositional analysis (Tuynenburg-Muys, 1975) and uses the wide availability and power of computers, often using dedicated microbial modelling software (Baranyi and

Tamplin, 2004). McKellar and Lu (2003) provide a useful overview to the topic of modelling microbial survival, growth and death in foods.

The WTO/SPS proposed the concept of an 'appropriate level of protection' (ALOP) defined as 'the level of protection deemed appropriate by the member (country) to establish a sanitary or phytosanitary measure to protect human, animal and plant life or health within its territory'. Subsequently, the CCFH developed consensus protocols for risk analysis of pathogens in food (eg, Anon, 2000a,b, 2002a) and produced a report on the 'Principles and Guidelines for the Conduct of Microbiological Risk Management' (Anon 2002b,c). A key output was the re-definition of the WTO/SPS definition, as 'ALOP refers to a level of protection of human health established for a food-borne pathogen'. However, there was little guidance on the nature of an ALOP or how it might be established. Several alternative approaches for setting an ALOP were debated including the concept of 'as-low-as reasonably achievable' (ALARA) based on the performance of the available risk management options.

All aspects of control require the definition of criteria for the 'disease burden' that public health can accept for a population, for instance, 'the number of cases per year per 100,000 head of population for a specific hazard in a specific food commodity'. But even this leaves much room for debate. Mossel and Drion (1979) proposed the concept of a 'lifetime tolerable risk' for botulinum and other toxins, but other objectives were more constrained in relation to both the time span and the defined population. So what is the tolerable risk to which any consumer should be exposed and over what timeframe? Is the 'population at risk' the total population, the most susceptible group(s) or only that proportion of the population who actually consume a particular food? Does ALOP include specific demographic groups of the population? Are related health concerns linked together (eg, all cases of *Salmonella* food poisoning) or considered only in relation to specific foods? What is the impact of alternative transmission routes, such as, transmission by food handlers; cross-contamination between foods due to poor handling and storage practices, and transmission of pathogens from food to consumers; or, even, person-to-person transmission? Is the risk timeframe a definable period (eg, a year) or a lifetime? Havelaar et al. (2004) proposed a definition for an ALOP as '*no more than x cases of acute gastroenteritis per 100,000 population per year associated with hazard Y and food Z*'. This ALOP concept provides a useful target metric for public health policy but is of limited usc in implementing food chain safety measures.

The ICMSF introduced the concept of 'food safety objectives' (FSO) that was adopted subsequently by CCFH as part of its microbiological risk management strategy. An FSO provides a metric to convert public health goals into parameters that can be controlled by food producers and monitored by government agencies. It is defined (ICMSF, 2002) as 'the maximum frequency and/or concentration of a microbial hazard in a food considered tolerable for consumer protection'. ICMSF (2002) notes that FSOs are 'typically expressions of concentrations of microorganisms or toxins at the moment of consumption'. Concentrations at earlier stages in the food chain are defined by performance criteria. Hence, an FSO seeks to take account of hazards arising in commercial processing and unpredictable effects associated with retail and domestic food storage and handling. By contrast, performance criteria relate to the requirement to control hazards at earlier stages of the food chain, that is, from primary (farm) sources to industrial production.

ICMSF (2002) used the following equation to describe the concept: $H_0 - \sum R + \sum I \leq \text{FSO}$. The terms in the equation (ie, the performance criteria) are as follows: the initial level of the specific hazard (H_0) associated with raw materials and ingredients, the cumulative decrease (or increase) in hazard level due to all processing factors (ΣR) and the cumulative increase in hazard as a consequence of post-process microbial growth (ΣI). The symbol $\leq$ implies that the cumulative effect should be less, or not greater, than the FSO expressed in terms of $\log_{10}$ units for a specific organism. Suppose, for instance, that the FSO for a specific pathogen in a food is considered to be not more than 0.1 organism/g at the time of consumption. Suppose further that the maximum initial contamination level is likely to be 100 organisms/g (2 $\log_{10}$ organisms/g) and the maximum reduction likely to be obtained by a combination of thermal and/or other processes is 3 $\log_{10}$ units. Then, to ensure that the FSO is not exceeded, the risk of re-growth between processing and consumption (ΣI) must be zero; substituting $H_0 = 2$, $\Sigma R = -3$ and FSO $= -1$, in the rewritten equation $\sum I = \text{FSO} + \sum R - H_0 = -1 + 3 - 2 = 0$. Thus conditions of storage and handling post-processing must be such as to prevent *any* growth of surviving organisms. However, if the initial level of contamination of the product (H_0) were 10 organisms/g (1 $\log_{10}$ organism/g), then a re-growth allowance would not be greater than 1 $\log_{10}$ unit: $\sum I = -1 + 3 - 1 = +1$.

Although such calculations are based on theoretical values that make no allowance for variations in microbial distribution within or between batches of food, they provide a simple way to demonstrate how an FSO can be used to assess risk for specific products and processes. Zwietering et al. (2010, 2016) describe how such calculations can be used to set and monitor an FSO and demonstrate how changes in process conditions would affect the acceptability of a process.

The FSO could be defined as 'the maximum likely level of hazard that is acceptable at the point of consumption' following the integration of several stages in food processing and distribution, based on knowledge of the microbial associations of foods, the processing hurdles (Leistner and Gould, 2001) and the likelihood of re-contamination and/or re-growth of organisms during subsequent storage and handling. Thus the FSO concept relates also to the use of the hazard analysis and critical control point (HACCP) concept for controlling the effectiveness of food processing operations.

For any manufacturing process, HACCP requires assessment of potential hazards and the identification and monitoring of control points (CCPs) that are critical for elimination or reduction of a hazard. Furthermore, the concept requires that each CCP will be monitored using simple, indirect methods to ensure the process is operating correctly and that the effectiveness of monitoring is verified by appropriate microbiological examination. HACCP is now required by law in most countries and forms part of good manufacturing practices (GMP) within a wider total quality management (TQM) approach that should be used within a business to ensure that all food products conform to criteria that define acceptable quality and safety standards.

These underlying themes imply that all persons responsible for production, distribution, sale and preparation of foods work together to ensure that potential hazards are identified and controlled effectively in order to minimise 'so far as is achievable' the risks to consumer health. This requires knowledge and experience of potential microbiological hazards and the likely effects of both acceptable and unacceptable practices at all stages from 'farm to fork'.

An FSO differs from a microbiological criterion (MC); it is not applicable to individual 'lots' and does not specify sampling plans. Rather an FSO is a statement of the level of control expected for a food processing operation that can be met by the proper application of other procedures and metrics including GMP, HACCP, performance, process, product and acceptance criteria (ICMSF, 2002). Ideally, an FSO should be quantitative and verifiable, though not necessarily by microbiological examination of food.

An FSO provides a means by which control authorities can communicate to industry what is expected for foods produced in properly managed operations, for instance, by specifying the frequency or concentration of a microbial hazard that should not be exceeded at the time of consumption. An FSO therefore provides a basis for the establishment of product criteria that can be used to assess whether an operation complies with a requirement to produce safe food.

FSOs should be established following *risk evaluation* by an expert panel using quantitative risk assessment. In all cases, the first step is the identification of hazards associated with specific foods based on epidemiological and other data. Next an *exposure assessment* is required to estimate the probable prevalence and levels of microbial contamination at the time of consumption. Such assessment requires information about the amount of product consumed by different categories of consumers, use of mathematical models that take account of the prevalence of the organism(s), the nature of the food processing operations, the probability for growth and survival of the relevant organisms in the food, both before and after processing, and the impact that food handling practices will have on the levels of organisms likely to be consumed (see, eg, Bassett et al., 2010; Jongenberger, 2012; Gonzales-Barron et al., 2014; Jongenberger et al., 2015).

Characterisation of the hazard (Anon, 2000b) requires assessment of the severity and duration of adverse effects resulting from individual exposure to a specific pathogen. A dose–response assessment provides a measure of the potential risk. The likelihood of exposure is dependent not only on the characteristics of specific strains of microbe and on the infective dose of the organisms, but also on the susceptibility of the host and the characteristics of the food that acts as the carrier for the organisms. *Risk characterisation* combines the information to produce a risk assessment that indicates the possible level of disease (eg, the number of cases per 100,000 people per year) likely to result from the given exposure. Risk characterisation needs to be validated by comparison with epidemiological and other data and should reflect the distribution of risk associated with the many facets of contamination, survival and growth of a specific organism in a food processed in a specific way. Bassett et al. (2010) provide examples (their Chapter 5) to illustrate how microbiological and epidemiological aspects can be taken into account in such an assessment.

By way of an example let us look at the worldwide situation of chicken meat contaminated with *Campylobacter* as the causal agent of a food-borne disease (Anon, 2009, 2013b) that is responsible for up to 9 million cases of illness per year in Europe alone. A recent UK report (Anon, 2015) showed that up to 80% of retail samples of chicken were contaminated with *Campylobacter* with up to 33% having more than 1000 cfu/g neck skin. In addition, *Campylobacter* were found on the outer packaging in up to 12.5% of samples tested, although the prevalence varied considerably. The latter point is important since it is believed that domestic cross-contamination is a major factor in many cases of the illness.

Earlier, Nauta et al. (2005) developed a risk assessment for *Campylobacter* exposure of the population of the Netherlands and concluded, *inter alia*, that it was not possible to provide any guarantee for provision of uncontaminated meat unless the chicken were irradiated. The European Food Safety Agency (Anon, 2011) estimated that cases of *Campylobacter*-associated illness across the EU created a disease burden costing €2.4 billion p.a. They suggested that a >50% reduction in prevalence of the disease could result from introduction of a MC requiring <1000 cfu *Campylobacter*/g of neck skin, but they also noted that at least 15% of all broilers would not comply with this criterion. Swart et al. (2013) advised that in the Netherlands process hygiene improvements, linked to establishment of an MC of <1000 cfu *Campylobacter*/g chicken neck skin, would reduce prevalence of the infection by two-thirds. Although the estimated cost to the industry would be €2 million/year, the forecast economic benefit to the country was €9 million/year by reducing the disease burden. The cost/benefit proposition is obvious – but even if the specific proposals were put in place, significant further improvements in farm practices, process hygiene and consumer education will still be required for many years to resolve the problem. However, Lee et al. (2015) have reported significant improvements in the *Campylobacter* disease burden in New Zealand through the mandatory introduction of MC and regulatory *Campylobacter performance targets* that are required of the poultry processing industry.

Whilst *Campylobacter* presents a major worldwide problem, the consequences to individuals of other food-borne diseases that occur at a lower frequency are equally important. The overall process of establishing an FSO for any one specific food–pathogen combination is difficult and challenging. It is clear from the FAO/WHO Expert Consultation (Anon, 2000b) that hazard and risk characterisation even for limited scenarios (eg, salmonellae in eggs; *Cronobacter sakazakii* in infant formulae) requires data of the highest quality and use of effective mathematical models to interpret the data (cf Bassett et al., 2010; Jongenberger et al., 2015). This is not to suggest that the approach is invalid or unworkable – rather it shows how little we really understand about the occurrence in foods of many microorganisms that are responsible for apparently commonplace causes of human food-borne disease.

Many approaches have been, and are still being, made to establish FSOs. One of the first published evaluations (Szabo et al., 2003) described the development of an FSO for control of *Listeria monocytogenes* in fresh lettuce that had been washed in chlorinated water prior to packaging in a gas-permeable film, followed by chilled storage. Other workers have sought to establish FSOs for salmonellae and *Listeria* in soft cheese and Shiga toxin–producing *E. coli* in raw steak. Stringer (2004) illustrated the difficulty of producing an effective FSO for control of *L. monocytogenes* in cold-smoked fish that is eaten raw, whilst rightly pointing to global success over many years in the control of *Clostridium botulinum* in low-acid canned products subjected to a 'botulinum cook'.

CODEX published a *Guide for National Food Safety Authorities* (Anon, 2006a) that explains the concept of food safety risk analysis. The report covers all aspects of risk assessment for foods including guidance on the four stages of a risk management procedure (Table 14.1). Microbiological risk assessment is based on the use of 'quantitative microbiological metrics' as a risk management option. 'Quantitative metrics' are defined as the *quantitative expressions that indicate a level of control at a specific step in a food safety risk management system … the term "metrics" is used as a collective for the new risk management terms of food safety objective (FSO), performance objective (PO) and performance criteria (PC), but it also*

Table 14.1 The Stages of Risk Assessment, Control and Management

Step	Process
1	**Preliminary risk management activities**
1.1	Identify and define the food safety issue
1.2	Develop a risk profile
1.3	Establish broad management goals
1.4	Decide whether a risk assessment is necessary
1.5	Establish a risk assessment policy
1.6	Commission the risk assessment
1.7	Review the results of the risk assessment
1.8	Rank the food safety issues and set priorities[a]
2	**Select risk management options**
2.1	Identify available risk management options
2.2	Evaluate identified risk management options
2.3	Select risk management option(s)
2.4	Identify a desired level of consumer health protection
2.5	Decide on preferred risk management option(s)
2.6	Deal with 'uncertainty' in quantitative risk and policy
3	**Implement risk management decisions**
4	**Monitor and review**

[a]*Applicable only if more than one risk is associated with a specific food.*
Based on Anon (2006a).

refers to existing microbiological criteria (MC). The report provides a case study for *L. monocytogenes* in ready-to-eat foods and recognises the desirability of using FSOs, POs and PCs in the development of risk-based MC; however, effective procedures for achieving these objectives are still under development.

Statistical methods play a major role in the development, use and interpretation of mathematical models of microbial contamination, growth and survival in relation to the epidemiology of food-borne disease. The reader is recommended to consult publications that consider this subject in more detail (Jouve, 1999; ICMSF, 2002, 2011; Szabo et al., 2003; Havelaar et al., 2004; Anon, 2002c, 2005a, 2006a, 2013b; Rieu et al., 2007; Bassett et al., 2010; Gonzales-Barron et al., 2013; Jongenberger et al., 2015; Zwietering et al., 2016). However, we do need to consider the implications that arise in the application of quantitative and qualitative data in food control situations.

It must be recognised that no amount of testing can ever *control* the safety and quality of manufactured foods. Rather, such testing indicates whether or not a production process, including all sources of contamination, was adequately controlled in terms of process conditions, process hygiene, pre- and post-process storage and distribution, etc. End-point testing in a production environment provides data for feedback control of a process. Quality assurance programmes such as those which seek to identify and control potentially hazardous stages of a process, including HACCP, are described, for instance, by Bauman (1974), Ito (1974), Peterson and Gunnerson (1974), ICMSF (1988), Wallace and Mortimer (1998), and Jouve (2000). In the HACCP system, end-point testing is used to validate and verify effective process controls: for raw material quality; time–temperature relationships for heating, cooling, freezing, etc.; process plant cleaning and disinfection; operator hygiene; and the many other factors critical to the production of foods under GMP.

A different aspect of food control relates to the assessment of actual or potential health risks associated with particular food commodities or products imported into a country and the assessment of 'quality' and 'safety' of foods in retail trade. For such purposes, it is not possible to 'control' the process or the post-process distribution and storage, although inspection of process plants, including those in exporting countries, is increasingly the 'norm'. Testing by enforcement authorities is targeted to assess whether foods on sale have the necessary qualities expected of them and/or whether they constitute a (potential) risk to the health of the consumer. Consequently, various qualitative and quantitative MC have been derived to provide guidance both for production personnel within industry and for enforcement authorities. Such criteria may or may not have legislative status but properly devised criteria can be key to ensuring compliance with GMP.

MICROBIOLOGICAL CRITERIA

MC can be defined (ICMSF, 2002) as 'limits for specific or general groups of microorganisms that can be applied in order to ensure that foods do not present a potential health hazard to the consumer and/or that foods are of a satisfactory quality for use in commerce'. This definition is deliberately vague since it encompasses many types of MC:

1. A *microbiological guideline* is used to provide manufacturers, and others, with an indication of the numbers of organisms that should not be exceeded if food is manufactured using GMP and stored in appropriate conditions during its 'normal' shelf life.
2. A *microbiological specification* defines the limits that would be considered appropriate for a particular food in a particular situation and may be used in contractual commercial agreements or may be recommended by national or international agencies as a means of improving/assessing the quality and safety of foods.
3. A *microbiological standard* is that part of national, or supranational, legislation that aims to define safety requirements and, in some cases, the quality of foods manufactured in, or imported into, a country.

Microbiological standards therefore have mandatory effect whilst guidelines and specifications do not. It is not intended to consider here the arguments for and against the use of legislative microbiological standards for foods since these have been well argued elsewhere (Thatcher, 1955; Wilson, 1955; Ingram, 1961; Baird-Parker, 1980; Mossel, 1982). It is important, however, to summarise the principles for establishment of 'microbiological reference values' (*sic* MC) described by Mossel (1982):

1. The number of criteria should be strictly limited so that a maximum number of samples can be examined for a given laboratory capacity.
2. The choice of criteria must be based on ecological considerations and be related to organisms of importance in public health and/or quality.
3. Criteria should be carefully formulated in *justifiable* quantitative terms.
4. Species, genera or groups of organisms to which the criteria are applied should be described in appropriate taxonomic terms and should be readily identifiable in the laboratory.
5. Test methods should be described in sufficient detail to permit their use in any reputable laboratory, and should be applied only after full validation including inter-laboratory trials.
6. Numerical values should be derived *only* as a result of adequate surveys to establish technological feasibility.

In 1979, the CCFH adopted these principles and agreed definitions for various types of MC that included the need for sampling plans and standardised methodology as a prerequisite for setting criteria (Anon, 1997). It is noteworthy that the revised CCFH definition of a MC (Anon, 2013a) now recognises that it is a risk management metric that can indicate either the acceptability of a product or the performance of a process or food safety control system at a specified point in the food chain. Hence, before setting an MC it is necessary to establish the intended purpose of the criterion. However, Mossel's principles '3' and '6' are clearly key matters for interpretation of laboratory data in relation to quantitative and qualitative MC.

HOW SHOULD MICROBIOLOGICAL CRITERIA BE SET?

Collection of data

Having decided upon the tests to be done and the methodology to be used, a survey should be made of products from several manufacturers who operate GMPs in order to determine what is technically feasible. As a prerequisite, an inspection should be made of the manufacturing processes together with microbiological examination of products so that any deficiencies in the process can be identified and corrected before the survey is done.

Samples, obtained at various times during the working day, should be stored and transported to the laboratory under appropriate conditions for examination using validated methods. The laboratory examination should be undertaken within a few hours of sample procurement. Sampling is repeated over a period of several days in one process facility and, ideally, also in other facilities producing the same generic product, until sufficient data (not less than 100 data sets) have been collected. A distribution curve is prepared of the $\log_{10}$ colony counts against the frequency of occurrence of counts, for example, the data for a cooked meat product shown in Fig. 14.1; the 95th percentile (φ) value (ie, the $\log_{10}$ count which is exceeded only by colony counts on 5% of the samples) is also shown on the graph (cf Fig. 2.1 for colony counts on sausages manufactured in two factories). In the case of the cooked meat data (Fig. 14.1) the 95th percentile is 3.5 $\log_{10}$ cfu/g indicating a well-controlled process. By contrast the data shown in Fig. 2.1 have a 95th percentile slightly below 7.0 log cfu/g, which is unacceptable even for a raw meat product. No attempt should

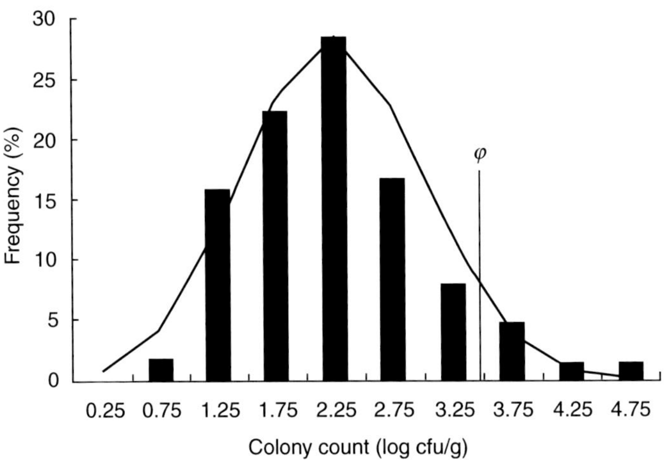

FIGURE 14.1 Frequency Distribution of Colony Counts on Cooked Meats, Overlaid With a 'Normal' Distribution Curve and Showing the 95% Percentile (φ) at 3.49 $\log_{10}$ cfu/g

Based on data of Kilsby et al. (1979)

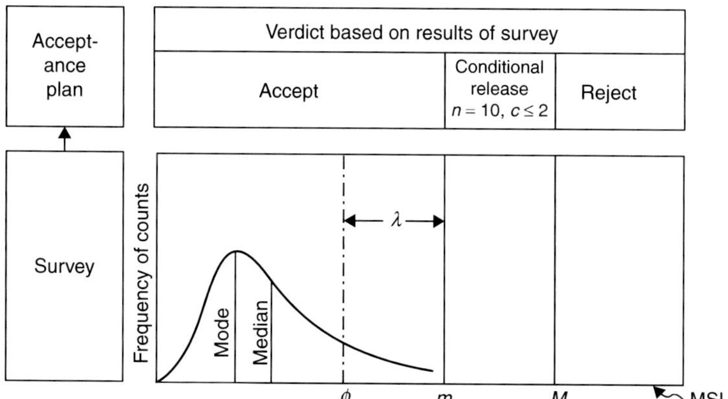

FIGURE 14.2 Distribution Plot of the Results of Microbiological Surveys Used to Derive Criteria for a Defined Food Product

M is an absolute limit (*M*) set in relation to the minimum spoilage level (MSL) or minimum infective dose (MID) for a pathogen; *m* is the acceptance target value set at a distance (λ) above the 95% percentile level (φ). Colony counts are plotted on the *x*-axis as $\log_{10}$ cfu/g or $\log_{10}$ cfu/mL

Modified from Mossel (1982)

be made to use such data to set critical values without first undertaking a branch and root investigation of the reasons for the high counts and the manufacturing processes must be improved before a survey is repeated.

Setting criteria limits

Fig. 14.2 is an idealistic curve that shows the mean, mode and the 95th percentile value (φ). For quantitative data, it is necessary to decide on the maximum number of organisms (the critical level: *M*) that should not be exceeded in any circumstances. This will normally be related to the minimum spoilage level (MSL) or, in the case of pathogens, to the minimum infective dose (MID) and is the maximum level that is acceptable for foods manufactured under GMP. It will usually be set at least one log cycle above the 95th percentile, provided that such value does not approach too closely the MSL or MID; Dahms and Hildebrandt (1998) recommended that *M* should be 1.5–2.5 SDs above a lower acceptability limit *m*.

The lower critical limit (*m*) is usually set very close to the 95th percentile value (φ) and below the maximum (*M*) such that the value of *m* represents counts that can be achieved most of the time in product manufactured under GMP. The tolerance (λ) between φ and *m* should be set to take account of the imprecision in microbiological measurement, the variability likely to occur in good-quality raw materials, the manufacturing process, etc. Fig. 14.2 illustrates the establishment of a 3-class sampling plan with $n = 10$, $c = 2$, based on the results of a survey, which requires not more than 2 of 10 samples tested to have counts $>m$ and $<M$ with none exceeding *M*.

In establishing criteria for 'point of sale' or import sampling (ie, by enforcement authorities), due attention is required not only to what can be achieved in GMP but also to the effects of subsequent storage and distribution. This requires understanding of the microbiological consequences of storage time and temperature and of the effects of the intrinsic properties of the food (ie, pH, a_w, presence of anti-microbial agents, etc.). In addition, from a public health viewpoint, the potential for consumer mishandling during storage, cooking and serving needs to be considered, together with the vulnerability of particular consumer groups.

To establish criteria for qualitative testing (i.e., quantal tests for pathogens), a survey of product manufactured under conditions of GMP must be done using realistic sample sizes (eg, at least 25 g) and numbers of replicate samples in order to determine the prevalence of the target organism in the product. As explained in Chapter 5 the value of quantal tests is dependent not only on the relative sensitivity and specificity of the test procedure but also on the number of samples examined. To obtain meaningful results, such surveys should examine at least 100 samples from each of a number of manufacturers. If the target organism is likely to be present in the test material, then the survey should be done using several quantities of each replicate sample, possibly using a Most Probable Number (MPN) test, in order to ascertain both the prevalence and level of potential contamination.

There is evidence (Anon, 2013c) that legislators may be influenced by the amount of media and social media information in response to a microbiological, or other, food scare and then seek to introduce new MC in the belief that a problem will be solved merely by setting criteria to define acceptable levels of specific organisms in a particular product. The revision to the European MC for foods (Anon, 2013d) is an example: criteria were set for Shiga-toxin-producing *E. coli* in sprouted seeds. Unfortunately these MC are so lax that effective control is unlikely. It is essential that MC are set only after appropriate risk assessment. Even then, the existence of the criteria may not be effective in controlling the safety of foods because safety and quality cannot be tested into a product – it is only through the use of appropriate performance objectives to improve the safety of manufactured products that improvements will be seen. A *Risk Manager's Guide* (Anon, 2016) to the statistical aspects of setting MC for foods provides much sensible advice. Anyone interested in understanding how to set up sampling plans for defined acceptability limits should read this guide.

Establishing sampling plans

Computer software packages for deriving and comparing the efficiency of statistically based acceptance sampling plans are freely available for download (ICMSF, 2009; Anon, 2016). A user-friendly package, including a capability to generate variables sampling plans, will be available in 2016 for download from FAO/WHO (Anon, 2016).

In setting any sampling plan, it is essential to recognise that the level of analysis which would be required to provide a high degree of protection to both the manufacturer and the consumer can rarely be achieved in practice (Chapter 5). Two-class attributes plans are used to assess quantal and some quantitative data for pathogens but three-class attributes sampling plans (see Chapter 5) are usually recommended for quantitative data. An attributes scheme normally requires a decision to accept or reject a lot based on grading of analytical results in relation to quantitative limit values (the acceptability and rejection criteria) and takes little note of microbial distribution or methodological imprecision except in so far as a tolerance may have been built into the limits. Of course, making a decision as to whether a specific colony count exceeds, or not, a pre-determined limit is simple to understand and requires neither a statistical appraisal of the data nor use of measurement uncertainty of counts on replicate samples. Yet, those responsible for food examination in 'accredited laboratories' are required to derive and report values for the measurement uncertainty of their results (Chapters 10 and 11). Hence, one must question the wisdom of using attributes schemes for MC based on quantitative data. When only a limited number of representative samples have been examined, the risk of making a wrong decision is high, so it is sensible to use all available analytical data to assess whether or not a set of results is compliant with a defined limit. I would argue that this justifies the wider adoption of variables sampling schemes for quantitative data, provided that knowledge exists of the microbial distribution in a specific food (Example 14.1).

EXAMPLE 14.1 COMPARISON OF ATTRIBUTES AND VARIABLES SAMPLING PLANS

Suppose a manufacturer wishes to ensure with a probability of 95% that the level of Enterobacteriaceae in a product does not exceed a limit of 3.0 log_{10} cfu/g. Tests on replicate samples of product show a typical average level of 1.6 log_{10} cfu/g with a standard deviation of 0.6. If he chooses to use a 2-class sampling plan, how many samples must be tested to give a 95% probability that he will not falsely reject product that contains <2 log_{10} cfu/g and less than a 5% probability of accepting product with a count greater than the limit value of 3 log_{10} cfu/g?

The answer can be found by using the companion spreadsheet that accompanies the FAO/WHO guidance on risk analysis (Anon, 2016). Use of the '2-class concentration' spreadsheet, for the criteria identified, shows that he must test at least 12 samples of each lot and accept up to 2 results which exceed the limit of 3 $\log_{10}$ cfu/g. For a plan of $n = 12$, $c = 2$, $m = 3.0$, the actual probability of accepting product with a count ≥ 3.0 $\log_{10}$ cfu/g is about 2% and the producer's risk of rejecting product with a count of 2.0 $\log_{10}$ cfu/g is 1.7% (probability of acceptance is 98.3%).

Whilst these probabilities are acceptable to him, the number of samples that need to be tested is too high. He knows that if he tests fewer samples using, for example, a 2-class plan with $n = 5$, $c = 0$, $m = 3.0$, the probability of rejecting a lot with a log count >3.0 remains high (96.9%) but the probability of accepting product with a log count of 2.0 falls to 78.3%; hence, he might wrongly reject almost 25% of his product.

Following discussion he considers two other options: using a three-class attributes plan or using a variables plan. A 3-class plan with, for example, $n = 5$, $c = 2$, $m = 2.0$, $M = 3.0$ for the typical average log count of 1.6 and standard deviation of 0.60, shows that the consumer's risk of accepting a lot with a log count >3.0 is <0.01%, but the producer's risk of rejecting a lot with a mean count of 2.0 $\log_{10}$ cfu/g is about 21%. This is still unacceptable. He next considers the option of using a variables sampling plan.

After trying various options he decides that, provided the variation in contamination level in his product does not change significantly, a variables plan with $n = 5$ and $m = 3.0$ provides a more acceptable option. Based on the Normal distribution of the log count with mean = 1.6 and SD = 0.6, the probability that the log count might exceed the limit value (3.0) is about 4% and gives a consumer's risk point of 96%. The calculated variables factor (k) is 1.015 giving a 92.7% probability of accepting product with a mean log count of 2.0, which is only slightly below his chosen probability of acceptance of 95%. Most importantly, the plan requires that product with a mean log count ≥2.4 should be rejected.

Before using this software, it is essential to read carefully the *Risk Manager's Guide* (Anon, 2016) and to look at the associated YouTube® videos (linked within the guide) that show effectively how to develop sampling plans. The worksheets include illustrations of the distribution of data based, on the assumption that log-transformed colony counts conform to a Normal distribution, operating characteristics (OC) curves based on both geometric mean (ie, log counts) and arithmetic mean counts. Of course, if there is evidence that the data distribution is not lognormal, then alternative methods to derive sampling plans are necessary. An example of an OC curve for probability of acceptance of a lot and the Normal distribution curve for these data are shown in Fig. 14.3 for the data used in this example.

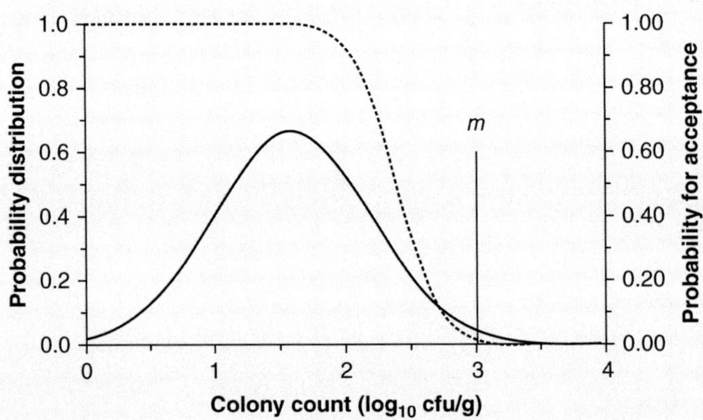

FIGURE 14.3

Probability distribution of hypothetical microbial counts (——) with a mean value of 1.6 $\log_{10}$ cfu/g and a SD of 0.60 and the probability for acceptance of lots (•••••) using a variables sampling plan with $n = 5$ and $m = 3.0$ $\log_{10}$ cfu/g.

The plots were derived using the companion spreadsheet to Anon (2016) kindly provided by Dr A. Kiermeier and published with permission of FAO and WHO.

Whichever scheme is adopted, it is essential to define the degree of 'risk' that is acceptable to the manufacturer and the consumer. Producer's risk has been defined previously (Chapter 5) as the 'risk of wrongly rejecting good product' (usually set at 5%) whilst the consumer's risk relates to 'wrongly accepting unsatisfactory product' (often set at 10%). Such risks can be quantified for both attribute and variable sampling schemes. However, a variables scheme uses the actual concentration data, thereby allowing the plan to 'yield similar discrimination between acceptable and unacceptable lots with fewer sample units (smaller *n*) than a similarly performing attributes plan' (Anon, 2016). This is because the variables scheme incorporates a tolerance that is related to both the number of samples to be examined and the likely variance in the distribution of the target organisms based on experience of analysis of replicate samples. A variables sampling plan is appropriate for use in well-controlled food manufacturing situations where the microbial distribution in food lots is known and reasonably constant (ICMSF, 2002). Variables plans have traditionally been considered to be inappropriate in official food control situations where knowledge of the microbial distribution may be unknown, for example, examination of samples taken at retail or at a port of entry. But does this situation always exist in practice? Anon (2016) notes that 'this type of sampling plan (*sic* a Variables plan) may be applicable to hygiene indicator organisms and for pathogens that are less likely to cause illness at low levels'.

Most legislation concerned with MC, including European legislation (Anon, 2005b) for food safety, uses the two-class attribute sampling plans for defined pathogens with three-class attribute sampling plans for most process hygiene criteria. However, the European MC for aerobic and Enterobacteriaceae colony counts of swabs from carcases of meat animals are required to comply with MC based on daily mean counts rather than on a set of individual counts. Gonzales-Barron et al. (2012, 2013) reported the effective level of control for this MC to be poor and proposed a variables sampling plan for Enterobacteriaceae on sheep carcases. The plan uses the gamma–Poisson, rather than the lognormal, distribution to produce optimised and statistically sound sampling plans that are claimed to be more efficient and discriminatory than attribute plans. Santos-Fernández et al. (2014) have also recommended use of a variables sampling plan for assessing food safety based on use of a sinh–arcsinh transformation of counts the distribution of which conforms to a Normal distribution for which variables sampling plans are used.

These are useful developments in the argument for using variables sampling plans that will hopefully become more widely accepted in the future. So far as I am aware, no national or international agency has adopted a variables scheme for microbiological examinations, although such schemes have been recommended for quantitative determination of contaminants such as mycotoxins (see Anon, 2006b). However, the mycotoxin sampling plan requires comminution and blending of a large number of representative samples before drawing replicate analytical samples in order to minimise 'between-sample' variation and ensure that test samples are representative of the entire 'lot'. In these analyses allowance is made for the measurement and sampling uncertainty. Because of the possible risks of cross-contamination, such procedures are not generally attempted for microbiological examinations, although Corry et al. (2010) showed the benefits in relation to estimation of microbiological sampling and measurement uncertainty.

THE RELEVANCE OF MEASUREMENT UNCERTAINTY TO MC

The CODEX Committee on Analysis and Sampling (Anon, 2004) recommended that 'Codex Commodity Committees concerned with commodity specifications and the relevant analytical methods should state, *inter alia*, what allowance is to be made for measurement uncertainty when deciding whether or not an analytical result falls within specification'. The proposal also noted that 'this requirement may not apply in situations where a direct health hazard is concerned, such as for food pathogens'. Whilst indicating a need to have a standardised approach to the use of uncertainty measurement in interpreting microbiological data, no specific approach was proposed.

Previously, the UK Accreditation Service (Anon, 2000c) had provided guidance on how laboratories might cite and interpret analytical data and the European DG SANCO (Anon, 2003a) assessed the problem associated with differing national interpretations of uncertainty in relation to compliance with defined limits. The DG SANCO report (Anon, 2003a) recommended that '… measurement uncertainty (should) be taken into account when assessing compliance with a specification'. The report continued, 'In practice, if we are considering a maximum value in legislation, the analyst will determine the analytical level and estimate the measurement uncertainty of that level, subtract the uncertainty from the reported concentration and use that value to assess compliance. Only if that value is greater than the legislation limit will the control analyst be sure *beyond reasonable doubt* that the sample concentration of the analyte is greater than that prescribed by legislation'. Note that this recommendation is dependent on expressing uncertainty on an additive scale, where the expanded uncertainty (U) can be subtracted from the estimated value.

Samples drawn from a 'lot' should, but may not, be truly representative of the 'lot' (Anon, 2002d) and the results of any examination will provide only an estimate of the true microbial population. As discussed previously, even in a well-controlled laboratory, quantitative microbiological methods are subject to many sources of error that often lead to substantial estimates of measurement uncertainty. The expanded repeatability measurement uncertainty (ie, the 95% CI bounds) of an analysis can frequently extend to ±5% of the mean colony count (ie, up to ±0.25 log cfu) from replicate tests on a single sample, and inter-sample and intra- and inter-laboratory reproducibility estimates give even wider estimates. Furthermore, the intrinsic errors associated with MPN and other tube dilution methods are even greater, and the 95% CIs are so wide that these methods should generally be restricted to use in industrial quality assessment except where one is looking for specific changes in an otherwise stable situation, for example, in potable water analysis.

The level of uncertainty attributable to sampling is but poorly recognised (see Chapter 11). For chemical and microbiological analyses of foods the uncertainty due to sampling is much greater than the measurement uncertainty. Hence, the use of criteria limits that take no account of such variation cannot be considered to be metrologically sound. For instance, if a mean log count is (say) 5.8 $\log_{10}$ cfu/g and the overall 95% measurement uncertainty is ±0.5 log cycles, then, on 19 occasions out of 20, the mean count indicates that the 'true' result lies within the log range 5.3–6.3, but a true result outside this range could be expected on 1 occasion in 20. An attributes scheme makes no allowance for such variation except in relation to the positioning of the control limits.

There are different ways in which measurement uncertainty estimates could be applied in relation to MC limits. For upper control limits, the suggestions of various bodies include the following:

1. subtracting the expanded 95% uncertainty value from a mean result before comparing the 'corrected' result with the limit, to ensure 'compliance without reasonable doubt' (Anon, 2003a);
2. applying a 'guard band' to the limit by adding the uncertainty value to the limit before comparing the actual result (essentially the same effect as in '1' but avoiding the necessity to 'correct' each result) (Anon, 2007);
3. applying the general statistical procedures for confidence intervals of a mean value estimate in relation to the 'true' value before comparison with the MC limit;
4. ignoring uncertainty estimates completely because wide estimates demonstrate poor laboratory technique – this is not necessarily true but it can be arguable.

Some of these options are illustrated in Example 14.2.

EXAMPLE 14.2 SHOULD MEASUREMENT UNCERTAINTY BE USED IN ASSESSING COMPLIANCE WITH A CRITERION?

Compliance of mean colony counts

EC legislation (Anon, 2005b) on microbiological criteria for swabs of meat carcases (other than pig for which different limits are set) uses a 2-class plan and sets criteria limits for Enterobacteriaceae (2.5 $\log_{10}$ cfu/cm^2) and aerobic colony count (5.0 $\log_{10}$ cfu/cm^2) that require compliance of the daily mean colony count with the absolute limit (*M*).

Suppose that replicate swabs of a carcase are examined daily for aerobic colony count and mean values together with estimates of expanded measurement uncertainty for the data are determined. Suppose, also, that the expanded (95% probability) reproducibility uncertainty is ±0.5 $\log_{10}$ cfu/cm^2 and the mean colony counts (as $\log_{10}$ cfu/cm^2) on four separate series of replicate examinations are A = 4.0 $\log_{10}$ cfu/cm^2, B = 4.6 $\log_{10}$ cfu/cm^2, C = 5.4 $\log_{10}$ cfu/cm^2 and D = 6.0 $\log_{10}$ cfu/cm^2.

Fig. 14.4 illustrates the upper and lower bounds of the uncertainty distribution limits of these four sets of mean colony counts in relation to the criterion limit value (M = 5.0 $\log_{10}$ cfu/cm^2).

Scenario A shows a mean colony count of 4.0 $\log_{10}$ cfu/cm^2 with the upper boundary of the 95% confidence interval (CI) at 4.5 $\log_{10}$ cfu/cm^2; here there can be no dispute – the test result is compliant with the limit of 5.0 $\log_{10}$ cfu/cm^2. *Scenario B* shows a mean colony count of 4.6 $\log_{10}$ cfu/cm^2 but the upper boundary of the 95% CI (5.1 $\log_{10}$ cfu/cm^2) slightly exceeds the criterion limit. *Scenario C* shows a mean colony count (5.4 $\log_{10}$ cfu/cm^2) that is above the criterion limit but the lower boundary of the 95% CI (4.9 $\log_{10}$ cfu/cm^2) is below the criterion limit – are these data compliant with the limit or not compliant? *Scenario D* shows both the mean colony count of 6.0 $\log_{10}$ cfu/cm^2 and the lower CI bound of 5.5 $\log_{10}$ cfu/cm^2 above the criterion limit – here there can be no dispute; the results do not comply with the criterion.

Whilst the decisions regarding *Scenarios A and D* are clear-cut, the situation is less clear for data in Scenarios B and C. Some might consider *Scenario B* to be acceptable, since the mean value

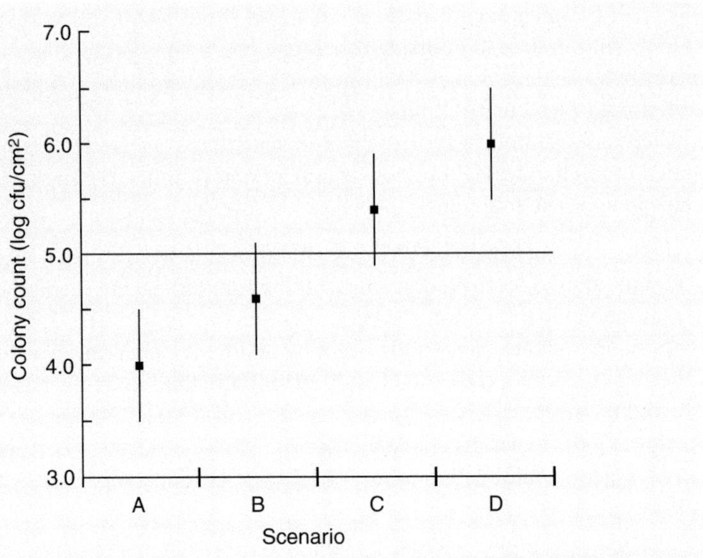

FIGURE 14.4

Implication of measurement uncertainty on mean values (■) in relation to a criterion limit of 5.0 $\log_{10}$ cfu/cm^2. Four scenarios show mean counts for A = 4.0 $\log_{10}$ cfu/cm^2, B = 4.6 $\log_{10}$ cfu/cm^2, C = 5.4 $\log_{10}$ cfu/cm^2 and D = 6.0 $\log_{10}$ cfu/cm^2 with upper and lower confidence intervals based on an expanded measurement uncertainty value of ±0.5 $\log_{10}$ cfu/cm^2

is below the criterion limit and, if one follows the proposal of DG SANCO (Anon, 2003a) to deduct the expanded uncertainty value from the mean value, the 'corrected' mean value [ie, the lower bound of the 95% confidence limit (CL)] is significantly below the limit value. Hence, it can be argued that the colony count is compliant with the limit. However, others might deem this to be unacceptable since the upper boundary of the 95% CL exceeds the criterion limit. For *Scenario C*, the mean value is above the criterion limit and might be interpreted as non-compliant; however, by deduction of the expanded uncertainty value the 'corrected' mean count is below the criterion limit, so the count would again be compliant with the limit. In both cases, it would be 'beyond reasonable doubt' that the test data comply with the criterion. However, this approach is not generally accepted at this time.

Compliance of individual colony counts

When a criterion requires that each individual result does not exceed a criterion limit, as in a normal two-class or three-class attributes sampling plan, the situation would be different. Suppose that, for the data from *Scenario C* above (mean count 5.4 $\log_{10}$ cfu/cm^2), the individual colony counts on 5 replicate samples were 5.0, 5.0, 5.1, 5.8 and 5.9 $\log_{10}$ cfu/cm^2. If the 2-class sampling plan was, for instance, $n = 5$, $c = 2$, $M = 5.0$, then 3 of the 5 counts would exceed the criterion limit value (5.0 $\log_{10}$ cfu/cm^2) so the data set would be deemed to be non-compliant. But if the expanded measurement uncertainty were to be subtracted from each individual value, the 'corrected' values would be 4.5, 4.5, 4.6, 5.3, and 5.4 $\log_{10}$ cfu/cm^2, respectively. Hence, the corrected mean value (4.9 $\log_{10}$ cfu/cm^2) would be below the limit value (and therefore compliant) and the individual colony counts would also be deemed to be compliant since only two of the individual corrected values would exceed the limit.

Guidelines have been published for expression and use of measurement uncertainty in chemical analyses (see, eg, Anon, 2007), but none exist for data on microbiological examination. The international microbiology community needs to derive guidelines on use of measurement uncertainty in tests for compliance with MC.

Eurachem (Anon, 2007) recommended the approach in which 'guard bands' are set around a criterion limit to show the level of tolerance that can be given for a particular form of analysis. The width of the guard band is determined by knowledge of acceptable limits for the measurement uncertainty of the test method and forms part of the decision process in setting criteria. The benefit of this approach is that there is a clear statement of intent with regard to the measurement uncertainty that is deemed acceptable for a specific form of analysis. Such an approach provides a greater transparency in setting and interpreting criteria.

It can be argued that a three-class sampling plan for MC implicitly incorporates the guard band concept since the lower limit (m) is the actual acceptance limit whilst the upper criterion limit (M) is an absolute that should never be exceeded. But the breadth of the middle band (ie, $M - m$) is often set not in relation to statistical principles but rather on a 'gut feel' of what should be achievable. Furthermore, there is a significant difference in the approaches, since most microbiological three-class plans are predicated on the interpretation of each of n results against the criteria, rather than the average value of the n colony counts.

This illustrates one of the issues concerning microbiological measurement uncertainty that still need to be addressed for microbiological examination of foods. It again raises the issue of whether it is justifiable to convert data into simple 'go/no go' values for use in attributes sampling schemes, or whether the time is now right to change to wider use of variables sampling schemes that takes account of all colony count data relevant to a set of replicate samples drawn from a 'lot'. Failure to make a clear decision on this approach will hinder future development of effective MC for foods.

Criteria based on presence or absence of a particular organism, or a group of organisms, are affected not only by the distribution of the target organism(s) within the food but also by the adequacy, or otherwise, of the test procedures. Hence, it is essential to use validated methods with a level of sensitivity and specificity suitable for the intended purpose. Even then, tests giving negative results on a number of individual replicate samples do not guarantee freedom from the organism in question; they merely indicate that the probability for occurrence of the organism in the 'lot' is below certain tolerances. The tolerance will be dependent on both the number of samples tested and the efficiency of the method used. However, in the same way that a scheme showing apparent absence of specific organisms does not guarantee total absence of the organisms from the product, the detection of one or more confirmed positive samples could also arise by chance. Hence, product of equivalent quality could be rejected on one occasion yet be accepted on another occasion if only a small number of samples were tested.

If one tests 10 samples of a lot having, say, 1% defectives (ie, 1% of 25-g sample units from a lot contain at least 1 detectable *Salmonella*), then on average, one would expect to detect salmonellae and therefore reject the lot (with $c = 0$) on 10 occasions out of 100, yet not detect them and accept the lot on 90 occasions (Table 5.1). However, if the true prevalence of defectives is only 0.1%, on average, then tests on 10 samples would expect to detect salmonellae only once in every 100 tests. The analyst is unable to differentiate between false-negative and true-negative results when carrying out quantal tests on real-life samples, although detection of confirmed positive results would confirm that the product is contaminated.

The situation becomes even less clear once we accept that the distribution of low-level contaminants in food products is not homogeneous – for instance, clusters of organisms may occur in small quantities of otherwise uncontaminated product. In reviewing a risk assessment model for *Cronobacter* and *Salmonella* spp. in powdered infant formulae (PIF) the WHO/FAO Joint Expert Meeting on Microbiological Risk Assessment (Anon, 2006c) assumed homogenous distribution of potential contaminants; however, the report noted that if clumping of organisms in PIF were significant, then the risk assessment strategy would not be effective and regretted a lack of data on microbial distribution in PIF. Subsequent detailed studies of the occurrence of *C. sakazakii* in PIF have shown that the contamination occurs as clusters of contaminants within a large quantity of otherwise uncontaminated material. The distribution of the organisms is described best by one or more compound Poisson or other distributions (Mussida et al., 2011, 2013; Jongenberger, 2012; Jongenberger et al., 2015). The operating characteristics (OC) curves associated with such models differ markedly from those used generally so that the probability for rejection of contaminated product differs markedly (Fig. 14.5). To improve the likelihood of detecting contaminants that occur at low frequency in an over-dispersed distribution

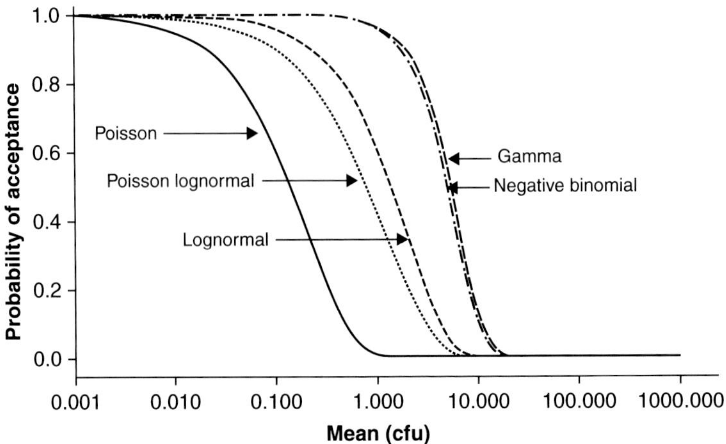

FIGURE 14.5 Operating Characteristics Curves for the Sampling Plan n = 5, c = 0 With Mean Acceptable Count of 1 cfu

The curves illustrate the differences in the probability of acceptance of a product assuming that different frequency distributions describe a significantly over-dispersed count of organisms (variance 1000 cfu) compared to random distribution of cells. The distribution curves are for Poisson (——), Poisson–lognormal (··········), lognormal (----), negative binomial (-·-·-), and Gamma distributions (– – –)

Adapted from Fig. 6.4 of Bassett et al. (2010) with permission from ILSI (Europe).

requires testing of large numbers of small (eg, 1 g) samples rather than small numbers of larger samples (Jongenberger et al., 2015).

CONCLUSIONS

The efficiency of a laboratory examination for pathogens or spoilage organisms is dependent on the number and size of samples examined and the distribution of organisms in a product. In establishing MC due cognisance is required of these effects. The potential cost of intensifying testing schemes needs to be balanced against the costs of unnecessarily rejecting valuable food materials and/or of increasing consumer's risk by accepting defective products. Acceptable quality levels, producer's risk and consumers' risk must be defined in relation to the likely distribution of organisms and must take account of the imprecision of microbiological methods. It is noteworthy that in 2013, CCFH established an electronic working group on 'Statistical and Mathematical Considerations for the Elaboration of Microbiological Criteria' (Anon, 2014). The recent publication by FAO/WHO (Anon, 2016) provides a clear and extended description of the issues and a framework for evaluation of different strategies on development and assessment of MC. This provides a more 'transparent' approach to the setting of MC including a generally understandable set of decision rules for MC. However, it is also essential to have internationally acceptable guidelines for use of measurement uncertainty in compliance assessment. In addition, Zwietering et al. (2016) have stressed the importance of understanding the nature of over-dispersed populations of pathogens in relation to food safety.

Effective microbiological control comes from use of GMP, HACCP and other control strategies at all stages of food production, distribution and storage, based on knowledge of the microbial ecology of particular foods under different process and storage conditions. Such control strategies also require effective introduction and monitoring of FSOs and FPOs. Lack of effective strategies is a matter of concern in relation to the control of food-borne disease organisms, such as the present-day widespread dissemination of *Campylobacter* in chicken. Although statistical process control (SPC) of microbiological data provides a vital additional aid to monitoring changes in a manufacturing process, end-point testing of manufactured foods is effective only as a means of retrospective monitoring of process, distribution and storage conditions. The distribution of organisms in foods and the statistical variation associated with methods of detection and enumeration lead to the conclusion that at present MC should be used primarily as guidelines and specifications, except for monitoring of high-level contamination with pathogens.

REFERENCES

Accum, F., 1820. A Treatise on Adulteration of Food and Culinary Poisons. Longman, Hurst, Rees, Orme & Brown, London.

Amos, A.J., 1960. Pure Food and Pure Food Legislation. Butterworths, London.

Anon, 1997. Principles for the Establishment and Application of Microbiological Criteria for Foods, third ed. Codex Alimentarius Guidelines, Report No. CXG 21e. WHO/FAO, Rome.

Anon, 2000a. The interactions between assessors and managers of microbiological hazards in food. Report of a WHO Expert Consultation, Kiel, Germany, 21–23 March 2000. WHO, Geneva.

Anon, 2000b. Joint FAO/WHO Expert Consultation on Risk Assessment of Microbiological Hazards in Foods. FAO Food and Nutrition Paper 71. WHO/FAO, Rome.

Anon, 2000c. The expression of uncertainty in testing. Laboratory of the Government Chemist, London.

Anon, 2002a. Regulation (EC) no 178/2002 of the European Parliament and of the Council of 28 January 2002 laying down the general principles and requirements of food law, establishing the European Food Safety Authority and laying down procedures in matters of food safety. Official J. Eur. Comm. L31, 1–24.

Anon, 2002b. Principles and guidelines for incorporating microbiological risk assessment in the development of food safety standards, guidelines and related texts. Report of a Joint FAO/WHO Consultation, Kiel, Germany, 18–22 March 2002. FAO/WHO/Institute for Hygiene and Food Safety of the Federal Dairy Research Centre, Kiel.

Anon, 2002c. Proposed draft principles and guidelines for the conduct of microbiological risk management. Joint FAO/WHO Food Standards Programme. Codex Committee on Food Hygiene 35th Session. CX/FH 03/7. FAO and WHO, Rome, Geneva.

Anon, 2002d. Guide on Sampling for Analysis of Foods. NMKL Procedure No. 12. NMKL, National Veterinary Institute, Oslo, Norway.

Anon, 2003a. The relationship between analytical results, the measurement uncertainty, recovery factors and the provisions in EU food and feed legislation. Report to the EU Standing Committee on the Food Chain and Animal Health Working Group Draft, 5 June.

Anon, 2003b. WHO/FAO Guidelines on Hazard Characterisation for Pathogens in Food and Water. WHO/FAO, Geneva, <http://apps.who.int/iris/bitstream/10665/42693/1/9241562374.pdf> (accessed 11.10.15).

Anon, 2004. The use of analytical results: sampling plans, relationship between the analytical results, the measurement uncertainty, recovery factors and provisions in Codex standards. CODEX Report ALINORM 04/27/23 Appendix VII. FAO/WHO, Geneva.

Anon, 2005a. Food Safety Management Systems – Requirements for Any Organisation in the Food Chain. ISO 22000:2005.

Anon, 2005b. Commission Regulation No. 2073/2005 of 15th November 2005 on microbiological criteria for foodstuffs. Official J. Eur. Union L338, 1–26.

Anon, 2006a. Food safety risk analysis: a guide for national food safety authorities. FAO Food and Nutrition Paper 87. WHO/FAO, Rome.

Anon, 2006b. Commission Regulation (EC) No 1881/2006 of 19 December 2006 setting maximum levels for certain contaminants in foodstuffs. Official J. Eur. Union L364, 5–24.

Anon, 2006c. Joint FAO/WHO Expert Consultation on *Enterobacter sakazakii* and *Salmonella* in Powdered Infant Formula, January 16–20, Rome. <http://www.who.int/foodsafety/publications/powdered-infant-formula/en/> (accessed 07.10.15).

Anon, 2007. Use of uncertainty information in compliance assessment. In: Ellison, S.L.R., Williams A. (Eds.), Eurachem/CITAC Guide. Eurachem, LGC, Teddington. <https://eurachem.org/index.php/publications/guides/uncertcompliance> (accessed 07.10.15).

Anon, 2009. Risk assessment of *Campylobacter* spp in broiler chickens. Microbiological Risk Assessment Series No. 11. WHO/FAO, Geneva. <http://www.who.int/foodsafety/publications/micro/MRA11_En.pdf> (accessed 11.10.15).

Anon, 2010. Preparatory study on food waste across EU27. Technical Report 2010-054. <http://ec.europa.eu/environment/archives/eussd/pdf/bio_foodwaste_report.pdf> (accessed 02.11.15).

Anon, 2011. EFSA Panel on Biological Hazards (BIOHAZ), 2011. Scientific opinion on *Campylobacter* in broiler meat production: control options and performance objectives and/or targets at different stages of the food chain. EFSA J. 9, 2105–2246, <http://www.efsa.europa.eu/sites/default/files/scientific_output/files/main_documents/2105.pdf> (accessed 11.10.15).

Anon, 2013a. Principles and guidelines for establishment and application of microbiological criteria related to foods. Codex Report CAC/GL 21-1997, Revision 1-2013. Codex Alimentarius Commission, FAO/WHO, Geneva.

Anon, 2013b. The global view of campylobacteriosis. Report of a WHO Expert Consultation, Utrecht, NL (ISBN 978 92 4 1564601).

Anon, 2013c. Perceptions and Communication of Food Risks and Benefits Across Europe. <http://resourcecentre.foodrisc.org/> (accessed 11.10.15).

Anon, 2013d. Regulation (EC) No. 209/2013 of 11 March 2013 Amending Regulation (EC) No. 2073/2005 as Regards Microbiological Criteria for Sprouts and the Sampling Rules for Poultry Carcases and Fresh Poultry Meat. <http://eur-lex.europa.eu/LexUriServ/LexUriServ.do?uri=OJ:L:2013:068:0019:0023:EN:PDF> (accessed 06.12.15).

Anon, 2014. Proposed draft annex on statistical and mathematical considerations to the principles and guidelines for the establishment and application of microbiological criteria related to foods. Codex Report CX/FH 14/46/6. Codex Alimentarius Commission, FAO/WHO, Geneva. <ftp://ftp.fao.org/CODEX/Meetings/CCFH/ccfh46/fh46_06e.pdf> (accessed 12.10.15).

Anon, 2015. A *microbiological* survey of *Campylobacter* contamination in fresh whole UK-produced chilled chickens at retail sale – February 2014 to February 2015. Food Standards Agency Report. <https://www.food.gov.uk/sites/default/files/full-campy-survey-report.pdf> (accessed 07.10.15).

Anon, 2016. FAO/WHO. The Statistical Aspects of Microbiological Criteria Related to Foods: A Risk Managers Guide. Microbiological Risk Assessment Series 24. FAO/WHO, Rome, Geneva, In press. <ftp://ftp.fao.org/codex/meetings/CCFH/CCFH46/FAO MC draft 140814a.pdf > (accessed 25.11.15).

Baird-Parker, A.C., 1980. Food microbiology in the 1980's. Food Technol. Aust. 32, 70–77.

Baranyi, J., Roberts, T.A., 1994. A dynamic approach to predicting microbial growth in food. Int. J. Food Microbiol. 23, 277–294.

Baranyi, J., Tamplin, M., 2004. Combase: a common database on microbial responses to food environments. J. Food Prot. 67, 1834–1840.

Bassett, J., Jackson, T., Jewell, K., Jongenburger, I., Zwietering, M., 2010. Impact of microbial distributions on food safety. International Life Science Institute Europe Report Series, Brussels, Belgium (ISBN 9789078637202).

Bauman, H.E., 1974. The HACCP concept and microbiological hazard categories. Food Technol. 28 (9), 30–34, 74.

Corry, J.E.L., Jarvis, B., Hedges, A.J., 2010. Minimising the between-sample variance in colony counts on foods. Food Microbiol. 27, 598–603.

Dahms, S., Hildebrandt, G., 1998. Some remarks on the design of three-class sampling plans. J. Food Prot. 61, 757–761.

Gonzales-Barron, U.A., Lenahan, M., Sheridan, J., Butler, F., 2012. Use of a Poisson–gamma model to assess the performance of the EC process hygiene criterion for *Enterobacteriaceae* on Irish sheep carcasses. Food Control 25, 172–183.

Gonzales-Barron, U.A., Zwietering, M.H., Butler, F., 2013. A novel derivation of a within-batch sampling plan based on a Poisson–gamma model characterising low microbial counts in foods. Int. J. Food Microbiol. 161, 84–96.

Gonzales-Barron, U.A., Cadaves, V.A.P., Butler, F., 2014. Statistical approaches for the design of sampling plans for microbiological monitoring of foods. In: Granato, D., Ares, G. (Eds.), Mathematical and Statistical Methods in Food Science and Technology. Wiley Blackwell, Chichester, West Sussex, UK.

Haas, C.N., 1983. Estimation of risk due to low doses of microorganisms, a comparison of alternative methodologies. Am. J. Epidemiol. 118, 573–582.

Haas, C.N., 2002. Conditional dose–response relationships for microorganisms: development and application. Risk Anal. 22, 455–463.

Havelaar, A.H., Nauta, M.J., Jansen, J.T., 2004. Fine-tuning food safety objectives and risk assessment. Int. J. Food Microbiol. 93, 11–29.

ICMSF, 1988. Microorganisms in Foods. 4. Application of the Hazard Analysis Critical Control Point (HACCP) System to Ensure Microbiological Safety and Quality. Blackwell, Oxford.

ICMSF, 2002. Microorganisms in Foods. 7. Microbiological Testing in Food Safety Management. Kluwer Academic/Plenum Publishers, New York.

ICMSF, 2009. <http://www.icmsf.org/main/software_downloads.html> (accessed 25.11.15).

ICMSF, 2011. Microorganisms in Foods. 8. Use of Data for Assessing Process Control and Product Acceptance. Springer, New York.

Ingram, M., 1961. Microbiological standards for foods. Food Technol. 15 (2), 4–16.

Ito, K., 1974. Microbiological critical control points in canned foods. Food Technol. 28, 46–48.

Jongenberger, I., 2012. Distributions of microorganisms in foods and their impact on food safety. Ph.D. Thesis. Wageningen University, Wageningen, NL.

Jongenberger, I., den Besten, H.M.W., Zwietering, M.W., 2015. Statistical aspects of food safety sampling. Ann. Rev. Food Sci. Technol. 6, 479–503.

Jouve, J.-L., 1999. Establishment of food safety objectives. Food Control 10, 303–305.

Jouve, J.-L., 2000. Good manufacturing practice, HACCP and quality systems. In: Lund, B.M., Baird-Parker, A.C., Gould, G.W. (Eds.), The Microbiological Safety and Quality of Food, vol. II. Aspen Publishers, Gaithersburg, MD, pp. 1627–1655 (Chapter 58).

Kilsy, D.C., Aspinall, L.N., Baird-Parker, A.C., 1979. A system for setting numerical microbiological specifications for foods. J. Appl. Bact. 46, 591–599.

Lee, J., et al., 2015. Example of a microbiological criterion (MC) for verifying the performance of a food safety control system: *Campylobacter* performance target at end of processing of broiler chickens. Food Control 58, 23–28.

Leistner, L., Gould, G.W., 2001. Hurdle Technology: Combination Treatment for Food Stability, Safety and Quality. Kluwer Academic/Plenum Publishers, New York.

McKellar, R.C., Lu, X. (Eds.), 2003. Modelling Microbial Responses in Food. CRC Press, Boca Raton, FL.

Mossel, D.A.A., 1982. Microbiology of Foods: The Ecological Essentials of Assurance and Assessment of Safety and Quality, third ed. University of Utrecht, Utrecht.

Mossel, D.A., Drion, E.F., 1979. Risk analysis. Its application to the protection of the consumer against food-transmitted diseases of microbial aetiology. Antonie van Leeuwenhoek 45, 321–323.

Mussida, A., Gonzales-Barron, U., Butler, F., 2011. Operating characteristic curves for single, double and multiple fraction defective sampling plans developed for *Cronobacter* in powder infant formula. Procedia Food Sci. 1, 979–986.

Mussida, A., Gonzales-Barron, U., Butler, F., 2013. Effectiveness of sampling plans by attributes based on mixture distributions characterizing microbial clustering in food. Food Control 34, 50–60.

Nauta, M.J., Jacob-Reitsma, W.F., Evers, E.G., van Pelt, W., Havelaar, A.H., 2005. Risk assessment of *Campylobacter* in the Netherlands via broiler meat and other routes. RIMV Report 250911006/2005. RIMVA, Bilthoven, Netherlands.

Peterson, A.C., Gunnerson, R.E., 1974. Microbiological critical control points in frozen foods. Food Technol. 28, 37–44.

Rieu, E., Duhem, K., Vindel, E., Sanaa, M., 2007. Food safety objectives should integrate the variability of the concentration of pathogen. Risk Anal. 27, 373–386.

Roberts, T.A., Jarvis, B., 1983. Predictive modelling of food safety with particular reference to *Clostridium botulinum* in model cured meat systems. In: Roberts, T.A., Skinner, F.A. (Eds.), Food Microbiology: Advances and Perspectives. Academic Press, London.

Santos-Fernández, E., Govindaraju, K., Jones, G., 2014. A new variables acceptance sampling plan for food safety. Food Control 44, 249–257.

Stringer, M., 2004. Food safety objectives – role in microbiological food safety management. International Life Science Institute Europe Report Series, Brussels, Belgium (ISBN 1-557881-175-9).

Swart, A.N., Mangen, M.-J.J., Havelaar, A.H., 2013. Microbiological criteria as a tool for controlling *Campylobacter* in the broiler meat chain. RIVM Report 330331008/2013. National Institute for Public Health and the Environment, Bilthoven, NL.

Szabo, E.A., Simons, L., Coventry, M.J., Cole, M.B., 2003. Assessment of control measures to achieve a food safety objective of less than 100 cfu of *Listeria monocytogenes* per gram at the point of consumption for fresh pre-cut iceberg lettuce. J. Food Prot. 66, 256–264.

Thatcher, F.S., 1955. Microbiological standards for foods: their function and limitations. J. Appl. Bacteriol. 18, 449.

Tuynenburg-Muys, G., 1975. Microbial safety and stability of food products. Antonie van Leeuwenhoek 41, 369–371.

Wallace, C.A., Mortimer, S., 1998. HACCP – A Practical Approach, second ed. Aspen Publishers, Gaithersburg, MD.

Whiting, R.C., 1995. Risk Analysis. A Quantitative Guide. Wiley, Chichester, UK.

Wilson, G.S., 1955. Symposium on food microbiology and public health: general conclusion. J. Appl. Bacteriol. 18, 629–630.

Zwietering, M.H., Stewart, C.M., Whiting, R.C., 2010. Validation of control measures in a food chain. Food Control 21 (12 Suppl.), 1716–1722.

Zwietering, M.H., Jacxsens, L., Membré, J.-M., Nauta, M., Peterz, M., 2016. Relevance of microbial finished product testing in food safety management. Food Control 69, 31–43.

Subject Index

G

H

I

K

L

M

N

O

S

T

U

Printed in the United States
By Bookmasters